dith Kohlenberger

v Formula For Cool

American Culture Studies | Volume 12

Judith Kohlenberger (Dr. phil.) is a post-doctoral researcher with a degree in English and American studies from the University of Vienna. Her research interests include gender and cultural theory, popular culture, North American film and television, cultural studies of science and science and technology studies, romanticism and postmodernism, and hemispheric approaches to the Americas. She works and lives in Vienna, Austria.

Judith Kohlenberger

The New Formula For Cool

Science, Technology, and the Popular in the American Imagination

[transcript]

Supported by a publishing grant from the Austrian Academy of Sciences.

Bibliographic information published by the Deutsche Nationalbibliothek
The Deutsche Nationalbibliothek lists this publication in the Deutsche Nationalbibliografie; detailed bibliographic data are available in the Internet at http://dnb.d-nb.de

Cover layout: Concept by Alexander W. Doms
Cover illustration: yod77 – fotolia.com
Proofread by Sarah Thomas
Printed in Germany
Print-ISBN 978-3-8376-3092-3
PDF-ISBN 978-3-8394-3092-7

Table of Contents

Acknowledgements

Keep cool but care.
THOMAS PYNCHON/*V.*

This study would not exist without the help, support, and inspiration from many professors, students, colleagues, mentors, friends and family. In those times of academic and personal struggle when I was in danger of losing my cool, it was them who provided me with motivation, confidence, and the strength to carry on.

First of all, I have to thank my supervisor, Astrid M. Fellner, for her unfailing support, intellectual wisdom, patient guidance, and enthusiastic encouragement. Throughout my academic career, she was so much more than a supervisor—she is a true mentor to me and, needless to say, my academic role model. It is thanks to her that I enjoyed working on this project from beginning to end, and that those brief moments of frustration and doubt were fleeting and soon forgotten indeed. Apart from her perceptive comments, numerous letters of recommendation, and the unrelenting willingness to listen to my ideas, it was her unwavering belief in my abilities and talents that helped me tremendously from the moment I embarked on this study. She met all of my suggestions, at times wild and terribly undeveloped, with an open mind and genuine interest, and I truly wish to thank her for it.

Secondly, the work on this study was immensely supported by the DOC fellowship that I was granted by the Austrian Academy of Sciences for the years 2012-2013. The two years, exclusively devoted to this study, may well have been the most productive in my life, at least in terms of pages written, talks given, and seminars attended. The annual grantee weekends organized by the Academy not only provided me with an intellectually open space to extend my academic horizon and forge rewarding interdisciplinary networks; they have also, perhaps even more substantially, enriched my life with new colleagues and friends, whose fresh perspective and critical feedback left an unmistakable mark on this study. Finally, the Academy facilitated the birth of this book by generously awarding a publishing grant.

In the wake of my DOC fellowship, the Department of English and American Studies at the University of Vienna readily took me aboard as an external staff member. I especially have to thank the then Head of the Department, Margarete Rubik, and Monika Wittmann for their quick, non-bureaucratic help and the institutional affiliation they enabled me to enjoy. In the same vein, the 'Abschlussstipendium' granted by the University of Vienna for the final months of my doctoral studies contributed immensely to the swift and successful completion of the thesis.

Special appreciation is extended to the members of my doctoral advisory board at the Dies Doctoralis as well as the board of examiners for my thesis defense: Susanne Reichl, Rudolf Weiss and Werner Huber from the Department of English and American Studies, Mitchell Ash from the Department of History, and Ina Hein from the Department of East Asian Studies at the University of Vienna. I thank them for their constructive comments and their encouragement in various stages of my work.

Susanne Reichl also has to be thanked for her monthly PhD group meetings, which contributed so much to the conceptualization of my study in those tricky early days. In all regards, her advice was sensible, accurate, and spot on. I wish to thank her and all participants of the group for their constructive criticism and constant encouragement.

At Saarland University, I am grateful to all the staff members who welcomed me with open arms and offered helpful feedback on my work in the research colloquium. Equally, the University of Graz gave me the opportunity to present my project at an early stage and to a large audience, for which I would like to express my sincere gratitude. Invaluable feedback was also offered by Lutz Musner and Jim McGuigan, then Senior Research Fellow, at the IFK Vienna. I am grateful to them for the time they generously granted me and my project.

I also wish to acknowledge the help of the Center for Doctoral Studies at the University of Vienna, in particular Bianca Lindorfer, Christian Kolowrat, and Lisette Schmidt. They have not only assisted me with acquiring funding for my project, but also provided treasured opportunities for networking and critical debates.

In terms of professional networks, my sincere thanks also go to the Austrian Association for American Studies, which I have come to consider my academic home. I especially wish to thank the organizers of the last annual meeting, above all the Association's president at the time, Ralph Poole from the Paris Lodron University Salzburg, for providing such an ideal, intellectually stimulating retreat before the finalization of my studies. I also want to thank its sister organization, Austria's

Young Americanists, whose members have given me the opportunity to serve on the board for several months and which I still feel deeply committed to.

There are many friends and colleagues whose support I wish to acknowledge. I am thankful to Tamara Radak and Lisa Edelbacher from the University of Vienna, Sabine Harrer from the IT University of Copenhagen, Mario Rader from the University of Graz, Georg Drennig from the University of Duisburg-Essen, Samuel Zwaan from the Georgia Institute of Technology, and Cornelia Klecker from the University of Innsbruck for their perceptive comments on earlier versions of the draft. My appreciation also goes to Timon Jakli and Hano Pipic from the University of Vienna for offering indispensable material and expert advice. Thank you for your friendship and for accepting nothing less than excellence from me.

In revising this study for publication, I need to thank my project manager Annika Linnemann from *transcript Verlag* for her unfailing patience and assistance with my exhaustive and at times surely bizarre questions. Many thanks also go to my editor Sarah Thomas, whose thoughtful commentary vastly improved my manuscript and whose constructive feedback encouraged me to believe that there is a book publication happening after all. All mistakes, needless to say, remain my own. During the time of revision, I could not have asked for a better cheerleading team than my co-workers at the Department for Socioeconomics at the Vienna University for Economics and Business. Annika, Armon, Asjad, Barbara, Bernhard, Erich, Florentin, Franzi, Heike, Hendrik, Karin, Kasia, Michael, Uschi, Vera—thank you for rooting for me, for offering friendship, enthusiasm and open ears, and for never declining a cup of coffee. Special thanks go to Florian Bast, Gil Rodman, and Netflix—though only present digitally, you never failed to brighten my spirits in hours of need.

I specifically wish to thank my long-term friend and colleague Susanne Hamscha, whose intellectual power and prudent advice have been inspiring beyond words. Her constructive recommendations have accompanied this study from beginning to end, and her vision has been nothing but encouraging.

Special thanks also go to Alexander Doms for his unwavering enthusiasm, his technical assistance in every possible (and impossible) regard, and for reminding me that there is a life beyond the written page. His readiness in designing the cover illustration for this book yet again demonstrated his unmatched resourcefulness, creative powers, and true friendship that continue to inspire me daily.

Above all, I deeply thank my parents and my sister for the unfailing support and encouragement I received throughout my studies. I am indebted to them for their absolute trust in my talents and their unconditional love, which has been my protective shield against occasional feelings of insecurity and doubt. Words cannot express how grateful I am to them for giving me the utmost confidence, taking the blows, and catching me whenever I was about to fall.

This book is dedicated to them.

Introduction

Images of Technoscience in the New Millennium

[A] sideshow can sometimes be the main event.
GEORGE LIPSITZ/*TIME PASSAGES*

"Our society has undergone a paradigm shift. In the information age, you and I are the alpha males," Dr. Leonard Hofstadter, experimental physicist and protagonist of the American hit sitcom *The Big Bang Theory* (2007-present), assures himself and his fellow-scientists during a fancy-dress party. The success of the show proves him right: Not only was the format soon syndicated all over the world, it has also inspired a wealth of popular cultural productions, from daytime shows to Hollywood movies, similarly exploring the hitherto shunned world of science and research. As if determined to prove Dr. Hofstadter's point, innovative infotainment shows, scientists as cult stars, and traditional television formats newly invested with science nowadays draw unprecedented numbers of viewers. Spectacular pictures from the Mars expedition grace a multitude of computer screensavers and dorm rooms, while the former derogative 'nerd' has been re-appropriated into an expression of teenage approval. The times when science and its devotees were represented by one likeable, yet hopelessly pathetic sidekick seem to be long gone. Today, the geeks have taken center stage as admirable heroes and witty protagonists. Contemporary popular culture appears to embrace wholeheartedly what the world of science has to offer - and vice versa. Individual disciplines have started to appropriate and make avid use of popular media. News of the first proton collision at CERN, one of the most prestigious scientific endeavors of our time, was spread on *Twitter*, and there is hardly a budding scientist who does not present his or her findings on an online blog, thus ridding the profession of its outdated image. In response, workshops for scientists teaching them how to become mediagenic and make their research appear exciting, hip, and zeitgeisty flourish. More serious assessments, such as *The Guardian*'s recent enthusiastic proclamation of "a golden age of science" (Cox et al. n.p.), construct scientific research as one of Western society's prime assets and a chief arbiter of cultural, rather than merely epistemic capital. Science, it appears, has finally discovered the formula for cool.

The present study is dedicated to the exploration of this perceived paradigm shift: While the ubiquitous notion of cool has long pervaded the realms of fashion, advertisement, and youth culture, it now also seems to invade the world of hard science, as depicted in mainstream film and television formats from the United States. Without doubt, coolness constitutes an omnipresent and simultaneously elusive quality of contemporary postmodern society and effectively functions as one of the foremost cultural sensibilities of our time. In correspondence with the triumphant evolution of cool from an attitude of 1960s US-American counterculture into a principal aesthetic norm of mainstream society, the past two decades have witnessed a substantial increase in studies on coolness from a multiplicity of angles and academic traditions. Far from being trivial or arbitrary, cool has been recognized as a dominant mode of affective comportment in the twentieth and early twenty-first century and accordingly been established as a viable field of research within such diverse disciplines as media theory, psychology, art history, and cultural studies. As a case in point, two recent edited collections in the field of American studies testify to the ongoing appeal of cool for academic investigation: *Is It 'Cause It's Cool? Affective Encounters with American Culture* (Fellner et al. 2014), which records the proceedings from the eponymous 2011 Annual Conference of the Austrian Association for American Studies, and *The Cultural Career of Coolness* (Haselstein et al. 2013), which emerged out of a research cluster on *Languages of Emotion* launched at the Freie Universität Berlin.

The indeterminate ontological status of cool and its resulting semantic flexibility seem to figure as prime motivations for the continuing popularity of the concept: Its cultural force derives from a universal applicability and simultaneous ontological vagueness. Underlying the term is, of course, a central ambiguity: On the one hand, cool has devolved into a universal, semiotically drained term of approval, a mere verbal tic, while its etymological roots, on the other hand, still evoke far-reaching connotations of coldness, dispassion, and emotional restraint. A myriad of studies have demonstrated that the concept of cool is, despite its obvious linguistic drainage, linked to a whole set of cultural associations, be it rebelliousness, irony, hedonism, masculinity, consumerism, youth, counterculture, or, of particular value for the present study, Americanness. Allegedly originating in a pose of affective detachment worn by African warriors in battle, the attitude is believed to have been imported to the United States via the slave trade, where it was cultivated and refined in the vibrant jazz and blues cultures of the 1920s and 1930s. Today, cool displays a noticeable proximity to the values of the modern information age, thus emerging, according to Alan Liu's seminal *The Laws of Cool* (2004), as "the techno-informatic vanishing point of contemporary aesthetics, psychology, morality, politics, spirituality, and everything. No more beauty, sublimity, tragedy, grace, or evil: only cool or not cool" (3). In view of this assessment, the cultural omnipresence of cool must be understood as not only impinging on the realms of advertisement, fashion, celebrity culture, and the American music industry, but also affecting the conditions of epistemic production and representation in the post-industrial information society, especially in the context of modern ICTs (information and communication technologies) and digital culture. The present study aims to address these crucial concerns by examining the depiction of 'cool technoscience' in seven selected films and television programs.

Embarking from preliminary academic accounts of the functionalities of cool, this study is based on the central hypothesis that recent popular cultural representations of (techno)science in mainstream American film and television are increasingly informed by a prominent focus on cool as an aesthetic and affective, rather than cognitive or ethical form of scientific legitimation. The growing emphasis in popular scientific imagery on cool is understood as a response to, or even indeed substitution of, former sources of scientific legitimation, which have been described as superseded in postmodern society. Cool thereby acts as a novel and popular form of legitimation, challenging and potentially replacing traditional cognitive and/or ethical justifications, as illustrated in the selected sample of popular films and television formats. The current penetration of science and technology into all aspects of life in the modern information society, from household gadgetry to digital fingerprints, both results from and contributes to these constructions of 'cool science' in the popular cultural fabric of the United States. Accordingly, the prevalence of codified knowledge and the rise of modern ICTs in today's post-industrial landscape are considered as providing the necessary socio-cultural backdrop for these cool representations of technoscience to emerge in contemporary film and television.

The central hypothesis of the study will be tested by examining seven recent audiovisual productions in American popular culture. This includes three recent feature films, the mainstream Hollywood productions *The Day After Tomorrow* (dir. Roland Emmerich, 2004) and *The Social Network* (dir. David Fincher, 2010), as well as Joshua Michael Stern's indie production *Jobs* (2013), and two television formats, CBS's crime drama show *CSI: Crime Scene Investigation* (2000-present) with its recently cancelled spin-offs set in Miami and New York City, as well as the domestic sitcom *The Big Bang Theory* (2007-present). The analytical focus is thus placed on popular audiovisual productions that are informed by varying notions of 'cool technoscience,' ranging from depictions of geek chic, spectacular experimental arrangements, and scientific entrepreneurship to media-saturated laboratories, cool-blooded researchers, and their sleek tool boxes. Notwithstanding the obvious aesthetic and generic diversity, the multifarious films and television series analyzed in this study are united by their common emphasis on cool as a novel, yet highly effective source for legitimating the cultural prestige, epistemological authority, and financial, ecological, and other resources enjoyed by contemporary technoscience.

Given this strong analytical focus on popular legitimatory discourses, the present study engages with leading explorations of the problem of scientific legitimation in postmodern society, in particular Jean-François Lyotard (*The Postmodern Condition,* 1979) and his philosophical adversary Jürgen Habermas (*Legitimation Crisis*, 1973), with whom he entered into an intense academic debate on issues of legitimation, consensus, and progress. Famously, Lyotard characterized the crisis of "the metanarrative apparatus of legitimation" (xxiv) as the defining feature of the postmodern age, suggesting that ethical and cognitive sources of metanarrative legitimation (i.e. the humanist understanding of science as a means to progress and liberation vs. the Hegelian notion of science for science's sake) have become entirely obsolete in today's knowledge economy. While Habermas remained skeptical of Lyotard's fervent proclamation of postmodernity and chose to view the present period as a continuation of the yet incomplete project of Enlightenment, his seminal analysis of the universal collapse of governing institutions concurs in the general assessment that legitimation crises constitute a distinctive problem of contemporary late-capitalist societies. Inspired by this rich philosophical dispute, the present study suggests a novel perspective on the multi-layered and heavily contested issue of legitimation: It proposes to understand the unprecedented emphasis in popular cultural representations of technoscience on the (undeniably imprecise and non-factual) notion of cool as replacing or, at the very least, responding to former discourses of scientific legitimation, which are generally regarded as dysfunctional in post-industrial society. Conversely, this means that the roots of what I perceive as a novel popular cultural attitude toward technoscience and its parameters (including equipment, practitioners, and resources) can be traced back to the much proclaimed postmodern crisis of legitimation.

Accordingly, the original contribution to knowledge made by this study can be located in its unique focus on popular American film and television as a matrix for contemporary discourses of scientific legitimation. As suggested by George Lipsitz's epigraph to this introduction, the sideshow staged by popular culture, in this case explored in the realm of mainstream audiovisual production, will be understood as the main event for heightened demands of legitimation in today's information society, of which the United States stand as one of the most prominent instances. Rather than treating popular culture as trivial, immaterial, and merely a minor prop on the center stage taken by 'high brow' cultural practices, the present study will direct the spotlight onto intellectually mundane and highly commercialized instances of mass cultural expression and demonstrate how they may abet the legitimatory needs of one of the noblest and most elitist realms of human activity, i.e. Western industrial science. The study thereby responds to perceptible research gaps in the involved areas of enquiry, as in-depth studies on the representation of contemporary discourses of science via the modes of popular culture, such as, in the present case, feature film and television, are comparably rare. The majority of scholarship in the field portrays a sociological angle, focusing on the quantitative analysis of science or processes of scientific popularization. Analyses adopting an interdisciplinary, interpretative approach to the study of technoscience and its representation in popular cultural production are only slowly emerging. Equally, the study of cool is a relatively new area of research, whose appropriation by scientific representations has so far not been examined. While interdisciplinary studies of cool in the context of fashion, music or the arts abound, the use and effects of cool beyond its traditional realms of thematic application have so far not been the subject of a comprehensive study. By advancing a conceptualization of cool as an affective and aesthetic source of popular scientific legitimation, the present study aims to adopt an innovative approach to engage with these research gaps and thus contribute to the ongoing dialogue between the scientific and the popular in contemporary American society.

The discursive focus on mainstream American film and television as central arenas of popular cultural production establishes the present study as an instance of cultural studies of science, an emergent, heterogeneous area of interdisciplinary research, which creatively absorbs scholarship from history, philosophy, sociology, anthropology, gender studies, and literary criticism. While it is indebted to related and established fields of research like science studies and science and technology studies (STS), its philosophical underpinnings differ substantially from sociological or historical perspectives. In his paradigmatic essay "What Are Cultural Studies of Scientific Knowledge?", Joseph Rouse explicates the analytical focus of the discipline as follows: "[C]ultural studies of scientific knowledge take as their object of investigation the traffic between the establishment of knowledge and those cultural practices and formations which philosophers of science have often regarded as 'external' to knowledge" (4). Examples of such 'external' or, in accordance with the above epigraph, 'sideshow' practices would be the analysis of advertisements in *Science* (Haraway 1989), the cultural and symbolic co-option between *Star Trek* and NASA (Penley 1997), the personal investedness of researchers exploring tobacco control (Reid 2000), or, in the present case, representations of cool technoscience in popular film and television.

Based on the philosophical premises of this interdisciplinary area of enquiry, the key research questions that are approached in the present study are the following:

- To what extent and in which particular ways are representations of technoscience in recent American film and television predicated on popular notions of cool? What role do different reifications of cool play in these representations? How are channels of popular cultural production, including major film genres and television formats, appropriated to convey a cool image of science?
- Secondly, the study addresses the immediate ramifications of this proposition: Which dominant image(s) of and discourses on science do these cool representations help to create? And how far, conversely, does this affect the nature, operational modes, and the effectuality of cool? The latter question will be discussed in view of the historical development of cool from a countercultural attitude into a phenomenon of modern mass society.
- What are the reasons for the prominent emphasis on cool in contemporary representations of science? Can these representations be understood as an (affirmative) reaction to a potential crisis of scientific legitimation?
- Finally, the study explores the epistemological implications of such cool representations for contemporary information societies, in particular the United States as one of their prime instances.

Naturally, any analysis of cultural significations that centers on their specific treatment of 'science' already implies a preconceived definition of the term. As soon as cultural productions are treated and analyzed as representations of science, specific forms of epistemology are identified as 'scientific' and hence authorized, demarcated, and attributed cultural and epistemic authority. This makes an objective, distanced analysis of 'this thing called science' a vain endeavor—a circumstance which cultural studies of science are not only acutely aware of, but choose to embrace as a potentially powerful accomplice. Preceding the investigation of the above research questions must, therefore, be the task of contriving a working definition of the subject under scrutiny, all the while remaining conscious of its necessary provisionality and epistemological arbitrariness: What is science and which cultural productions qualify as representations of scientific matters? Which characters, if we look at American feature film and television, represent veritable scientists, as opposed to engineers or medical doctors? How have the images of science and technology been merged?

With the last question, the present study joins the wealth of interdisciplinary scholarship on 'technoscience,' a term originally coined in the field of bioethics to describe the tightening relationship between science and high-end technology, which have become virtually synonymous in many contexts.[1] Most prominently, the idea of technoscience as a defining feature of contemporary information culture was

1 See chapter 3 of this study for a closer discussion of the term and its relation to the concept of the information society.

developed by the seminal works of Bruno Latour and Donna Haraway, which this study is explicitly indebted to. The latter has perceptively demonstrated that today's omnipresence of technoscience "extravagantly exceeds the distinction between science and technology" (Haraway, *Modest_Witness* 3) and must be regarded as "a mutation in historical narrative" (Haraway, *Modest_Witness* 3), marking a new cultural epoch. The character of scientific practice and the concomitant knowledge it produces have been fundamentally transformed by the growing demands of the industry, politics, and the military. Technologized, mediatized information is omnipresent in modern information societies like the United States and has become virtually inescapable: The penetration of all aspects of life with technoscience, be it in the form of personal computers, mobile phones, or, even more consequentially, biotechnologies, makes a neat separation into formerly distinct domains—the political, the personal, the economic—nearly impossible. It goes without saying that this circumstance also impinges on representations in popular culture, where the boundaries between the images of science and technology, of pure and applied research, are habitually blurred or even indeed conflated.

In setting its own and unavoidably subjective borders between technoscience and related segments of professional activity, the present study chooses to rely on what may be called an intradiegetic definition of the utilized terms: As soon as the subject matter of a given film or television series is referenced as 'science' by the text or paratext, it was treated as eligible for a closer inspection of the employed representational strategies and scenarios. This pragmatic approach appears to be most feasible and promising for the purposes of this study, since the focus of analysis is not placed on the socio-political dynamics accompanying the attribution of the labels 'science' or 'technology' to popular entertainment formats in film and television, but the entwinement of these cultural formations with notions of cool. Throughout the analysis, special attention is granted to the credo of cultural studies of science as a discipline that commits itself to retaining a basic awareness of its necessarily biased position within the scientific landscape (Rouse, "Cultural Studies" 6). Consequently, a major aim of the present volume is to remain self-conscious and reflexive about its own political, epistemic, and personal investment in and engagement with the cultural significations it sets out to study.

STRUCTURE AND METHODOLOGY: A ROAD MAP

The present study is broadly divided into a theoretical and an analytical part. The former is dedicated to the discussion of central philosophical frameworks which my argument responds to, which will be introduced by brief literature reviews of the most relevant studies in the field. As the present study is heavily interdisciplinary in nature, it draws from a wide range of concepts and theories in the fields of American studies, cultural studies, science studies, media theory, as well as film and television studies. Previous research on cool has yielded some pertinent results in the fields of sociology, anthropology, and art history, which are equally incorporated for reference. An overview of relevant insights gained in these areas of enquiry will prepare the ground for my own take on the popular cultural representation of technoscience and its focus on cool as a viable new form of popular scientific legitimation.

The theoretical part of this study will elucidate the three main theoretical strands on which my argument is based. First of all, this involves theoretical approaches from recent research into the production, operational modes, and functionalities of cool, which chapter 2 on "The Conquest of Cool: From American Counterculture to Global Dominance" will survey in detail. Exploring the defining parameters of the concept, whose pivotal cultural position depends on its very elusiveness and mutability, will constitute a major goal of the initial chapter. The second strand of theory that this study builds upon is the shifting premises of knowledge production in the information society; studies of this will be reviewed in chapter 3 "The American Information Society and the Crisis of Scientific Legitimation." In addition to the influential dispute between Lyotard and Habermas, the chapter also engages with recent sociological, historical, and ethico-political perspectives on scientific legitimation. Despite divergences in methodology and outlook, the majority of studies concur in the assessment that discussions of scientific legitimation have increased exponentially with the advent of the modern information society. Closely related to studies on scientific legitimation is the second body of scholarship which the chapter will discuss, namely academic explorations of technoscience and its effects on social, economic, and political relations in the post-industrial information society. Finally, chapter 4 entitled "Nerd Alert: Science and the Popular" explores contemporary takes on scientific popularization and dominant models of the relationship between popular culture and science, before advancing my own approach to the study of science and/ in/ as popular culture. To varying degrees, all three areas of academic enquiry reveal perceptible research gaps, which the present study aims to address accordingly.

Following the theoretical part of this study, its analytical part will explore instances of two television genres (the crime drama and the sitcom) and two feature film genres (the disaster movie and the biopic). The selected examples easily rank amongst the most popular genres of the early twenty-first-century, so that each text stands as only the most prominent and/or feasible instance of a wide range of productions with similar aesthetics, narrative styles, and cinematography. The particular choice of primary material allows me to attend adequately to unique aspects in each text's representation of science, such as the aestheticization of crime-detection technology or the idealization of science as a heroic agent for saving the world. The common focus on the visual dimension ensures structural homogeneity of the primary sources, as does the regional restriction on popular cultural productions from the United States. Additionally, audiovisual media's modes of production can be regarded as more heavily capitalizing on notions of cool than written sources, since cool clearly emerges, as the theoretical part of this study will show, as a visual effect with a strong Americocentric orientation. In correspondence with W.J.T. Mitchell's seminal *Picture Theory* (1994), the chosen texts are treated as essentially heterogeneous media productions: Despite their evident foregrounding of nonlinguistic aspects, the analysis also takes into account the verbal dimension. Since the present study is dedicated to the examination of a recent trend in popular cultural imagery, it centers exclusively on contemporary examples, starting in the year 2000. References to earlier representations of technoscience in the realms of feature film and television are included where necessary for the argument. Despite these occasional and indispensable backward glances, the aim is not, however, to provide a comparative study: The analytical focus is firmly placed on what I perceive as a recently emergent trend in popular cultural imagery of technoscience.

In accordance with the constructionist approach in cultural theory, the scrutinized representations are understood as 'texts' in the broadest sense, i.e. sites of meaning production and negotiation, which will be subject to close reading and critical evaluation. This entails that the visual depictions under analysis are understood as exceeding the realm of mere reflection: Following Stuart Hall's seminal work on cultural representation (1997), the scrutinized examples are regarded as not only reflecting, but producing knowledge and meaning through language, image, and discourse, thereby actively contributing to a specific construction of dominant societal ideas. Consequently, texts are treated as open to revision and active interpretation, thus paying tribute to "cultural studies' sensitivity to differences and contested meanings and identities" (Rouse, "Cultural Studies" 6) in order to counter universalizing tendencies which may propose a false unity of social categories in lieu of their historical and local contingency. The goal of such a methodological approach is anything but the causal explanation of the epistemological consequences of the cultural formations under scrutiny: When exploring discourses and images of technoscience circulated in popular culture, the present study strives to recognize its own epistemic engagement with the objects of analysis, as every cultural critique is ultimately part of the culture it purports to observe. Recognizing one's own ultimately subjective stance toward the subject matter and its consequences for the ensuing interpretation is a key principle of analysis, as is the fundamental methodological premise that there is, provocatively speaking, no fixed meaning to be explained for once and for all. Equally, the focus of the study's analytical part will by no means be placed on the respective texts' purpose as intended by their authors/producers, such as their goal of presenting science in a more or less favorable light. The discipline of cultural studies has shown early on in the academic debate that these personal motivations are ultimately inaccessible. These are the central aspects in which the methodological approach adopted by the present study notably differs from the more empirically oriented science studies.

Another major difference to contemporary sociological explorations is the present study's eschewal of quantitative enquiry: Since my analysis examines the cultural work performed by popular cultural representations of science, rather than the public responses they evoke or their concrete effects on scientific practice, it is based on close readings of the material instead of empirical analysis. The careful and sustained interpretation of the primary sources allows me to attend adequately and individually to each of the above outlined research questions and conduct a thorough testing of the central hypothesis. In approaching each of the primary texts, my close reading is guided by the following consecutive steps:

- How is science represented in the text? What (narrative) function(s) does it fulfill?
- Which popular notions of cool is the representation of science in the respective text based on? What notions does it mobilize and reinforce or challenge and subvert?
- How are these considerations borne out in a close reading of a signature scene from the text?

These three steps in analyzing a text correspond to the individual structure of the analytical chapters. Accordingly, each analytical chapter starts with a brief overview of the particular image(s) of science advanced in the text and concludes with a close reading of a specific scene or episode, which is intended to combine and consolidate the initial reflections. The aim of the final close reading is to crystallize the key argument of each chapter through recapitulating its central themes and applying them to a pertinent and coherent example from the text.

The first analytical chapter, "'Cool Forensics and the Spectacle of Technoscience in *CSI: Crime Scene Investigation*, *CSI: Miami*, and *CSI: NY*," explores the representations of technoscience in the internationally successful crime drama series and its equally popular spin-offs. The *CSI* universe will be understood as displaying an unprecedented focus on forensic science, especially DNA processing, for crime detection. The curious figure of the scientist-detective, a profession specifically invented for the purposes of the show, and the repeatedly stressed centrality of physical evidence are vital factors in *CSI*'s particular representations of science. The tight amalgamation of science with high-end technology, from the chrome-and-steel laboratory to impressive computer simulations, fulfills a crucial aesthetic function. The close reading of one of the show's signature lab scenes will elucidate the way in which science, the epitome of consolidated knowledge, can be represented as cool in the context of the classic police procedural: The focus is shifted from the knowledge that science produces to a situation which communicates no relevant information at all. The audience is thus greatly encouraged to appreciate, or even indeed tolerate, the heightened narrative attention to tedious scientific processes because of their extreme stylization and visual coolness.

The second analytical chapter entitled "Geek Cool and the Comedy of Science in *The Big Bang Theory*" examines the only text in my sample which employs scientific representations for a humorous effect. Since comedy might be treated as either undermining or affirming the presented situation, thus fundamentally bearing upon the analysis in question, the chapter also includes a brief overview of basic (television) humor theories. Consequently, I will argue that science comes to assume the role of what postcolonial theory classified as 'the Other,' since it repeatedly serves to provoke comic, yet highly consequential deviations from the norm. In analyzing the particular notions of 'geek cool' advanced by the show's protagonists, differentiations between conflicting types of mainstream and alternative cool will prove highly useful. Similar to how technoscientific expertise in *CSI* allows the scientist-detectives to emerge as superior and cool, it is *because of*, rather than despite the association with science that the characters in *The Big Bang Theory* display typical contrarian traits of coolness. The close reading will focus on the first season of the show and illustrate the trope of social divergence, which acts as a vital factor in the display of alternative forms of cool as well as the basis for the show's comedy.

"Only the Cool Survive: The Science of Disaster in *The Day After Tomorrow*" is the third chapter of my analytical part and deals with the genre of the disaster movie. As a prime representative, the Hollywood blockbuster *The Day After Tomorrow* depicts a (man-made) natural catastrophe causing the potential eradication of all life on the planet, which can only by mitigated by Western industrial science and technology. The text features an intricate set of recurring tropes and aesthetics that typically inform the treatment of science in the genre. Hence, I will argue that science (applied sensibly in the hands of a young, adventurous, and cool scientist) acts as the miraculous *deus ex machina* that rescues humanity from the imminent apocalypse. The cognitive value of science emerges as a vital factor in making science and its practitioners appear cool: It is only by virtue of scientific knowledge about the disaster that the characters, above all the all-powerful male hero, are endowed with a higher chance of survival and may hence face the crisis in cool blood. On the formal level, this is augmented by the restricted use of facial close-ups and a high degree of digitally produced special effects, which replace the human agent as the site of affective audience involvement. The scene chosen for close reading will reflect how the core scientific virtues of reason, rationality, and literacy act as the saving grace, with the story's hero as their living embodiment.

The last analytical chapter of this study, chapter 8 entitled "From Nerds to iCons: Consumer Cool and the Rise of the Scientist-Entrepreneur in *The Social Network* and *Jobs*," explores the recent biopics about two well-known public figures in the computer sciences, Apple tycoon Steve Jobs and Facebook founder Mark Zuckerberg. Both films mobilize notions of a countercultural cool that is fundamentally hinged on the social and aesthetic value of modern ICTs. Through its promotion of self-creation, perseverance, and individualism, the genre of the biopic offers a viable vehicle for perpetuating the myth of the American Dream and the self-made man, both of which are employed as key narrative strategies in the movies under scrutiny. However, the analysis will demonstrate that the biopics strive to present their protagonists as scientific innovators and corporate rebels, rather than shrewd entrepreneurs, by projecting the dissenting aura of cool on the popular electronic brands they embody. Resorting to seminal studies on the uses of cool in business culture, the final chapter will argue that the cinematic representations of the tech founders perpetuate the pervasive ideology of contemporary 'cool capitalism' (McGuigan 2009) to legitimate commercial exploitation and technoscientific supremacy. The final close reading focuses on the first and only scene published before *Jobs*'s official release date, which depicts the protagonist as sagaciously recognizing the transformative potential of personal technology—not only for corporate business ventures, but, much more consequentially, for its fabrication of consumer cool.

Finally, the study's conclusion will combine and contrast the findings gained in the close readings of the primary sources and take into account the variability of technoscientific representation across genres and televisual media. This shall allow for more wide-ranging, yet nuanced statements on the production and circulation of cool in cultural significations of technoscience. The individual results from the analytical part of this study will thus serve as a well-founded basis for the final critical evaluation of my initial research questions and augment the verification of the central hypothesis. Additionally, the concluding chapter will place the discussion into the larger context of the academic study of popular culture and examine the inferences that can be drawn from the preceding analyses.

As this brief road map for the ensuing pages intends to show, a major aim of the present study will be to determine whether the growing emphasis of popular scientific imagery on cool may indeed be regarded as a response to the obsolescence of former sources of legitimation. The analysis of the above listed, multifarious films and television productions shall help to shed light on current processes of interaction between science and popular culture, both of which constitute pivotal sources for change in post-industrial American society. Through this informed assessment, the study seeks to make a valuable contribution to contemporary cultural studies scholarship on the discursive nature of (scientific) representation and demonstrate that sometimes indeed, the analytical spotlight is best placed on the supposed sideshow.

The Conquest of Cool

From American Counterculture to Global Dominance

> Attempting to capture cool is a trap.
> *WIRED*/NOVEMBER 2001[1]

In July 2006, the progressive business magazine *Fast Company*, self-acclaimed place of innovation 'where ideas and people meet,' issued an elaborate chart with some of the most frequently used expressions among adolescents from the 1950s to the present (Conley 2006). Among them were rare gems like 'groovy,' 'boss,' or 'gnarly,' nowadays virtually extinct and alien to the contemporary teenager's vocabulary, but also more recent colloquialisms including 'sweet' and 'sick,' each one of them with a maximum life span of two decades. There was only one adjective, right at the very top of the graph and highlighted in bright red, which, according to the company's careful market research, had never gone quite out of fashion: cool.

In contrast to the present study, *Fast Company*'s market researchers were hardly interested in a cultural history of the pervasive idiom, but much more keen on its commercial exploitation, and sure enough, the fancy graph was accompanied by a short commentary on cool's vital significance for marketing enterprises in the twenty-first century—appropriately entitled "A Craving for Cool." Cool, according to this article, has become a major concern for companies worldwide, so much indeed

1 See pages ix-xv in the November 2011 issue of *Wired* magazine entitled "A Special Advertising Section: The Phenomenon of Cool."

that big enterprises (*Fast Company*'s prime example being, of all brands and trademarks, Hasbro's toy line My Little Pony) have begun to outsource cool, i.e. enlist smaller, closer-to-the-ground agencies to detect ideas, innovations, and new trends, thus revealing, by some mercantile sorcery, the future route of cool. Exactly ten years earlier, author and journalist Malcolm Gladwell already dubbed this activity "The Coolhunt" (1997). Coolhunting, conducted by eternal teenagers who answer to the glamorous names of DeeDee or Baysie, means exciting field work in Brooklyn, Berlin, or Tokyo, with the aim of predicting which sneakers design will sell best next season.[2] It is a billion-dollar business, proving that regardless of all the products cool might yet become attached to, there is one thing that it will surely never be: irrelevant.

"WE KNOW IT WHEN WE SEE IT": THE (NEARLY) IMPOSSIBLE TASK OF DEFINING COOL

It thus not only appears logical, but indeed unavoidable that academic study discovered cool, an intrinsic part of contemporary popular cultural expression, as a rewarding field of research, far more extensive than Gladwell's coolhunters would have ever dared to imagine.[3] Cool, as one recent monograph has chosen to phrase it, "touches upon the domain of the individual and the collective, the aesthetic and the mental, the social and the political, but also economic dimensions, i.e. media and market, as well as gender, nationality, and race. The concerned field of research could not be vaster" (Geiger, Schröder, and Söll 7).[4] Especially the last three decades have witnessed an enormous increase in studies on coolness from a multiplici-

2 Gladwell's article, published in *The New Yorker,* originates from the author's research upon accompanying two cool hunters on their quest for the ultimate new trend. A PBS *Frontline* documentary, called "The Merchants of Cool" (2001, dir. Barak Goodman), complements the journalistic take on coolhunting as a new means of trend scouting that emerged in the early 1990s. See also Nick Southgate (2004), who dissects the concept from a marketing perspective and suggests that "coolhunting now offered insight into the only consumers that truly mattered, the cool" (453).

3 This is especially true for the last three decades, which witnessed a considerable increase in cultural studies scholarship on cool, thus paralleling the concept's sedimentation as an integral part of contemporary (youth) culture.

4 Original wording: "Cool als Kulturtechnik berührt die Bereiche des Individuellen und Kollektiven, des Ästhetischen und des Psychischen, des Sozialen und Politischen, aber auch die Dimensionen der Ökonomie, d.h. der Medien und des Marktes, ebenso wie die der Geschlechter, Nationalitäten und Hautfarben. Das betroffene Forschungsfeld könnte also nicht weiter gefasst sein." All translations mine.

ty of angles and academic disciplines, including psychology, sociology, art history, and media studies, thus paralleling a myriad of popular articles, documentaries, opinion pieces, trend bibles, and fashion manuals, of which the above chart is only a minor example.

The dazzling list, by no means exhaustive, already demonstrates one central characteristic of contemporary scholarly engagement with the amorphous concept: Cool might well rule academic study today, but it does so in vastly different shades and colorings, which may hardly be subsumed under the same heading. Depending on the respective discipline, cool is simultaneously treated as a personal attitude (Pountain and Robins), an aesthetic principle of knowledge work (Liu), an emotional mantle of indifference (Stearns), an elaborate charade for self-defense (Harris), a cultural strategy (Geiger, Schröder, and Söll), an almost empty signifier (*OED*), a formerly countercultural value now appropriated by corporate interests (Frank), a rhetoric (Rice), an African-American male pose (Majors and Mancini Billson), one of the most popular contemporary compliments (Poschardt), as well as, plainly but tellingly, 'It' (Roach). Several, if not almost all of these meanings seem to surface in the *Fast Company*'s "Craving for Cool," which unwittingly merges linguistic, ontological, and conceptual concerns to create one giant bubble of cool.

While cool, as these and other popular usages suggest, might in fact derive its cultural pervasiveness from its very elusiveness and structural flexibility, employing it as a parameter in academic study makes a working definition indispensable, although this may ultimately remain an irresolvable paradox. "We know it when we see it" (67), John Weir writes in the *Rolling Stone* magazine, while Gladwell concluded his ethnography of coolhunters with the damning verdict that coolness "cannot be observed at all, because the act of discovering cool causes cool to take flight" (n.p.). Bent on the allegedly impossible, I shall nonetheless embark on the Herculean task of pinpointing this perhaps most elusive entity. The following pages are devoted to a concise and eclectic survey of some of the most relevant studies of coolness, with the intention of elucidating the changing uses of cool and its structural definitions, which shall, eventually, prepare the ground for the particular conceptualization of cool that the present study will be predicated on.

A closer look at the etymological roots of the term might serve as a helpful point of departure. Cool, as the *Oxford English Dictionary* relates, is, first of all, an adjective indicating low temperature, especially when agreeable and refreshing. In its figurative usage, it may be applied to fabrics, landscapes, or medical treatment, again in a predominantly positive sense. Its use as a personal attribute is a metaphorical extension, which the *OED* defines as "not affected by passion or emotion, dispassionate; controlled, deliberate, not hasty; calm, composed" (s.v. cool *adjective*). It dates back to as far as the Old English epic *Beowulf* (composed sometime between 1700 and 1100 A.D.), the oldest surviving manuscript of vernacular European literature. Several centuries later, William Shakespeare also employed the word in this sense, as these lines from *Hamlet* testify: "O gentle son / upon the heat and flame of thy distemper / sprinkle cool patience" (1599-1602: 3.4.122-124). When discussing more contemporary uses, however, even the *OED* seems to be hard put, suggesting a wild array of near-synonyms, such as clever, stylish, shrewd, sexually attractive, admirable, fashionable, up-to-date, and classy. As both *Fast Company*'s coolhunters and *OED* researchers suspect, cool is all that and much more.[5]

The online resource *Urban Dictionary*, in this case perhaps more up to the challenge, lists more than 160 definitions of the term, the most popular (by common vote) being: "The best way to say something is neat-o, awesome, or swell" (*Urban Dictionary*, s.v. cool), which references the term's suitability as an elaborate term of approval or even indeed an empty conversation filler, used abundantly by teenage speakers. This weakened usage has almost completely lost its former connotations of physical and/or emotional coldness, detachment, and nonchalance, but testifies to the global and so far unparalleled success of the term as an all-purpose attribute. Indeed, it seems that cool has had to undergo severe linguistic drainage in order to achieve its current popularity, resulting in an almost limitless applicability. In that sense, its semiotic force might be reduced to that of a positive epithet, a "verbal tic" (Gioia, "Death of the Cool" 1), or a mere interjection.[6] Not only, as Gabriele Mentges observes, is the range of practices which cool may nowadays be applied to almost infinite, but it also produces, in a kind of eternal loop, ever new meanings: "[T]he unrestrained use of cool strengthens its connotational variety and semantic reach and generates new meanings and lexical fields" (26).[7]

It is with regard to this hermeneutic circle of cool that the *Fast Company*'s chart reveals its deeper truth: Cool, as Dick Pountain's and David Robins's comprehen-

5 As a case in point, most studies concerned with the historical dimensions of cool agree on its connotations of rebelliousness, defiance, and subtle irony that define a cool character, attributes which remain conspicuously absent in the *OED* definitions (see, for instance, Pountain and Robins 30).

6 See, for instance, the collocations 'cool cat' and 'cool costumer' or cool as an expression of consent, signifying little more than 'Ok! Alright!' See *OED*, s.v. cool, for further uses.

7 Original wording: "Allerdings werden durch diese offene Verwendung die Konnotationsvielfalt und Bedeutungsreichweite von Cool gestärkt und neue Sinngebungen und Bedeutungsfelder generiert." The example Mentges gives is the British Labor Party's 1990s slogan 'Cool Britannia,' which created a connection between future, consumerism, and technology, all of them united through the virtue of cool (26). While several studies choose to assess this trend as inherently negative (see, for instance, Dietrichsen 243), Mentges refrains from dismissing the concept has utterly drained; rather, she understands cool's multifacetedness as symptomatic of a complex time and age (35).

sive study *Cool Rules* argues, must be understood as "a meta-code that cuts across all the many subspecies of teenage argot" (30). This claim is supported by several linguistic studies, such as Teresa Labov's (1992), which concludes that cool and its connotations are understood in practically all teenage cultures on the North-American continent. The obvious ubiquity and social penetration of the term seem to be tightly linked to its lexical elasticity, or, as Diedrich Dietrichsen puts it, "maximum degree of drainage" (243).[8] Nonetheless, Pountain/Robins agree with most other etymological analyses that cool is far from being the fancy teenage equivalent of 'good' without deeper implications. It comes with a rich cluster of associations and a clear agenda. "Using the word," as Peter Stearns maintains, "is part of the process of conveying the right impression" (1). Marcel Danesi's *Cool: The Signs and Meanings of Adolescence* (1994) supports this argument, as it shows that coolness (in this case employed by 1990s teenagers) depends on a variety of semiotic codes, including body image, social cognition, and language. A vital part of coolness is the use of the right idiolect, whose adolescent varieties Danesi calls "pubilect" (96). The word and the deed, bluntly put, complement one another.[9] Accordingly, the linguistic complexity of cool already implicates the mutability of its ideational scope.

MADE IN THE USA? CULTURAL ORIGINS OF COOL

While contemporary scholarship, according to the above survey, seems to struggle with grasping cool's vast semiotics, locating its conceptual rather than etymological origins appears to be a much easier task. A variety of historical studies trace the roots of modern-day cool back to African or African-American cultures respectively. One of the earliest in this vein is "An Aesthetic of Cool" (1973) by the eminent art critic Robert Farris Thompson, as well as his subsequent monographs *African Art in Motion* (1979) and *Flash of the Spirit* (1984). According to Thompson, coolness in body, mind, and spirit is one of the most important religious virtues, an "all-embracing aesthetic attitude" ("Aesthetic" 41) pervading West and Central African

8 Original wording: "Coolness ist ein Begriff, der heutzutage einen maximalen Grad an Entleerung erreicht hat."

9 Mentges's conceptual history supports this idea: "The step from the term cool as mere adjective to the noun coolness must be seen as a qualitative differentiation: Coolness is not only the proper noun, but always already denotes a disposition of qualities with conceptual status" (19). Original wording: "Dabei ist der Schritt von dem Wort cool als einfachem Adjektiv hin zur substantivierten Form der Coolness als qualitative Unterscheidung zu verstehen: Coolness ist nicht nur das einfache Nomen, sondern bezeichnet bereits eine Disposition von Eigenschaften mit konzeptionellem Status."

societies in a myriad of idioms and rites. One of the most prominent is the concept of *itutu* of the Yoruba tribe, an attitude of detachment worn by warriors in battle and associated with the color blue. In *Cool Pose*, Richard Majors and Janet Mancini Billson support Thompson's argument of African philosophy as a forerunner of today's cool and stress its significance as an emotional mask: "Coolness means poise under pressure and the ability to maintain detachment, even during tense encounters" (2).[10] The attitude was allegedly imported via the slave trade to the North American continent, where it served the plantation workers as a "survival strategy" (Majors and Mancini Billson xi) against the horrors of slavery. In times of physical harassment and exploitation, keeping a cool pose allowed them to preserve their spiritual integrity and conceal their contempt for the white owners, potentially dangerous emotions inviting harsh punishment (Pountain and Robins 38).

Thompson, however, also maintains that African cool "is worn not only in time of stress, but also of pleasure" ("Aesthetic" 41), a usage which seems to resurface in the centrality of the cool pose for blues and jazz musicians. Several studies have explored the emergence of coolness as a (predominantly male and African-American) attitude in the jazz club milieu of the early twentieth century, including Lewis MacAdams's *Birth of the Cool* (2001), Ted Gioia's *The Birth and Death of the Cool* (2009), Marlene Kim Connor's *What is Cool? Understanding Black Manhood in America* (1995), Pountain/Robins's aforementioned *Cool Rules* (2000), as well as Joel Dinerstein's forthcoming *The Mask of Cool: Jazz, Film Noir, and Existentialism in Postwar America.* Marcel Danesi's *Cool: The Signs and Meanings of Adolescence* (1994) locates the origins of jazz musicians' cool stance in the habit of opening the windows to cool the air in the heated and smoke-filled jazz clubs. With reference to Tony Thorne's *Dictionary of Contemporary Slang* (1990), he argues that "the slow and smooth jazz style that was typical of that late-night scene came to be called 'cool'" (37).[11] Indeed, the particular musical style widely known as Cool Jazz was incubated by trumpeter Miles Davis, who recorded the famous compilation album *Birth of the Cool* (1957).[12] Together with Malcolm X, another key figure of early twentieth-century black aesthetics, Davis epitomized the invulnerable pose that Majors and Mancini Billson define as vital for the performance of

10 Majors and Mancini Billson emphasize the survival of cool in the fetishism and talismans of African voodoo, a usage which is supposed to date back to as far as the third century B.C. (57).

11 Pountain and Robins remain skeptical toward this myth of origins, to which they attribute a "slightly bogus feel" (32). They do, however, support Danesi's argument of 1930s jazz club culture as a main incubator of modern-day coolness.

12 In contrast to the preceding, very fast and energetic Bebop, Cool Jazz is slower, subdued, and calm. See, for instance, Ted Gioia's elucidative chapter on "Cool Jazz and West Coast Jazz" (332-342) in *The Oxford Companion to Jazz* (ed. Bill Kirchner, 2000).

black masculinity; according to George Elliott Clarke' "Cool Politics," X's and Davis's "public postures comprised [...] the essence of cool" (par. 15).[13]

In contrast to these explorations of cool as an African-American attitude, fewer studies have attempted to unearth European influences on coolness, such as the aristocratic disdain of *sprezzatura* maintained by Italian courtiers during the Renaissance (Pountain and Robins 53-55) or the virtues of apatheia and ataraxia promoted by Greek stoicism (Sommer 30-44). Similarly, the Romantic archetype of the cynical and emotionally aloof Byronic hero seems to resurface in many contemporary cool characters in fiction and film, such as Batman or *House, M.D.* The nineteenth-century dandy, on the other hand, reflects the centrality of decadent self-stylization through fashion, art, and physical comportment (Geiger, Schröder, and Söll 13).[14] Various avant-garde movements of European modernism, including Dadaism (Pountain and Robins 56), 1920s architecture of the *Neue Sachlichkeit* (Söll 149-151), or Bertolt Brecht's cynical characters in "In the Jungle of Cities" (1927) and *The Threepenny Opera* (1928) (Hofmann 127-137), also display styles and demeanors we might nowadays interpret as cool. Gerald Schröder, on the other hand, argues for an understanding of coolness "as a product of the Cold War, at least when it comes to its popularization as a concept" (183).[15] All of these philosophical trajectories ought, however, best be understood as influences on, rather than direct predecessors of contemporary notions of coolness. The Greek and Italian, Romantic and Victorian versions of emotional detachment and proud aloofness all reflect some aspects of twenty-first-century cool, while several crucial others remain notably absent.

Whatever the historical origins of the term, it seems safe to assume that coolness in the sense in which it is widely understood and used today emerged as a by-product of the rise of youth culture from the 195os onwards. 1950s actors and singers, such as James Dean, Marlon Brando, and Elvis Presley, were immortalized as timeless icons of coolness. In these formative years, cool became a substantial part

13 For this aspect of cool, see also bell hooks's *We real cool: Black men and masculinity* (2004). Studies like hooks's, Majors and Mancini Billson's (1992), and Connor's (1995) make us aware of cool's gendered nature, in addition to its racial dimension. Connor, for instance, goes as far as claiming that "for black men: Cool essentially defines manhood" (19). Theories on the reasons for women's alleged lack of coolness abound. See Pountain and Robins 136-142.

14 A more comprehensive overview of potential European and other influences on contemporary cool would by far exceed the limits of the present study. More information can be obtained in Pountain and Robins (52-70) and Geiger, Schröder, and Söll (105-163).

15 Original wording: "Somit kann Coolness in gewisser Weise als Produkt des Kalten Krieges beschrieben werden, zumindest was die Popularisierung des Konzepts in dieser Zeit anbelangt."

of the avant-garde art scenes in Europe and North America, as Herbert Gold argues in *Bohemia: Digging the Roots of Cool* (1996). Notorious personalities like William S. Burroughs and Allen Ginsberg cultivated coolness as part of their nonconformist lifestyle, disaffected by the demands of respectable middle class society. It is in the time of Dean's *Rebel Without a Cause* (1955) and the beatnik's unabashed attack on the loathed bourgeoisie that coolness gained its particular flavor of rebelliousness and individuality, a quality that sets its owner apart from the masses. Several youth sub cultures, be it the hippies of the 1970s or the punks of the 1980s, made cool an integral part of their philosophy, although they frequently chose to rebaptize it. Indeed, while the ubiquitous expression 'cool' might, contrary to the findings visualized in *Fast Company*'s fancy graph, have gone in and out of fashion as the preferred term of approval fairly frequently, its appeal as a carefully cultivated personality trait seems to be have thoroughly consolidated over time. According to Ted Gioia, the rapid and unprecedented rise of cool after 1945 is closely related to the "growing importance of the concept of lifestyle" ("Death of the Cool" 44), a luxury which teenagers of an earlier and less affluent generation could not afford to worry about.

Consequently and despite its alleged origins in sub- and countercultural milieus, cool was quickly adopted by mainstream society as a particularly attractive form of self-fashioning. The cultivation of a cool and supposedly individualistic lifestyle was greatly abetted by several other emerging phenomena of postwar society, as Thomas Frank's insightful study *The Conquest of Cool* (1998) perceptively shows. Especially the rising business and consumer culture of the 1960s, epitomized by Madison Avenue's booming advertising agencies, helped to enhance the co-option of the cool rebellion of alternative youth cultures by the capitalist mainstream. Frank discusses intriguing examples, such as Pepsi's fictional youth movement The Pepsi Generation (170), which reflects a style of advertising that "positioned products as bearers of nonconformity, escape, resistance, difference, carnival, and even deviance" (133), all of which might well be classified as essentially cool qualities. Drawing on Frank's landmark study, Jim McGuigan's *Cool Capitalism* (2009) explores how the incorporation of cool legitimizes contemporary corporate interests. Nonchalant disaffection is turned into hedonistic acceptance of the values promoted by the capitalist market. Cool, as *Fast Company* is acutely aware of almost half a century after the emergence of a postwar consumer ethos, has become one of the most celebrated cash cows of the West's free market economy.

Kalle Lasn's *Culture Jam: The Uncooling of America* (1999) and Daniel Harris's essay "Coolness" (1999) similarly identify contemporary coolness with conspicuous consumption and commercial exploitation. The attitude formerly fostered by rebels, criminals, slaves, and underdogs has fast devolved from a survival into a marketing strategy, a "heavily manipulative corporate ethos" (Lasn xiii-xiv), which only seemingly promotes rebellion, individuality, and nonconformity. The kind of cool employed by modern advertising succeeds in transforming "shopping [...] [into] nothing less than a subversive form of civil disobedience" (Harris 47). Pountain and Robins even go as far as concluding, perhaps a tint too bleakly, that "[c]ool as an ethic is exquisitely suited to a life of consumption" (166). The history of cool since its rapid infusion into 1950s youth culture is one of commercial co-option, commodification, and semantic diffusion, which might have just reached its climax in the twenty-first-century information society and the latter's obsession with cool electronic gadgetry. A conspicuous feature of contemporary cool thus seems to be a certain degree of "depersonalization" (Gioia, "Death of the Cool" 4) or reification, resulting in its increased use as an attribute of inanimate objects rather than a personal character trait.

CONTEMPORARY COOL: DIRECTIONS, TRAJECTORIES, AND DEAD ENDS

So far, this survey has focused on (largely diachronic) explorations of coolness as an individual attitude or strategy, deliberately or unconsciously employed to convey an impression of aloofness, emotional detachment, and hedonistic self-indulgence, frequently in order to hide boiling and potentially destructive feelings underneath. The majority of scholarship does indeed exploit these dimensions of coolness, which is, as Pountain and Robins's study postulates, "not inherent in objects but in people" (21). Paralleling these explorations is a range of accounts focusing on more contemporary uses of cool as a cultural, rather than personal phenomenon that may be understood as symptomatic of an era dominated by technology and indirect modes of communication and social interaction. A particularly instructive study is Poschardt's *Cool* (1999), which harks back to the notion's etymological roots denoting low temperature. By analogy, the contemporary pervasiveness of coolness is indicative of the social temperature of today's civilization, which Poschardt illustrates with acute analyses of popular cultural productions, ranging from cool movies to comic heroes combating mutated polar bears. In these renderings, cool is understood as "the attempt to counter the ice ages of our existence with affirmative strategies" (Poschardt 10),[16] a definition which seems to echo African notions of cool composure. Poschardt locates the origins of this markedly aesthetic attitude in the budding mass society of the early twentieth century; its full bloom, however, coincides with the social alienation experienced in the late twentieth century.

In a similar vein, Peter Stearns's *American Cool* (1992) constructs the notion as deeply entwined with the affective milieu of the 1990s. The rise of cool to a domi-

16 Original wording: "der Versuch, den Kältepassagen der Existenz affirmative Strategien entgegenzusetzen."

nant sensibility of contemporary American society entails a decisive break with earlier Victorian mentality, which can be characterized by the heightened value it attributed to (the expression of) passion. Today, emotional restraint and disaffection rank among our society's most cherished virtues. Coolness is thus not an affirmative attitude or welcome strategy for survival, but becomes the prevailing (and in Stearns's opinion problematic) mode of comportment in a culture embarrassed by emotional excess. While Poschardt, however, chooses to view cool as a global phenomenon, ranging from Russian opera to French novels, Stearns maintains that "[t]he concept is distinctly American, and it permeates every aspect of American culture" (1). Quite contrary to this programmatic definition, Stearns's analysis focuses on a very specific segment of American culture, namely the white middle class, thus neglecting the very stratum of society which, according to the historical accounts discussed in the previous section, functioned as the chief incubator of this coolness.

A somewhat underexplored area of research is the use of cool as an aesthetic phenomenon. A prime example of scholarship in this field is the aforementioned collection of essays edited by Annette Geiger, Gerald Schröder, and Änne Söll *Coolness: Zur Ästhetik einer kulturellen Strategie und Attitüde* (2010), which includes a range of insightful contributions examining coolness in particular contexts, such as literature, painting, photography, film, advertisement, and popular culture. Cool is here understood as a central category of the twentieth and twenty-first century, which the collection aims to analyze by investigating its aesthetic dimension. The rationale for this approach is connected to the fact that coolness "is always about performances - of body and fashion, habit and attitude, style and changes in style and many more" (Geiger, Schröder, and Söll 8).[17] The collection's focus is clearly on the modernist period as the first age of mass media and technical reproduction, frequently described in contemporary literature as cold and thus "particularly relevant [...] for the development of cool strategies" (Geiger, Schröder, and Söll 11).[18] In contrast, coolness after 1945 is understood as the postmodern equivalent, less informed by existential threat than demands of individualization and self-stylization (Geiger, Schröder, and Söll 15-16). Methodologically, the collection resorts to Marshall McLuhan's distinction between hot and cool media. In his landmark *Understanding Media* (1964), McLuhan establishes cartoons and television as examples of the latter, since in these media "so little is given and so much has to be filled in by the listener [or viewer]. On the other hand, hot media do not leave so much to be filled in or completed by the audience. Hot media are, therefore, low in

17 Original wording: "bei [Coolness] handelt es sich stets um Inszenierungen—von Körper und Mode, von Habitus und Attitüde, von Stil und Stilbruch und dergleichen mehr."

18 Original wording: "so relevant [...] für die Herausbildung von coolen Strategien."

participation, and cool media are high in participation" (1964: 36).[19] Accordingly, the majority of contributions focus on medial-aesthetic substantiations of cool.

A comparable approach is taken by Jeff Rice, whose exploration of cool as a writing style also draws on McLuhan as well as Jean Baudrillard's modification of this distinction in *Symbolic Exchange and Death* (1976). According to Baudrillard, it is the "extermination [...] of the real of production and the real of signification" (*Symbolic Exchange* 7) that must be understood as cool; coolness is thus a form of expression or discourse which exemplifies the way in which "signs are exchanged against each other rather than against the real" (*Symbolic Exchange* 7), i.e. they are no longer associated with referents and have become indeterminate. The already discussed lexical drainage of cool thus serves as a starting point for Rice's analysis of a particularly cool electronic rhetoric and writing style, which is produced by merging various discourses, practices, and theories. Hence, he urges his readers to think of cool "as a juxtapositional practice" ("Writing" 224), "creating associations and emotional responses out of the combination of unlike words and images" ("Writing" 233). The corporate appropriation of cool, discussed by Frank, McGuigan, and others, is primarily understood as a rhetorical one: Coolness is not only a lifestyle, an emotional mask, or a strategy of comportment, but a language, generated by new connections in the conjunction of consumerism, youth culture, fashion, and the web. Such a cool verbal aesthetic, as Rice maintains, can be produced through working with hypertext, a form of writing that allows for a high degree of juxtaposition. Again, his definition constructs cool as a central characteristic of the information age, symptomatic of "a world where everything is interconnected" ("What is Cool?" par. 4) and tightly linked to the ever evolving electronic culture.

Alan Liu's *Laws of Cool* (2004) supports this line of argument by establishing cool as "the nascent, everyday aesthetics of knowledge work" (8) and a fundamental part of contemporary information society. His study connects the aforementioned strands of research by advancing an understanding of cool as the leading currency of the information age with a significant aesthetic dimension, which he closely analyzes in several pertinent examples. A vital characteristic of this new aesthetic, which has virtually replaced all other ethic, aesthetic, and cognitive categories, is unknowability, or, in Liu's words, "the ethos of the unknown" (9) potentially spread by cool practices. This harks back to the already discussed lexical mutability and elusiveness of cool, which can never be accurately known and defined. Cool may provide spaces of imprecision and obscurity in a culture obsessed with information transmission, manipulation, and distribution, thus impeding, parodying, or

19 Instances of hot media would be radio, photography, and printed matter. For a historical example, see Lewis MacAdams's exploration of hot and cool media in relation to the pivotal 1962 CBS broadcast from the White House (220-221).

simply refusing exact knowledge (9).[20] In that sense, cool is the dominant social imperative (or 'law,' as Liu's title suggests) of information society, while simultaneously committing acts of destruction against what is most valued in it, so that it "at once mimes and critiques knowledge work" (Liu 327), a definition which echoes Rice's understanding of juxtaposition and contradiction as prerequisites for cool. Examples are, above all, digital practices, such as cool graphics on the web. New technologies are thus identified as central to contemporary cool,[21] which thrives on the very mainstream consumer gadgetry, from game consoles to high-speed processors, it had hitherto been debarred from or had rebelled against.

WHAT IS COOL? A SUMMARY

Despite its obvious heterogeneity, the above survey may nonetheless reveal several basic and commonly agreed upon premises about the subject under scrutiny. First of all, it suggests that the uses of cool can be grouped into the following broad categories: cool is an aesthetic category; a personal attitude, pose, or strategy, tightly linked to a very specific (yet potentially changing) set of behavioral standards and character traits; a dominant cultural discourse of postmodern information society, illustrative of a consumer ethos hinged on technology, digital expression, and pseudo-rebellious attitude; finally, a virtually empty signifier, expressing admiration, consent, or agreement, applicable in almost any social interaction. The above discussed analyses all concur in one or a combination of two of these categories, while the occasional side glance to any of the other definitions is the rule rather than the exception. Already the exact nature of cool's ontological status, thus, seems to evoke highly convergent answers from its researchers.

Of course, this may well have to be understood as stemming from the concept's inherent vagueness, which interestingly leads to the first conclusion the majority of scholarship seems to arrive at when striving to unravel its underlying dynamics and defining characteristics: What appears to define cool, paradoxically, is its mutability and "playful fluidity" (Gioia, "Death of the Cool" 4), in other words its very resistance to definition, a supposition in which virtually all the studies discussed above concur. The preceding survey indeed implies that the term's omnipresence today has happened not despite but because of its structural flexibility, which allows for ever changing and evolving areas of application: "The diffuse semantics of the notion obviously act as necessary conditions for its very popularity and force,

20 According to Liu, "traits of efficient information [such as relevance, clarity, and utility] are parodied, misapplied, radicalized, and otherwise used against the grain" (186).

21 See Mentges (26), who observes a tight connotational web of future, consumerism, and technology.

while making it notably complex" (Geiger, Schröder, and Söll 7).[22] Rice calls cool "allusive and indefinable, its meanings perpetually shifting" ("What is Cool?" par. 3), while Mentges even suggests that the only means of tackling the obscure workings of cool is an elaborate chain of associations that will allow the positioning of oneself in this vast discursive field (33). Indeed, contemporary cool seems to act almost like a free-floating signifier that may be attributed to diverse scenarios, hardly subject to limitations: It can (simultaneously or consecutively) be applied to fashion, music, and advertisement, to jazz, punk, and hip hop, to cowboys, movie stars, and prison inmates, to rebellion, outsiderdom, and mass consumerism, as the above discussed examples have shown. It seems, in short, impossible (nor indeed desirable for those thriving on it) to define cool, which paradoxically constitutes a major part of its definition.

Figuratively speaking, cool could thus be classified as exuding an inescapable aura of unknowability and obscurity, which is, for better or worse, extended to its bearer: A cool character is always already surrounded by an air of mystery and secrecy, convincingly conveying the impression of possessing some deeper knowledge not disclosed to the ordinary onlooker. The above survey implies that cool sharply distinguishes between those who are familiar with and those who will forever remain ignorant of it: "Coolness is an aesthetic of insignia, […] of the passwords and Masonic handshakes through which card-carrying initiates gain entrance to the clubhouse" (Harris 48). Consequently, cool in the twenty-first century is necessarily entwined with power, a social imperative to acquire literacy in deciphering and employing its codes; it relies on a basic fluency in the relevant styles and demeanors, whose adoption equals "a process of becoming literate, the acquisition of a cultural technique" (Geiger, Schröder, and Söll 16).[23] In the information age, this fluency extends to the command of digital devices, but also, more substantially, to a basic understanding of as well as an ability to participate in the economic exploitation of knowledge. Eventually, this once again results in a paradoxical situation: While being cool equals being conversant in a particular style, the insider knowledge from which it derives will never be divulged or articulated. Liu aptly illustrates this understanding of cool as advanced in his study by discussing cool online: "When such sites [i.e. the coolest web pages by common vote] are most seriously, deeply cool—no information is forthcoming. Cool is the aporia of information" (179). Opaqueness, impenetrability, or downright resistance to (the transmission of) information should hence be taken note of as vital ingredients of the amorphous concept. In logical reverse, this also means that clarity, comprehensibil-

22 Original wording: "Gerade die diffuse Semantik des Begriffs bedingt offenbar seine Beliebtheit und Stärke, macht ihn jedoch auch ausgesprochen komplex."

23 Original wording: "ein[…] Prozess der Alphabetisierung, […][das] Einüben einer Kulturtechnik."

ity, directness, and the unrestricted diffusion of information, so central to contemporary cubicle culture, are impedimental rather than conducive to the truly cool pose.

A second principle of cool that emerges from the wealth of scholarship is closely related to its apparent ambiguity. The impression of obscurity and resistance necessarily maintained by coolness seems to stem from the fact that a cool character is compelled to encompass, smoothly and without evident effort, contradictory notions simultaneously. An important concept here is that of the ironic mask, which several astute analyses have pinpointed as a necessary prerequisite for any cool persona. Already its early uses in African tribal rituals establish the cool face in the presence of danger as the proficient embodiment of mutually exclusive affects, showcasing emotional composure while suffering from great agitation and distress (Thompson, "Aesthetic" 41). Today, one may discern this paradox already underlying the very production of any genuinely cool attitude: While cool, in its most basic sense, may be defined as the art of indifference, one must invest quite a considerable amount of thought, time, and money in order to accomplish this high degree of seeming complacency. Indeed, contemporary cool is perhaps best characterized by its ironic (and contrived) reversal of polarities, reminiscent of its original African uses, which shows itself, for instance, by "feigning boredom in the face of danger, or amusement in the face of insult" (Pountain and Robins 26). Substituting or at least contrasting one sentiment with its very opposite seems vital for a cool pose; a cool character, by analogy, must be prepared to hoard conflicting, if not seemingly incompatible qualities at the same time.

Several studies have been dedicated to the exploration of this "ambiguous sensibility [...] that lies beneath Cool" (Pountain and Robins 243), such as Majors and Mancini Billson and bell hooks, who both suggest that cool as a dominant form of social expression among African-American men may result in hyper-masculine behavioral patterns, ultimately damaging for the emotional set-up of its bearer and his relationship to fellow human beings (Majors and Mancini Billson xi; hooks, *we real cool* 111). In a wholly different context, Liu also maintains that cool necessarily contains the potential for both creative and destructive tendencies, describing it as "a paradoxical ethos" (72), which is "[a]t once wise to it all and profoundly ignorant, simultaneously arrogant and vulnerable, and [...] both strangely resistant to and enthralled by the dominating information of postindustrial life" (78). The globalized capitalist market, which postmodern society is anchored in, additionally emphasizes the contradictory sensibility cool is predicated on: Commercially co-opted cool enhances conformity and the adoption of a consumerist life style, while still, however faintly, promising rebellion and detachment, as if one was exempt from the temptations of square advertisement. As Frank's, McGuigan's, and Lasn's perceptive studies of contemporary corporate culture show, cool is the magic potion luring us into buying a mass-produced pair of sneakers for accentuating our individuality: "[C]oolness cleverly disguises [uniformity] as its opposite, nonconformity" (Harris 49). Juxtaposition of conflicting or indeed contrary ideas ought to be regarded as vital for cool. Rice's elucidative work proficiently illustrates how juxtaposition functions as a cool practice by analyzing the process of iconization which all the cool protagonists of twentieth-century popular culture, be it James Dean, Michael Jackson, or Elvis Presley, have undergone: The detached, incessant reproduction of the figures through a variety of forms and media results in a collage of representation, an unlikely arrangement, which only serves to enhance the solidification of the respective celebrity as iconic and cool ("Writing" 230).[24]

The above list of cool icons coincidentally reflects another central characteristic of cool which most academic investigations seem to agree on: Cool is a distinctly US-American phenomenon, which establishes it as an integral and noteworthy field of research for American studies. Not only does its modern-day usage date back to African-American musical innovations and the concomitant jazz age, it is also US-American rather than Western European society which first gave rise to the emergence of distinct youth cultures and the concept of lifestyle (Gioia, "Death of the Cool" 23-42). Diverse channels, including rock'n'roll music, fashion, and teenage jargon, as well as easily accessible media like MTV, have served, through the decades, as sources "for the mimetic spread abroad of North American teen models of behavior" (Danesi 46). The consumption of American products, from jeans to television, was eagerly seized on as a pathway into the very nucleus of cool production. "Americans [...]," as art critic and cultural theorist Dave Hickey coolly asserts in the late 1990s, "are best at cool" (n.p.). The abundant use of cool for advancing corporate interests, from Coca Cola to the iPhone, solidifies cool as a US-American concept, so much indeed that one might be inclined to view it as "nothing more than US cultural imperialism: [...] it is simply American popular culture exported around the world" (Pountain and Robins 23). From Hollywood stars to rock music to must-have electronic devices out of Silicon Valley, almost any reification of cool bears some observable relation to the United States as its mother country. Television and film productions like the ones analyzed in the present study may equally be understood as capitalizing on the general appeal of American popular culture, the epitome of cool, for drawing an international audience.

---

24 Rice here refers to Greil Marcus's eminent *Dead Elvis* (1991) and his analysis of the singer as the ultimate reified celebrity, who keeps reappearing through images, texts, and other media expressions in various ways. The "cultural joining of Elvis in an Energizer commercial with Elvis the alarm clock with Elvis the singer" (Rice, "Writing" 230) creates a discourse of juxtaposition that solidifies his iconicity, which Rice understands as deeply entwined with cool.

Not only has cool therefore "seized a central place in the American imagination" (Stearns 1),[25] but it is also, conversely, the concept of America that continues to loom large whenever cool is evoked.[26] Not coincidentally, thus, the majority of scholarship focuses on US-American society, be it Stearns's analysis of dominant sensibilities in the American white middle class, Majors and Mancini Billson's account of African American masculinity, or Frank's, McGuigan's, and Lasn's explorations of business culture.[27] However, processes of globalization and informatization have clearly affected cool and caused a perceptible adjustment, if not indeed wholesale weakening of its originally Americocentric perspective; as a dominant concept of consumer culture, cool has become intelligible almost worldwide.

Today's global appeal of cool suggests another premise which most analyses so far have defined as central to the study of cool, namely that this elusive and imprecise notion constitutes a fundamental part of today's zeitgeist: "Despite the vagueness of its definition, coolness must be regarded as a central category of the twentieth and twenty-first century, which has […] significantly shaped the cultural self-conception" (Geiger, Schröder, and Söll 7).[28] The wealth of academic disciplines and research areas involved testifies not only to the pervasiveness and indeed inescapability of cool within academia, but, more importantly, to its momentousness in

---

25 This is also supported by Mentges, who shows that internationally, the term cool has still not acquired the cultural force and pervasiveness it enjoys in (mainstream) American society: "In German, cool remains in the domain of jargon […], explicitly referencing youth culture" (25), while these restrictions do not exist in its US-American usage. Original wording: "Im Deutschen verhaart Cool eher im Bereich des Jargons […], explizit auf die Jugendkultur bezogen."

26 The centrality of cool for the American imagination was recently illustrated by the 2008 presidential elections. "Does Obama make it 'cool' to be American?" *Fox News* asked after the country's reputation had been seriously tarnished by George W. Bush's presidency. While cool here is apparently used in its weakened sense, merely indicating approval and endorsement of the election results, it is still interesting to note that even conservative mainstream news stations choose to use 'cool' before any other adjective. Cf. *Fox News*, aired on 9 November 2008, accessed 14 January 2012, <http://video.foxnews.com/v/3917474/americas-coolagain>.

27 A major exception is Pountain and Robins's *Cool Rules*, which is not only one of the first monographs to study cool in its entirety, but also from a notably British perspective. Similarly, the edited collection by Geiger, Schröder, and Söll as well as Poschardt's study of cool as an affirmative strategy discuss diverse national scenarios, with a special focus on (European) modernism.

28 Original wording: "Coolness muss trotz aller Vagheit in der Definition als eine zentrale Kategorie des 20. und 21. Jahrhunderts betrachtet werden, die das kulturelle Selbstverständnis […] maßgeblich geprägt hat."

contemporary American society and beyond. Research on cool has convincingly demonstrated that a deeper understanding of the evasive concept can help to illuminate dominant and emerging cultural narratives, trends, and discourses. As one of the major affective and aesthetic categories of the present, cool captures "the interpersonal climate in modern mass societies" (Poschardt 9),[29] while its semantic fluidity mirrors the growing complexity and changeability of our social, economic, and political relations. In so far, cool "gives the epoch a name" (Mentges 35)[30] and emerges as "the cultural dominant of our time" (Liu 76). The ever tightening web of corporate exploitation, youth culture, consumerism, digital practices, and new technologies has endowed cool with a wealth of new associations and markedly contributes to its proliferation in the new millennium. As several studies observe, new media and electronic culture are central components of the twenty-first-century understanding of cool (see, for instance, Mentges 26), making it the key asset of the digital citizen. In that sense, cool comes to denote the general temperature of social life in the mediatized age, which is governed by indirect modes of communication.[31] It is thus only consistent to conclude that cool is "in the process of becoming the dominant type of relation between people in Western societies, a new secular virtue" (Pountain and Robins 19).

There are, however, also studies which heavily dispute this prediction and rather suggest that cool is past its prime. Annette Geiger, for instance, observes a "postmodern rise in temperature" (87),[32] initiated as early as 1968 by the flower power movement of the hippies[33] and now culminating in the emotional backlash after 9/11 and the credit crunch.[34] Barack Obama's election as the first US president of

29 Original wording: "das zwischenmenschliche Klima in modernen Massengesellschaften."

30 Original wording: "gibt der Epoche einen Namen."

31 In an inspiring special edition on the first century of the new millennium, the German newspaper supplement *Zeit*-Magazin suggests that "the engines of coldness are now glacial white and called iPhone and iBook [sic] and iPod" (Amend, Illies, and Stelzer par. 1). Original wording: "die Kältemaschinen sind diesmal gletscherweiß und heißen iPhone und iBook und iPod."

32 Original wording: "postmoderne[r] Temperaturanstieg."

33 In contrast to Geiger, Pountain and Robins interpret the hippies as simply crafting their own version of cool. 1960s and early 1970s youth culture looked, thought, and acted very differently from the preceding beat generation, "but beneath such changes of image persists that blend of hedonism and narcissism that makes up Cool" (Pountain and Robins 72).

34 9/11 is regularly evoked as a caesura in the social climate, both in America and beyond. *The New York Times Book Review,* for instance, reports that "[s]ince 9/11, American readers have shown an appetite for simple tales told with becoming directness" (Turow par. 1). See also Gioia, "Death of the Cool" 3. In contrast, the German newspaper sup-

color is supposed to support this hypothesis, despite cool's centrality for African-American masculine behavioral codes.[35] Cool is 'out,' both as an aesthetic and affective category, and the new generation has well recognized that it "had degenerated into fashion attitudes" (Geiger 90).[36] A new social awareness among the young, such as the Occupy movement or MTV's campaigns for climate protection, suggests that coolness gives way to political commitment and charity, thus transcending former ironic detachment and aloofness.[37] Geiger's hypothesis is prominently supported by Ted Gioia (2009), who vehemently announces the death of cool as part of a wholesale cultural paradigm shift starting in the 1990s. The values that cool allegedly stands for, such as irony, sarcasm, emotional restraint, narcissism, and artifice, are nowadays viewed with suspicion, while sincerity, authenticity, and simplicity are fast becoming the preferred attitudes for both personal interaction and public discourse. Following this supposition, the growing appeal of social media must not be understood as enhancing cool by enabling an unprecedented degree of narcissistic self-stylization and individualization (allowing virtually everyone to be an artist, and, by analogy, exceptional), but ought to be interpreted as an expression of the users' desire to "experience the creativity of real people in real-life situations" (Gioia, "Death of the Cool" 16). The alleged death of cool is also reflected by popular culture, for example with the emergence of television formats "celebrating nerdiness" (Gioia, "Death of the Cool" 8), which portray the fictional lives of the utterly uncool. According to this cultural reading, postmodern society has irretrievably lost both the desire and the capacity for cool, alienated, and consumerist ways of living, and, much to the contrary, embraces passion and social commit-

plement Zeit-Magazin understands the period between 2001 and 2008 as a particularly cool one. "[A] new warm period" (Amend, Illies, and Stelzer par. 5, original wording: "eine neue Wärmeperiode") only starts with the election of president Obama, the global collapse of the financial system, and increased political efforts to stop the progressing climate change.

35 The styles and symbols which John Street (2006), for instance, defines as vital for the transformation of politicians into veritable pop culture icons can be detected in the public appearance of Barack Obama, who portrays several key characteristics of the charismatic political persona. See also Mentges 17-18.

36 Original wording: "die coolen Attitüden [...] waren zu Mode-Attributen degeneriert."

37 Pountain and Robins argue that political commitment must necessarily be hot about its cause, rather than cool: "Cool is never directly political, and politics, almost by definition, can never be Cool. To get anywhere in politics you need to care passionately about something" (171). Conversely, John Street's perceptive exploration of (celebrity) politicians' self-representation defines cool as "one of the possible aspects of contemporary political style" (369).

ment, holds no interest in style, attitude, or individuality, and successfully manages to resist the lure of false icons and brands.

THE FORMULA FOR COOL: TECHNOSCIENCE, INFORMATION AESTHETICS, AND THE RISE OF THE NERDS

The above synopsis of Geiger's and Gioia's argument, admittedly subjected to a tinge of irony, possibly already suggests that the present study does not support their point of view on the supposed demise of cool. A mere look at the preferred pastime activities in the twenty-first century should suffice to inspire a serious revision of their original hypothesis.[38] When Gioia lauds the return to authenticity and sincerity, one cannot help but wonder what exactly could actually be viewed as 'real' in the age of reality shows, online dating, and Photoshop. While simplicity, as Gioia correctly observes, might well emerge as a new ideal in advertisement and marketing, cultural practices intent on perfectly camouflaging the real by paradoxically enhancing the impression of authenticity flourish. As Frank's and other perceptive studies on the commoditization of cool have shown, coolness acts as a governing principle of postmodern existence, a corporate ethos, and pervasive cultural discourse that came to stay. In the age of iPhones and iPads, our new "engines of coldness [...] [in] glacial white" (Amend, Illies, and Stelzer par. 1), the cool appeal of electronic gadgets, digital media, and, more generally, progress in information technologies is irresistible. "[T]o be cool in the twentieth century," as Rice argues, "one has to be connected to electronic culture" ("Writing" 222). With YouTube, Twitter, and Facebook, the iGeneration, as sociologists nickname contemporary adolescents,[39] is offered highly efficient channels to cultivate its narcissism, hedonism, and ironic detachment, defined by Pountain and Robins as the "three core personality traits" (26) of a cool character. Its entanglement with electronic culture, in which cool continues to reign as a "paradoxical ethos" (Liu 72), guarantees that the appeal of cool as a both emotionally and physically detached way of living is unbroken.

38 According to a recent study by the British communications regulator Ofcom, children and adolescents spend more time surfing the web than watching television and would accordingly rather forego the latter (Arthur par. 1).

39 See, for instance, Larry D. Rosen's *Rewired: Understanding the iGeneration and the Way They Learn* (2010), in which he describes the iGeneration as the legitimate heirs of the preceding Net Generation (or, more commonly, Generation Y or Millennials). Rosen defines the members of the iGeneration as "multitaskers, social networkers, electronic communicators and the first to rush to any new technology" (*Rewired* 2).

In contrast to Geiger's and Gioia's account, I will thus maintain that cool has neither lost its appeal nor its pervasiveness in the information culture of the early twenty-first century. Rather than proclaiming the total death of cool, I would propose the emergence of an updated version of it, a new cool which may well have shed some of its (originally sub and counter cultural) connotations of emotional detachment and ironic aloofness in the face of adversity, but continues to constitute an integral part of the cultural sensibility in the electronic age. This understanding of cool draws heavily on Liu's account of knowledge work aesthetics, which defines cool as the underlying principle of production in the digital age. While twenty-first-century cool is heavily commoditized and commercially exploited, its associations with rebellion, toughness, and narcissistic self-stylization remain intact: Even though the promise of nonconformity may have well devolved into a nothing more than a "flamboyant charade" (Harris 40), it still represents, however distorted, cool's strongest asset. And it is exactly this individuality, expressed via new, easily accessible, and highly customizable channels of self-fashioning such as social media, which is admired as the foremost secular virtue in the age of mass production and artificial enhancement.[40] Indeed, the term iGeneration not only references the overwhelming percentage of Apple fan boys and girls among contemporary teenagers, but reminds us that (not only) adolescents today invest an increasing amount of attention, time, and money in the creation, representation, and celebration of the self (L. Rosen, "Teaching" 3). Cool as the perhaps most "self-contained and individualistic attitude" (Pountain and Robins 23) perfectly hits the mark, so much so that it seems safe to assume it "could even prove to be the new mode of individualism, an adaptation to life in post-industrial consumer democracies" (Pountain and Robins 13).

The growing individualization of the self and the according personalization of one's surroundings, from shopping recommendations to the workplace, not only constitute a major connotational field of cool today, but are evidently tightly linked to today's technoscientific progress. I would hence strongly contest Gioia's interpretation of contemporary popular culture, which prefers to understand the increased representation of tech-savvy characters, from engineers to physicists to lab technicians—the former "nerds, geeks, and dweebs" (Gioia, "Death of the Cool" 8)—as evidence for the demise of cool. Media celebrities like Steve Jobs and his biopic actor Ashton Kutcher (chapter 8) have impressively demonstrated that in to-

40 This elevated status of individuality might indeed be linked to a growing desire for authenticity and sincerity, as Gioia argues. These traits have, however, always been valued as essential characteristics of the cool persona, since only a truly self-possessed, calm, and straightforward character can enjoy composure and emotional control (Thompson, "Aesthetic" 41). Clarke (1998) skillfully dissects this cool performance of authenticity in Malcolm X and Miles Davis.

day's information culture, one is not deemed cool despite, but in fact because of one's mastery of technological gadgetry and digital practices. Sociological studies of adolescent behavior support this reading.[41] Contrary to Gioia and Geiger, who cite diverse recent television formats and advertisement campaigns as evidence for an extensive uncooling of American society, the present study sees the new cool of technoscience also reflected in popular cultural productions. Traditional television formats and film genres, from the prime-time police procedural to the disaster movie, have become invested with scientific contents to an unprecedented degree. Technologically and scientifically savvy characters like forensic analyst Horatio Caine (*CSI: Miami*), theoretical physicist Sheldon Cooper (*The Big Bang Theory*), and paleoclimatologist Jack Hall (*The Day after Tomorrow*) will, miracle-like, solve the story's central conflict and emerge as the nonchalant and highly popular protagonists of the respective production. In that respect, we are indeed witnessing a "revenge of the nerds" ("Death of the Cool", 15), as Gioia observes, only with the slight modification that these scientists have shifted their roles from former antagonists to new gurus of (an updated) cool. Rather than mirroring the death of cool, the cultural productions analyzed in the analytical part of this study suggest that its semantic field has once again redrawn its boundaries to align with the changed premises and demands of the contemporary information society. Conversely, this also means, in a nutshell, that science might have eventually discovered the formula for cool.

Contrary to scenarios of the twenty-first century as a post-cool end time, the present study argues that cool not only remains an irresistible cultural force in contemporary information societies, but that it has also notably adjusted its focus. In the first decades of the new century, cool not only continues to operate as an underlying principle of American society in the fields of advertisement, corporate ideology, and fashion, to name just a few; much more so, it also increasingly informs popular representations of science, technology, and concomitant practices, tools, and actors, especially so in the realms of television and feature film. The present study understands these currently prevailing popular images of science, from the popular physicist to the adventurous climatologist, from tech entrepreneurs to geeky criminologists, as prime instances of the changing semantics of cool, which seems to have shifted from the tough and rebellious to the nerdy and science-savvy. In that respect, these ostensibly 'cool science' representations in typical channels of popular culture generate an up to now unparalleled degree of interaction of technoscience with the styles, modes, and techniques of American popular culture. The emphasis on popular cultural productions is crucial here: It is imperative to note that while science might well be deemed cool in a technologized encounter with popular

41 See L. Rosen (2010) and Montgomery (2007) as two out of a wealth of studies on the adolescent craving for electronic devices.

culture, such as in the aforementioned Hollywood movies or TV shows, it can remain rather uncool in a more formalized context. This again testifies to the contradictory nature of cool, which conjures a central paradox: Personalized friend suggestions on Facebook, online gaming, and Google search, all of them based on and controlled by algorithms,[42] are highly treasured by the average American teenager, whereas the study of algorithms at high school is a pain in the proverbial. While the focus of the present study is placed on popular cultural *representations* of science, which appear to be fairly homogenous in their appropriations of cool, the plurality of meanings inherent in our modern-day notion of technoscience must nonetheless be heeded as even further complicating the ambiguous effects of cool.

Evidently, these considerations suggest several central issues to be explored in greater depth, above all the key question of how cool representations of scientific contents affect cultural discourses of natural sciences and technology within postmodern information societies, in particular the United States as one of their chief representatives. I argue that the new coolness of technoscience has noticeable effects on the status of scientific knowledge and practice, and might, in fact, have to be understood as responding to, complementing, or even indeed substituting former sources of scientific legitimation as incubated in popular cultural texts. In a culture obsessed with the retrieval, manipulation, and storage of information, the supposed virtue of cool acts as both a replacement of and a challenge to traditional discourses of scientific legitimation, which postmodern philosophy established as dysfunctional, obsolete and, accordingly, a major symptom of the characteristic demise of metanarratives in post-industrial society. In order to tackle these concerns proficiently and penetrate both the reasons and the effects of a new scientific cool, a deeper understanding of 'this thing called science' (Alan F. Chalmers) and its reciprocal influence on popular culture will prove indispensable. The following two chapters are hence dedicated to the exploration of the concept of the information society and processes of scientific legitimation and popularization, as well as a brief discussion of this study's methodological and (inter)disciplinary foundation.

42 Customized suggestions on the Internet, which the average consumer experiences as automatic and intuitive, depend on elaborate algorithms, allowing the respective digital device to learn from the habits of its user. See Slavin (2011) and Uricchio (2004).

The American Information Society and the Crisis of Scientific Legitimation

> Science and technology revolutionize our lives, but memory, tradition and myth frame our response.
>
> ARTHUR M. SCHLESINGER/
> *THE CYCLES OF AMERICAN HISTORY*

According to the *OECD Guide to Measuring the Information Society*, reissued in 2011 and intended as "a standard reference for statisticians, analysts and policy makers" (3), the emergence of modern-day information culture is closely associated with major and unprecedented progress in the field of Information and Communication Technologies (ICTs). As the defining feature of the contemporary information society according to the OECD, ICTs touch on the political, personal, cultural, and aesthetic aspects of contemporary life, and must be taken into account as fundamental economic forces. The supply, demand, and use of as well as infrastructure for ICTs in a given country, ranging from cash registers to personal computers, serve as core indicators for determining its level of transformation into an information society, which can be observed, to varying degrees, in both developed and developing countries. The share of information services in the total economy is thus substantially higher than in the preceding industrial age, which marks a fundamental change in the economic (and, consequently, social and political) foundations of a country. These paradigmatic developments in post-industrial American society are considered by the present study as providing the necessary socio-cultural backdrop for cool representations of technoscience to emerge.

While ranking among the most authoritative, the *OECD Guide* is only one out of a wealth of official publications intent on grasping the constitutive mechanisms of what sociologists, policy makers, and cultural theorists since the early 1970s have endowed with a range of labels, including, among the most prominent, those of the knowledge, information, post-industrial, or network society: The penetration of all aspects of contemporary American life, be it the political, personal, social, economic, or, as the present study demonstrates, aesthetic domain, with fast evolving high technology. Knowledge is no longer only a matter of professional growth and social status, but, perhaps even more importantly, a considerable economic factor determining social relations, the distribution of wealth, and aesthetic production. Paralleling the infiltration and manipulation of political, economic, and social conditions by IT services, we are also faced with fundamental changes of cultural codes and styles. As the literature review in the preceding chapter has shown, the semantic field of cool is equally informed by the changing demands of a digital lifestyle and cannot be meaningfully explored without a deeper understanding of its social, economic, and epistemic conditions. Accordingly, the present study regards the permeation of all aspects of life in the twenty-first-century information society by technoscience as the requisite socio-cultural condition that propels the popular cultural production of cool science. At the same time, these productions not only reflect the penetration of the previously discrete aesthetic domain by techno-scientific imagery, but, in return, also contribute to this very process by enhancing the progressing interaction of science with the aesthetics, modes, and strategies of popular culture.

This development is tightly connected to the changed status of science within contemporary information societies. While science, especially in its combination with high-end technology, has not only become widely accepted in, but indeed indispensable for politics, economics, media, and our everyday lives, the public demands on science have equally increased exponentially, leading to a veritable crisis of legitimation concerning scientific practice and its production of risks. The present chapter demonstrates how the focus on cool in contemporary popular representations of science must be understood as a reaction to this crisis as they challenge or indeed substitute earlier (modernist) discourses of legitimation that are no longer effective in the postmodern age. For this purpose, I engage with, among others, the work of Jean-François Lyotard, whose philosophical discussion of the problem of legitimation in post-industrial society has proven paradigmatic for future analyses. Embarking from these perspectives, I argue that cool as an aesthetic and affective, rather than cognitive or ethical form of (scientific) legitimation has become highly viable in contemporary American popular culture, a circumstance facilitated through the growing interdependence of formerly discrete areas of activity prompted by the rise of postmodern knowledge culture. The emergence of cool is thus tightly linked to and highly dependent on the emergence of the information society, whose characteristics and effects on scientific legitimacy are explored in this chapter.

THE KNOWLEDGE/ INFORMATION/ POST-INDUSTRIAL/ NETWORK SOCIETY: A CRITICAL OVERVIEW

While the *OECD Guide* declares that "[t]here is little doubt that information and communication technology (ICT) has promoted profound economic and social change *over the past decade or so*" (12, my emphasis), academic, rather than economic analyses of this transformation date its beginning to as far back as the early 1960s. Among the first to observe the growing importance of codified knowledge and techno-scientific progress for societal development were Fritz Machlup, whose eminent study *The production and distribution of knowledge in the United States* (1962) introduced the concept of the knowledge industry, and Robert E. Lane (1966), who wrote of a "knowledgeable society" (650), which is fundamentally rooted "in epistemology and the logic of enquiry" (650). One year earlier, in 1961, Helmut Schelsky already argued

> "that the all-encompassing scientization of our existence necessitates a repositioning of science within the relationship between humans and the world, that this scientization of our lives indeed generates a novel form of relationship between humans and the world. [...] a new cultural epoch of mankind." (6)[1]

Even earlier analyses, such as Max Weber's 1918 lecture "Science as a Vocation" ("Wissenschaft als Beruf," published in 1922), must also be taken into account as valuable indications of a changing paradigm. Since these first tentative studies, a range of labels for the current type of knowledge-based culture have been created. Especially the terms knowledge society and information society have, both in professionalized and public discourses, frequently been used interchangeably and

1 Original wording: "[...], daß die umfassende Verwissenschaftlichung unseres Daseins eine neue Zuordnung der Wissenschaft im Verhältnis Mensch und Welt erforderlich macht, ja, daß diese Verwissenschaftlichung unseres Lebens ein neuartiges Verhältnis von Mensch und Welt selbst entstehen läßt. [...] einer neuen Kulturschwelle der Menschheit." Schelsky's publication sparked off a vigorous debate on technocracy throughout the 1960s and 1970s, which was supposed to mark the end of political ideology and the simultaneous emergence of a scientifically-technologically ruled state. Cf. Reinharth 90, and Weingart, *Stunde* 19-21.

largely synonymously, depending on the epistemological focus of the respective study.[2]

Daniel Bell, another early protagonist in the defining years of this new phase of society, additionally introduced the attribute 'post-industrial'[3] to refer to comprehensive societal changes. In his seminal *The Coming of Post-Industrial Society* (1973), which he himself characterized as an "essay in forecasting" (3), Bell notes that a decisive property of this "interstitial time" (37), setting it notably apart from the earlier modernist period, is the heightened and consistently increasing importance of science. Bell especially emphasized the centrality of *theoretical* knowledge (20) and the growing significance of technological innovation and progress on the one hand and political decision-making and policy formation on the other, a definition which has continued to influence sociological and economic models of the information age until today. Scientifically produced knowledge thus acts as the "axial principle" (14) of post-industrial America, a development which Bell assesses as inherently positive, since it will prepare the ground for planned and controlled social, economic, and technological progress.[4] Bell's notion of the present time as a post-industrial one was quickly adopted and, in return, influenced by several influential theoreticians, including André Gorz, Kenneth Keniston and Paul Goodman, Zbigniew Brzezinski, and Alain Touraine.[5] In addition to theoreticians

2 While Anglo-American critique employs both terms, with a predilection for the notion 'information society,' the concept of the knowledge society ("Wissensgesellschaft") is more pervasive in the German academic community. German sources quoted in the present study will thus habitually utilize the latter expression. See also Weingart, *Stunde* 325-326, for a discussion of the diverging terms.

3 Bell briefly also discusses the concepts 'knowledge society,' 'information society,' and 'professional society,' all of which he judges as "somewhat apt in describing salient aspects of what is emerging" (37). Instead of limiting his enquiry to scientific and technological aspects, however, he also includes economic considerations, which the label 'post-industrial' best accounts for.

4 His study also notes a shift in professional occupation from the industrial to the tertiary (or 'post-industrial') sector, especially to the fields of education and research; consequently, it is the scientific personnel and the university which will increasingly act as key productive forces, replacing manual labor and the manufacture of material goods. While Bell emphatically asserted that "the ethos of science is the emerging ethos of post-industrial society" (386) and assessed the growth of the scientific sector as inherently positive, a number of other studies reacted markedly less euphorically. Already ten years earlier, Derek de Solla Price (1963), for instance, considered the exponential growth of academia problematic and highly alarming for the future state of science and research.

5 Touraine's *Return of the Actor: Social Theory in Postindustrial Society* (1988) defined the attribute 'post-industrial' in similar terms to Bell; their conceptualizations, however,

who adopted, yet notably modified Bell's terminology, the characterization of society after 1945 as one that exceeds its formerly industrial basis also inspired the development of other concepts, such as Alvin Toffler's notion of the Third Wave society (1980).[6] Bell's early characterization of the changing mechanisms of the social fabric has thus proven to be a highly influential and visionary early work in the development of information society models.

In response to Bell's and similar neoliberal analyses, socialist accounts of knowledge-based production also emerged, such as the groundbreaking volume *Civilization at the Crossroads* (1966), compiled and published by a conglomerate of researchers headed by Radovan Richta. Richta predicted a fundamental repositioning of the human within society and vis-à-vis industry, technology, and science, since the latter would become deeply entwined with human existence: It "serves [...] as a source generating new types of human endeavour, as an initiator and producer of wants. That is to say, it is a productive force that can create new demands, conflicts and outlooks" (213). In seeming unison with Bell, however, Richta trusts in the self-regulatory mechanisms of science and technology for controlling their own consequences and fulfilling humanistic hopes. Inspired by early socialist analyses like Richta's, contemporary neo-Marxist accounts of the knowledge society habitually stress the reliance of contemporary capitalism on information technologies and digital networks. In the wake of this, several more or less influential concepts like information capitalism (Morris-Suzuki 1997), virtual capitalism (Dawson and Foster 1998), digital capitalism (Schiller 2000), high-tech capitalism (Haug 2003), or transnational network capitalism (Fuchs 2008) were advanced. According to these labels, information technologies and the growing dependence of human existence on (the distribution of) scientific knowledge are understood as restructuring the dynamics of global capitalism, which continues as the dominant economic principle well into the twenty-first century. Knowledge is hence seen as functioning according to the same underlying logics as do material commodities, i.e. it enhances structural inequalities between those who have access to information (so-called dig-

differ with regard to the material basis of society, which Touraine saw deeply rooted in escalating production, reinvestment, and industrialization rather than science-based activity.

6 Toffler focused on the succession of American culture from agrarian to industrial to post-industrial society, and equally stressed the importance of knowledge-based production: "In a Third Wave economy, the central resource—a single word broadly encompassing data, information, images, symbols, culture, ideology, and values—is *actionable* knowledge (Dyson, Gilder, Keyworth, and Toffler, par.2, emphasis in the original). In an earlier publication (*Future Shock*, 1970), Toffler used the term super-industrial society with a similar meaning: Again, knowledge is identified as the key factor for propelling production, replacing material work, and accelerating cultural change.

ital citizens) and those who are excluded from it. Accordingly, new technologies are not neutral, but embedded in a global network governed by the economic exploitation of information.[7]

In view of the global interconnectedness of informational flows, the term 'network society' has been advanced. Particularly influential in this regard was Jan van Dijk, whose book *The Network Society* (1991) coined the term that was later elaborated and popularized in Manuel Castells's *The Rise of the Network Society* (1996), part of his trilogy *The Information Age* (1996-1998). Castells and van Dijk primarily draw on James Martin's concept of *The Wired Society* (1978) as a society connected by mass communication, as well as several other explorations of the growing interaction of electronic culture with human life, such as Starr Roxanne Hiltz' and Murray Turoff's *The Network Nation* (1978), and Paul Craven's and Barry Wellman's concept of the network city and networked individualism (Craven and Wellman 1973, Wellman 2001). While terms like 'information society' focus on the expansion of scientific knowledge as a pivotal source for change in society, the model of the network society proposes a range of political, economic, and social transformations provoked by the spread of information technologies[8] via a network of global flows. The organization via cultural and natural networks, i.e. the connection of discrete elements by links and relationships, has thoroughly pervaded the deep structure of society.[9] The network society remains capitalist, but power is now increasingly measured in terms of absence or presence in a given network and the ability to control its switches, rather than via supremacy in social action. By proposing that neither exploitation nor social inequality based on (networked) labor are ef-

7 In their controversial publication *Empire* (2000), Michael Hardt and Antonio Negri, for instance, focus on the contemporary dominance of immaterial labor, essentially knowledge- and communication-based, which creates new spaces of social inequality. While *Empire* is not, first and foremost, concerned with the cultural effects of science and technology in the tradition of Bell or Richta, its argumentation does reveal the pervasive influence of early information society discourses.

8 These information technologies include, but are certainly not limited to the Internet, as several groundbreaking studies in this field already emerged well before the advance of the World Wide Web, such as Hiltz' and Turoff's *The Network Nation* (1978).

9 According to Castells, "[n]etworks constitute the new social morphology of our societies and the diffusion of networking logic substantially modifies the operation and outcomes in processes of production, experience, power, and culture" (500). While networks have always existed as a form of social organization, new information technologies stimulate a much deeper and yet unparalleled penetration of the social tissue with underlying networking dynamics. Castells calls this new spatial logic a "*space of flows*" (408, italics in the original), which is in the process of substituting the traditional experience of a historically rooted "*space of places*" (Castells 409, italics in the original).

fectively surmounted by today's evident increase in information-based businesses, these noticeably less euphoric accounts put Bell's and Richta's utopian visions of a society guided by scientific objectivity and rationality into perspective.

The brief survey of information society models already reflects, despite their evident disparity, several vital characteristics and constitutive mechanisms with which most studies seem to concur. Above all, the majority of accounts stress the increasing significance and centrality of theoretical, codified knowledge, in contrast to the predominance of labor and material property in the preceding industrial era. The attributes 'theoretical' and 'codified' are essential here, since almost all civilizations, from ancient Egypt to the industrial Victorian era, did, in some way or another, depend on knowledge and its transmission from one generation to the next; information-based human action, as Gernot Böhme and Nico Stehr explicate in their influential monograph *The Knowledge Society* (1986), must hence be regarded as an "anthropological constant" (8) rather than an emerging trend. It is only with the growing *social*, *economic*, and *political significance* of *scientific* knowledge and the concomitant expansion of information production and storage that scholars observe a fundamental change in the way in which societies function, which is why some of them propose less ambiguous designations than that of the 'knowledge' or 'information' society to describe the range of changes that were initiated in the postwar era.[10]

Secondly, most studies characterize, in one way or another, the new, post-industrial state of society as one that is "based on the penetration of all its spheres of life by scientific knowledge" (Böhme and Stehr 8). Its distinctive features include the scientization of social life, the professionalization of the workforce, the replacement of other forms of knowledge by science, the emergence of science as an immediate economic resource, the differentiation of a new social class defined by its scientific activity, and an according transformation of the political sector.[11]

10 A number of studies also aim to decontextualize and historicize the concept of the knowledge society and detect processes of comprehensive scientization as early as the end of the nineteenth century. This rather broad understanding of a knowledge-based culture, advanced, for instance, by Peter Burke (2000), Margit Szöllösi-Janze (2004), and Jakob Vogel (2004), however, notably disregards the centrality of ICTs and their effects on scientific practice as key factors for the emergence of information societies after the Second World War.

11 Additionally, Stehr (1994) differentiates three different, historically contingent types of knowledge to illustrate the changing functions of theoretical knowledge in the information age: interpretative or orientative knowledge ("Deutungs- oder Orientierungswissen") of the Enlightenment era, productive knowledge ("Produktivwissen") of the nineteenth century, and contemporary practical knowledge ("Handlungswissen"), which endows its user with the immediate capacity to act upon it (Stehr 215).

Knowledge as the key factor for social change replaces labor as the foremost constituent of society, and, similar to the production of commodities in the nineteenth century, undergoes considerable growth and differentiation (Böhme and Stehr 23).[12] Consequently, science as the place where established, codified knowledge is created is no longer characterized by its epistemic singularity, but has acquired enormous social, political, and economic relevance. The growing informatization and scientization of society must certainly be understood as a global phenomenon transcending the political borders of nation-states (cf. the *OECD Guide*), although the specific processes involved may "function quite differently in different cultures" (Szöllösi-Janze 313).[13]

In this respect, most models seem to tacitly support Thomas Hughes's famous metaphor of "The Seamless Web" (1986) of society, science, technology, industry, "etcetera, etcetera" (Hughes 281), which references the penetration of all aspects of life in post-industrial society with science and technology to an unprecedented degree, and the mutual interdependence between these formerly distinct areas of activity. The idea of the web emphasizes the inextricability of seemingly independent domains, none of which assumes a superior position over the others (Felt, "Wechselwirkungen" 52). In a similar vein, Peter Weingart's eminent *Die Stunde der Wahrheit* (2001, *The Moment of Truth*) refers to this growing amalgamation of cultural domains as the "*scientization* of society" (*Stunde* 17, italics in the original),[14] and, conversely, the "*politicization*, *economization* and *mediatization* of science" (*Stunde* 18-19, italics in the original).[15] In sociological terms, this has often been visualized via the metaphor of boundary work, which is concerned with the rhetorical demarcation of science from related societal domains and 'pseudo-science.' As

12 This does not, however, entail that the average citizen may now profit from an unmatched amount of knowledge and information, as the sometimes misleading term '*knowledge* society' might suggest. Quite the contrary: The growing quantity of knowledge always also includes a growing awareness of yet underexplored areas of human activity, a knowledge of our lack of knowledge (Weingart, *Stunde* 20). In Jerome Ravetz' words, this engenders a new form of "ignorance created by science" (100). This argument had already emerged in Richta's *Civilization at the Crossroads*, in which he predicted that "for every want satisfied and every advance in knowledge, it [i.e. science] breeds a multitude of new questions, a spate of human dissatisfaction" (214).12 Systems theory as advanced by Niklas Luhmann also stresses the paradoxical increase of ignorance, or non-knowledge, due to scientific progress. See, for instance, Luhmann's essay "Ökologie des Nichtwissens" (1992).

13 Original wording: "derartige Prozesse in den einzelnen Kulturen ganz unterschiedlich ablaufen."

14 Original wording: "*Verwissenschaftlichung* der Gesellschaft."

15 Original wording: "*Politisierung*, *Ökonomisierung* und *Medialisierung* der Wissenschaft"

Thomas Gieryn (1983, 1995, and 1999) and others suggest, the socially contingent boundaries between formerly distinct areas of cultural activity become increasingly blurred, since scientifically generated knowledge has become relevant in virtually all sections of society.[16] Mitchell Ash's (2002, 2004, and 2007) conceptualization of science and political, economic, and educational endeavors as mutual resources equally supports this line of thought. According to Ash, the growing malleability of the boundaries between these levels due to current processes of informatization can be regarded as a stronger "mobilization of resources" ("Ressourcen" 352),[17] i.e. a heightened degree of mutual stimulation and enrichment. His perspective largely falls in line with Weingart's, stressing the increasingly reciprocal, rather than one-sided influence between science and other fields of cultural activity.[18]

Conceptualizations of the information society in terms of a blurring of boundaries, a reciprocal mobilization of resources, or a web of cultural activities, all of which are popular topoi within contemporary science studies scholarship,[19] enter into discourse with the influential work of Bruno Latour, whose theory of transcended categories ranks among the most notorious. In *Science in Action* (1987) and *We*

16 The concept of boundary work seems to emerge in Barnes (1974) and has prominently been elaborated by Gieryn (1983, 1995, and 1999). Since then, a variety of (sociological) studies have been devoted to boundary work in general (Shapin 1982) or the empirical analysis of particular historical episodes (Shapin and Schaffer 1985, Jasanoff 1990). While boundary work, more specifically, is used to refer to "the discursive attribution of selected qualities to scientists, scientific methods, and scientific claims for the purpose of drawing a rhetorical boundary between science and some less authoritative residual non-science" (Gieryn, *Credibility* 4-5), it also provides a feasible framework to conceive of the information society as a gradual blurring of boundaries between formerly distinct areas of cultural activity, such as science and politics. Conversely, the space of science itself is in turn invaded by neighboring areas and has lost the socially excluded position it has hitherto enjoyed.

17 Original wording: "Ressourcenmobilisierungen."

18 For completion's sake, it should be noted that Weingart prefers not to conceptualize the stronger interaction between science and politics, economy, or the media in terms of a blurring or greater malleability of boundaries: "The amalgamation of science and politics does not, however, abolish the border between them, but rather reinforces it" (*Stunde* 141). Original wording: "Die Vermischung von Wissenschaft und Politik führt aber nicht etwa zur Aufhebung der Grenze zwischen ihnen, sondern zu deren Bekräftigung."

19 See Weingart, *Stunde* 21-22, and Rip 141 for an exploration of the pervasive metaphor. The title of Nikolow's and Schirrmacher's edited collection *Wissenschaft und Öffentlichkeit als Ressourcen füreinander* (2007, *Science and the Public as Mutual Resources*), borrowed from Ash's largely eponymous article included in the monograph, takes the same line.

have never been modern (1993), Latour observed a wholesale abolition of the distinction between different professional fields and academic disciplines, as well as, more fundamentally, a progressing transcendence of the (presumably modernist) separation between natural, social, scientific, and political phenomena, between nature and culture. The contestation of established binaries such as subject/object, body/mind, modern/postmodern is understood as constitutive of contemporary society. While public discourses, such as the one surrounding the HIV pandemic (*Modern* 1-3), already seam- and effortlessly mix political, medico-scientific, and social concerns, Latour recommends to "ratify in public what is already happening" (*Modern* 144), namely a rejection of disciplinary boundaries, studied by specialists and attributed to discrete areas of activity and research. Following Latour, one must thus conceive of the contemporary information society, which, as the above considerations have shown, witnesses an unprecedented degree of interaction between formerly distinct cultural domains, as one which finally transcends dualisms and illegitimate categorizations.[20] Latour's work, however, implies a much more radical blurring of boundaries than proposed by most other models; it is traditional scientific practice itself, especially the clear division of labor between natural sciences, concerned with non-human nature, and the humanities, limited to the study of the social independent of material circumstances, which is fundamentally challenged by the changing premises of information culture.

20 While the modern era is commonly problematized as having introduced and being fundamentally based on these distinctions, Latour claims that they have, in actual fact, never existed, which is why 'we have never been modern.' Accordingly, he calls the actual web of human and nonhuman phenomena, of material and social conflicts which constitute our existence the "Gordian knot" (*Modern* 3). This knot is artificially broken "into as many segments as there are pure disciplines. By all means, they seem to say, let us not mix up knowledge, interest, justice and power. Let us not mix up heaven and earth, the global stage and the local scene, the human and the nonhuman. 'But these imbroglios do the mixing,' you'll say, 'they weave our world together!' 'Act as if they don't exist,' the analysts reply" (*Modern* 3). By demonstrating how the supposedly discrete entities of nature and culture have always interacted, but were constructed as dualisms in the course of the seventeenth century by early scholars like Thomas Hobbes and Robert Boyle, Latour argues for a more holistic understanding of society as well as science. Latour's argument has been supported and elaborated by Donna Haraway, whose work is discussed in the following section. See also Pickering (1994) and Crawford (1994) for critical discussions of Latour's argument.

SCIENCE IN THE INFORMATION SOCIETY

Judging by the above overview of models, a key factor in the emergence of the information society is the transformation of institutionalized science. This is also noted by Böhme and Stehr, who conclude from their comparison between the Information Revolution initiating today's knowledge society and the Industrial Revolution of the Victorian era "that the dynamics of the production of knowledge are subject to the same forces which propel the productions of commodities: the increased differentiation and independence of production corresponds to the growth and differentiation of science" (23). Following this line of argumentation, it would be the surplus value generated by post-industrial science, paralleling the expansion of industrial capital in the course of the nineteenth century, which serves as the driving mechanism behind the rise of knowledge at the expense of material commodities. This, however, also entails several fundamental changes affecting the nature, practice, resources, and responsibilities of science itself, including its public representation and legitimation. It is these extensive transformations, I argue, that prepare the ground for popular cultural representations of cool science to emerge. Accordingly, it is imperative at this point to explore the elusive notions of 'science,' its relation to 'technoscience,' and their significations in the information society.

Several models investigating the information age indeed focus on science as its most prominent constituent. One of these models was introduced by Silvio Funtowicz and Jerome Ravetz (1991), who explore the concept of post-normal science[21] to denote a novel form of science where "acts are uncertain, values in dispute, stakes high and decisions urgent" (Funtowicz and Ravetz, "Post-Normal Science" 1). The suggested remedy to increase certainty and objectivity is the extension of the peer community to include localized, lay knowledge provided by "all

21 The term 'post-normal science' draws on Thomas Kuhn's eminent *The Structure of Scientific Revolutions* (1962) and his notion of 'normal science,' which refers to the routine, everyday practice of science in between socially constructed paradigm shifts. According to Kuhn, an alleged scientific discovery is initiated when during phases of normal science, more and more abnormalities and contradictions accumulate, which will lead to a crisis and, if it cannot be resolved, a scientific revolution, i.e. a comprehensive paradigm shift introducing a new set of beliefs on which a given discipline will be founded. Once these changes are accepted as the new rules for the practice of science, the revolution is complete and normal science sets in again. Normal science is backed by broad agreement and relies on a set of established, commonly accepted principles, thus amounting, in Kuhn's words, to little more than "puzzle-solving" (35) within the currently dominant paradigm and a highly restricted range of potential action. By analogy, post-normal science denotes situations of extraordinary scientific challenge which necessitate an extension of known and proven methods.

those with a desire to participate in the resolution of the issue" (Funtowicz and Ravetz, "Post-Normal Science" 7) to guarantee a more democratic constitution of scientific findings that corresponds with current postmodern tendencies.[22] A similar path is taken by the concept of mode 2, advanced in the seminal publication *The New Production of Knowledge* (1994) by a collective of researchers, including Michael Gibbons, Helga Nowotny, and Peter Scott. In contrast to mode 1 knowledge, which is constituted in traditional, long-term research within disciplinary boundaries, mode 2 denotes scientific production that is interdisciplinary, applied, and context-based, organized in short-time individual research projects and teams, whose members bring in a variety of academic backgrounds to focus on a very specific (real life) problem.[23] John Ziman's notion of post-academic science, introduced in his eminent *Real Science* (2000), touches upon similar concerns and equally stresses the influence of social, political, and economic developments on scientific practice within the modern information society. The concept of the triple helix, developed by Loet Leydesdorff and Henry Etzkowitz (1996), explores this reciprocal reassignment of functions between the three actors of state, industry, and academia. While science is increasingly paying tribute to the demands of a free market economy in which codified knowledge and academic education are valuable commodities, industrial research conducted in big, global companies is simultaneously being scientized. As a result, the borders between public and private research, academic and industrial science are constantly shifting and the relationship between science, industry, and the government is becoming tighter, as reflected by the substantial increase in collaborations between state universities, their PR departments, and cooperate businesses.[24] In order to facilitate the cooperation of scientists from different

---

22 Funtowicz and Ravetz specifically discuss policy issues of risk and the environment, such as problems of sustainable development, to argue for a greater involvement of all concerned parties, both scientists and laypeople, in the implementation and solidification of scientific findings. Such "extended peer communities" ("Post-Normal Science" 7) of local laypeople would include, for instance, citizens' juries, focus groups, or consensus conferences, communities which are already actively practiced and ensure a democratization of science. See also Funtowicz' and Ravetz's chapter in Robert Costanza's *Ecological Economics: The Science and Management of Sustainability* (1991).

23 Mode 2 not only affects the actual production of scientific knowledge, but also its practitioners (from individual to group work), issues of quality control and peer review, as well as the organization of research institutions. It is understood as a historically emerging concept, starting in the middle of the twentieth century and thus strongly related to the advance of new technologies.

24 It should, however, be noted that these forms of collaboration tend to be restricted to very specific disciplines and research traditions, such as molecular biology, pharmacy, or IT research. See Weingart, *Stunde* 230-231, as well as Terry Shinn's essay "The Triple He-

disciplines with each other as well as with engineers, politicians, and industry-based researchers, effective borderlands and contact spaces must be established, which Peter Galison and David Stump (1996) have referred to as 'trading zones.' The interaction between formerly discrete areas of activity involves the creation of a new and commonly accepted set of ground rules, a distinct lingua franca, to enhance processes of mutual exchange.[25]

Borderlands and blurred boundaries appear especially poignant with regard to the tightening relationship between science and high-end technology, which have become practically synonymous in many contexts[26] and are frequently blended into the buzz word 'technoscience.' Originally coined in the field of bioethics by Belgian philosopher Gilbert Hottois,[27] the concept of technoscience has variously been theorized as a defining feature of Western scientific culture for the greater part of the twentieth century and is attributed to a variety of interrelated phenomena. It was prominently developed by the eminent work of Donna Haraway and Bruno Latour, who in turn seem to be influenced by Martin Heidegger's seminal *Die Technik und die Kehre* (1962). More recent directions include the edited collections by Don Ihde and George E. Marcus (1995) and Evan Selinger (2003). Above all, the neologism 'technoscience' is employed to illustrate the changed character of scientific practice, which has become acutely dependent on high technology gadgets, expensive equipment, and an according (inter)national industry to conduct, organize, and finance research. These extensive, international research projects are known as 'Big Science,' a notion popularized by Derek de Solla Price's *Big Science, Little Science* (1963).[28] The latter cautioned against the uncontrolled growth of science and

lix and New Production of Knowledge" (2002) for a detailed discussion and critique of the triple helix model.

25 Galison's and Stump's *The Disunity of Science* (1996) draws on linguistic concepts of language exchange in a colonial context, which works through the creation of trading zones, pidgins, and creoles. Scientific collaboration across disciplinary boundaries is conceived of in terms of an exchange of commodities, a trading situation, which requires a common language. Again, this supports an understanding of knowledge as a material resource in the information age.

26 As Marcel LaFollette sums up the results of recent surveys among the American public: "[T]he images of science and technology have either merged or are continually confused" (177).

27 See especially Hottois's more recent publications in English, such as "Technoscience: Nihilistic Power versus a New Ethical Consciousness" (1987).

28 While Price's book introduced the concept of Big Science to a broader public, the term had already appeared two years earlier in Alvin M. Weinberg's "Impact of Large-Scale Science on the United States," published in 1961 in *Science*. Weinberg's use of the term strikes markedly more positive chords than Price's: "When history looks at the 20th centu-

sparked off a heated debate on science funding, risk management, and research ethics among scientists and policy makers alike.[29] In the United States, one of the first and perhaps still most notorious Big Science endeavors was the Manhattan State Project (1942-1946), dedicated to the construction of the atomic bomb, which employed about two thousand scientists and engineers for several years (Felt, Nowotny, and Taschwer 198).[30] Contemporary examples of international Big Science collaborations are the Human Genome Project, founded in 1990 by the U.S. Department of Energy and the U.S. National Institutes of Health, the Fermi National Accelerator Laboratory ('Fermilab') near Chicago, and its European counterpart, the CERN in Geneva. In addition to their dependence on high-end technology, these projects also illustrate the growing differentiation, specialization, institutionalization, and trans- or interdisciplinary nature of science over the last decades (Weingart, *Stunde* 342-350).[31]

Concepts like 'technoscience' and 'Big Science' point to the fact that the underlying dynamics of contemporary information-based society must not only be understood in terms of a widespread scientization of society, but, perhaps even more fun-

---

ry, she will see science and technology as its theme; she will find in the monuments of Big Science—the huge rockets, the high-energy accelerators, the high-flux research reactors — symbols of our time just as surely as she finds in Notre Dame a symbol of the Middle Ages. [...] We build our monuments in the name of scientific truth, they built theirs in the name of religious truth; we use our Big Science to add to our country's prestige, they used their churches for their cities' prestige" (161).

29 Price predicted that if expenditure on science continued in the same rate as before, "we should have two scientists for every man, woman, child, and dog in the population, and we should spend on them twice as much money as we had. Scientific doomsday is therefore less than a century distant" (*Big Science* 19). While Price's prophesy has so far not proven true, scientific progress is indeed increasingly propelled by large-scale, government-funded projects.

30 The Manhattan State Project also serves as a prime example of the tightening relationship between science and the military, another constitutive feature of information societies, especially the United States. According to Felt, Nowotny, and Taschwer, 40% of the worldwide expenditure on science since 1945 has been spent on military research projects (198). If this seems an unlikely percentage, one must only think of several vital innovations of the twentieth century, such as the World Wide Web, which originate in military research. See also J. Weber 125.

31 This is also observed by Holton and Blanpied (1976), who claim that the institution of Big Science projects did not remained without consequences, but has had fundamental repercussions on scientific practice and public policy, including funding and the relationship between scientific colleagues. See also Galison's and Hevly's *Big Science: The Growth of Large-Scale Research* (1992).

damentally, a *technologization* of both society and science. The penetration of all aspects of American life with technology, be it in the form of computer systems, mobile phones, or artificial intelligence, makes a neat separation into formerly distinct domains—the political, the personal, the economic, and, more substantially, between pure science and applied technology—nearly impossible. Accordingly, Paul Forman (2007) suggests viewing technology, rather than science, as the primary constitutive mechanism of society after 1980. Indeed, he proposes the primacy of technology over science, or, more profoundly, the subsumption of science under the much more inclusive concept of technology, as a key "demarcator of postmodernity from modernity" (Forman 1). While modernity used to prioritize science, technology is nowadays the broader and more widely used concept, which may also, but not always, include scientific dimensions. Forman's argument concurs with Donna Haraway's work on technoscientific imaginary, which perceptively demonstrates that today's omnipresence of technoscience "extravagantly exceeds the distinction between science and technology" (*Modest_Witness* 3) and must be regarded as "a mutation in historical narrative" (*Modest_Witness* 3), the beginning of a new cultural epoch.[32]

Not surprisingly, the comprehensive transformation of traditional science into modern-day technoscience has also had wide-reaching consequences for its institutionalization and production. Since the 1980s, the practical turn in science studies has created heightened awareness for science's fundamental and increasing dependence on exclusive and cost-intensive equipment, which makes scientific knowledge locally and circumstantially contingent, highly contextualized, and influenced by

32 Haraway's studies not only demonstrate that technologized information has become ubiquitous and inescapable in knowledge-based cultures, but, more fundamentally, that it has converted its user into digestible data, for instance with the transformation of the human body into a decipherable (and manipulable) genetic code. Most prominently, this is elaborated in Haraway's influential "A Cyborg Manifesto" (published in 1985 in the *Socialist Review* and updated in a 1991 essay), in which she argues, in seeming convergence with Latour, that the growing technologization of human action "challenges these dualisms in intriguing ways" (177) and is responsible for a wholesale confusion of elemental categories that shape contemporary existence, such as the distinctions between self/other, civilized/primitive, reality/appearance, body/soul. The "Manifesto" is also considered a founding text of postfeminism, famous for its declaration that "there is nothing about being female that naturally binds women together into a unified category" (155). In order to transcend essentialist identity conceptions, feminists should consider themselves cyborgs, replacing identity with affinity. Haraway's vision of the destabilizing effects of technoscience on the central binaries of Western thought is realized in practices like Artificial Life, robotics, and biotechnologies, which present an 'implosion' of nature and culture (J. Weber 114-115).

the specific circumstances of its production within a given, localized laboratory.[33] So-called laboratory studies thus assert that "nothing epistemologically special is happening" (Knorr Cetina, "Laboratory" 151) when scientific knowledge is produced, a deliberately provocative formulation that has provoked intense debate.[34] In their influential *Laboratory Life: The Social Construction of Scientific Facts* (1979), Bruno Latour and Steven Woolgar investigate the agency of nonhuman objects and delineate in detail how the obscure process of research depends on routine lab practices as much as on matters of negotiation and the transcription[35] of material substance into figures, graphs, or otherwise exploitable knowledge via large and expensive lab equipment. Its results can hardly be reproduced and, consequently, regulated and critiqued by the scientific peer community due to the vast dimensions of the respective experiment:

"We are in fact faced with the problem that there are frequently only one or very few specimens worldwide of very large, complex, and expensive instruments. [...] This entails, first of

33 While the growing establishment of high-end research in local laboratories has drawn increasing scholarly attention to the fact that the process of scientific knowledge production is neither universal nor standardized, laboratory studies are well aware that the constitution of scientific facts has always been shaped by its particular context (Schirrmacher and Nikolow 11). The dependence of scientific production on collective, locally specific action and the social and cultural constructedness of a given scientific fact are neither a new idea that only emerged with modern science studies nor an exclusive feature of the information society. As early as 1935, Ludwig Fleck introduced these considerations with his notions of the 'thought style' ("Denkstil") and the 'thought collective' ("Denkkollektiv"), which are understood as fundamentally shaping scientific findings. Cf. Fleck's *Die Entstehung und Entwicklung einer wissenschaftlichen Tatsache* (1935, reprinted 1980) and Ash's discussion of it ("Ressourcen" 354-355).

34 Knorr Cetina draws on Richard Rorty's famous formulation that "no interesting epistemological difference" (qtd. in Knorr Cetina, "Laboratory" 151) can be identified between different social pursuits, such as between that of knowledge and that of power. Science is just as constructed as any other culturally contingent action.

35 Latour and Woolgar emphasize the centrality of writing or transcription for scientific research and describe scientists in a laboratory as "compulsive and manic writers [...] who spend the greatest part of their day coding, marking, altering, correcting, reading, and writing" (48-49). Following this postulation, several influential studies have since demonstrated the importance of narrative strategies, rhetorical devices, symbols, pictures, and metaphors for the scientific production of knowledge, thus pointing to the necessarily narrative construction of seemingly natural facts. Cf. Knorr Cetina's *The Manufacture of Knowledge* (1981), Evelyn Fox Keller's *Secrets of Life—Secrets of Death* (1992) and *Refiguring Life* (1995), as well as Christina Brandt's *Metapher und Experiment* (2004).

all, that such experiments are practically not reproducible and that scientific progress becomes a social negotiation process, which is restricted both locally and with regard to the group of people that will be granted access." (Felt, Nowotny, and Taschwer 53)[36]

In its basic form, this development had already been noted by Bell, who predicted "the primacy of theory over empiricism" (20) as a constitutive feature of the emerging post-industrial society. The growing dependence on machinery and collective analytic work in a big research conglomerate has thus eroded the social preeminence of the individual scientist, which might be understood as yet another defining feature of science in knowledge-based cultures (Daston and Galison 347-349, Szöllösi-Janze 282-283, and Stehr 350-420).

Another distinctive characteristic of science in the information society is the increase in risk potential via ecologically detrimental, ethically problematic, or failed research. Ulrich Beck (1986) proposed the term "risk society" ("Risikogesellschaft") to refer to the increasing willingness within technoscientific research to hazard its consequences. Indeed, the emergence of the information society coincides with a heightened discourse of risk in the 1970s, which effected an "erosion of faith in science" (Holton and Blanpied xxi). A famous example which illustrated to the public that Big Science is inevitably linked to a degree of uncontrollable danger is the 1979 Three Mile Island accident. As perhaps the most consequential nuclear meltdown in the United States' power plant history, the accident became notorious not only for the damages done to local residents and the environment, but also to the credibility of science due to unclear, highly contradictory, and evidently biased official communications to the American public.[37] Together with counterparts

36 Original wording: "In der Tat ist man mit dem Problem konfrontiert, daß es von sehr großen, komplexen und teuren Instrumenten meist nur eines oder zumindest ganz wenige Geräte weltweit gibt. […] Dies bedeutet zunächst, daß Experimente dieser Art praktisch nicht mehr wiederholbar sind und daß wissenschaftliche Erkenntnis mehr denn je ein sozialer Aushandlungsprozeß wird, der noch dazu lokal und vom Personenkreis her beschränkt ist." See also J. Weber 128-129.

37 The "mistakes, oversights, and misjudgments" (J. Walker 3) accompanying the accident at the Three Mile Island power plant severely challenged people's faith in science. CBS anchorman Walter Cronkite introduced his coverage on Three Mile Island as "the most confused day in the history of news media" (qtd. in Wilkes 1005), which could not only be attributed to a lack of knowledge of nuclear physics among science journalists (Wilkes 1005), but, more fundamentally, to the experts' obvious attempts at obscuring the actual dimensions of the accident and supporting economic and/or political interests, rather than those of the American people. This is nowadays understood as having done massive and irredeemable damage to the public image of science in the United States (Weingart, "Moment" 703). J. Samuel Walker's *Three Mile Island: A Nuclear Crisis in Historical*

in Europe and Asia, it stands as a symbol of how "by the 1970s belief in progress was tempered by growing awareness of risks" (Nelkin 445). In philosophical terms, it uncannily appeared to support dooming verdicts on the future course of science's legitimacy, which surfaced in the same decade. [38]

THE CRISIS OF SCIENTIFIC LEGITIMATION

Paralleling the comprehensive transformation and expansion of the scientific sector, public demands on science to account for its authority have increased dramatically over the last decades. Indeed, the heightened discourse surrounding issues of scientific (de)legitimation can be characterized as yet another defining feature of contemporary information societies, although issues of legitimation and consent are, of course, intrinsic to scientific practice from its ancient beginnings (Lingis 1). However, the contemporary close entwinement of science with other areas of cultural activity has noticeably increased calls for a feasible justification of its expenditures, outcomes, and risks in recent decades. In the 1980s, Arie Rip astutely observed that "[c]ollegiate controls and legitimate authority in the professional domain, all painstakingly won during the 19th and early 20th century, are eroded now that the structure of industrial society is changing" (143). This erosion of legitimation not only results in a variety of repercussions on the practice of science, but also informs its productive relationship with politics, economy, and, particularly noteworthy for the present study, popular culture. Accordingly, this book understands popular representations of cool science as deeply entwined with issues of scientific legitimation, authority, and public esteem. Indeed, I propose to view these cool representations as offering an alternative form of legitimation suited to the demands of modern infor-

Perspective (2004) offers a comprehensive account of the incident in its political and historical context.

38 While natural sciences have undergone a myriad of fundamental transformations due to the changing premises of scientific practice in the information age, the humanities seem to remain largely unaffected. Indeed, one might be inclined to conclude that the rapid advancement of natural sciences happens to the disadvantage of the humanities and social sciences (Ash, "Wissenschaftswandlungen" 95). On the other hand, the concept of the information society is tightly linked to the academic solidification of science studies and science and technology studies (STS), which emerged in the wake of public debates on science funding, risk management, and extensive scientific growth (Edge 6-7). The meta-knowledge provided by these disciplines becomes increasingly important for science practitioners, policy makers, and the general public alike, since the rising social and financial expenditures for Big Science projects necessitate a profound understanding of their potentials and risks. See also Reinhardt (2010).

mation culture. The representations of science explored in the analytical part of this study are regarded as highly dependent on, if not indeed arising from the crisis of legitimation that has characterized scientific discourse since the postwar era. The present section examines dominant ethico-philosophical and sociological conceptualizations of scientific legitimation and its concomitant crisis by referring to paradigmatic studies in the field.

In *The Postmodern Condition* (1979), widely appreciated as a foundational theorization of contemporary culture on par with Bell's notion of the post-industrial society, Jean-François Lyotard famously characterizes the widespread legitimation crisis of science as a defining feature of the postmodern age. According to his seminal publication, the fact that we have become incredulous toward "the metanarrative apparatus of legitimation" (Lyotard xxiv), a quintessential feature of modernity, and "no longer expect salvation to rise from [it]" (Lyotard xxiv) constitutes a vital characteristic of postmodernity. The postmodern, a term which Lyotard was the first to adapt from architecture and art criticism and introduce to general philosophy, is thus defined, in its most simplistic terms, as "incredulity toward metanarratives" (xxiv). The superseded metanarratives in question include, above all, the modernist notion of science as a means to progress and liberation, which is grounded in humanist thought and the French Revolution, and the Hegelian tradition of science for science's sake, in which "knowledge first finds legitimacy in itself" (Lyotard 34).

According to Lyotard, the postmodern crisis of legitimation must, first and foremost, be understood as a comprehensive backlash against the almost messianic image of science in the preceding modernist period. Especially during the jazz age, as the culmination of America's Progressive Era[39] in the 1920s has often, and at times rather derogatively, been dubbed, science perfectly suited the period's zeit-

39 The Progressive Era, dating from the late nineteenth century up to 1930, was a period of social and political activism, which witnessed a number of fundamental reforms, including the introduction of women's suffrage, the passage of the Federal Trade Commission Act, and the temperance movement. According movements in the fine arts, such as precisionist painting and the growing popularization of photography and motion pictures for supposedly truthful and efficient representations of reality, further enhanced the period's ideals of efficiency and political as well as scientific progress, especially in the fields of medicine, education, and engineering. Social, individual, and national improvement constituted the era's top priorities, which notably influenced the way science was perceived and institutionalized in public discourse. See Thurs (98-99) for a close analysis of the Progressive Era's effects on science as well as Robert Nisbet's *History of the Idea of Progress* (1980) and J. Leonard Bates's *The United States, 1898-1928: Progressivism and a Society in Transition* (1976) for a general introduction to the idea of progress in early twentieth-century America.

geist and came to embody its ideals of meticulousness, efficiency, and progress. What is more, some scientists even indeed "portrayed science and scientific thinking as important aspects of American democracy and bulwarks against the spread of totalitarianism" (Thurs 95).[40] This "rosy halo" (LaFollette 159) that surrounded science was buttressed by the enormous popularity of individual researchers, including the emigrant media celebrity Albert Einstein.[41] The unparalleled and, more often than not, unquestioned authority which science enjoyed during the modernist era was justified by its alleged beneficial effects on human existence. The reason for the eventual demise of this pervasive narrative of scientific progress lies in post-industrial scientific and technological developments, above all in the fields of ICT, and the concomitant erosion of former hierarchical structures, primarily in the context of the traditional university. It was radically enhanced by the advent of the Second World War and its atrocities realized through major 'breakthroughs' in the fields of physics and biochemistry. Lyotard hence maintains that science, in fact, destroyed its former sources of legitimation, including the myth of its own potency to thoroughly explain the physical world. Due to a growing insistence on the eventual applicability of scientific research, knowledge "ceases to be an end in itself" (Lyotard 5), but must answer to specific needs; it has become the foremost economic commodity. Humanist narratives of legitimation are thus replaced with those of performativity, efficiency, and usefulness: "[S]ince performativity increases the ability to produce proof, it also increases the ability to be right" (Lyotard 46).[42]

The invocation of the performative function of science reflects how the concept of the legitimation crisis is closely linked to the more general crisis of representation in the postmodern age, i.e. the loss of belief in the faithful reproduction of an accessible objectivity outside of us. Consequently, the new, postmodern legitimation of scientific work "is not to produce an adequate model or replication of some

40 For detailed information on the political implications of science during and after the jazz age, see Peter Kuznick's *Beyond the Laboratory: Scientists as Political Activists in 1930s America* (1987).

41 Einstein's rapid fame in the United States "has been puzzling to his biographers and to other commentators" (Missner 267) ever since. When considered in the framework of Lyotard's study, this mystery is, however, easily dispelled: The American public's veneration for the scientist perfectly exemplifies the cultural force of legitimation myths at the beginning of the last century, especially the humanist narrative of science as a means to national emancipation (Lyotard 31-37, Jameson, "Foreword" ix-x).

42 Similar to Judith Butler's later conceptualization of performative (gender) identity (*Gender Trouble*, 1990), Lyotard here draws on the work of linguists John L. Austin (1962) and John R. Searle (1969), who introduced the performative as a special form of speech act that not only describes, but produces the subject. See Austin's *How To Do Things With Words* (1962) and Searle's *Speech Acts* (1969).

outside reality, but rather simply to produce *more* work, to generate new and fresh scientific *énoncés* or statements" (ix, italics in the original), as Fredric Jameson argues in the foreword to *The Postmodern Condition*. In connection to this premise, which seems to echo several tenets of Paul Feyerabend's and Thomas Kuhn's analyses of scientific progress,[43] Lyotard advances a radical vision of science in the postmodern age:

> "Postmodern science—by concerning itself with such things as undecidables, the limits of precise control, conflicts characterized by incomplete information, '*fracta*' catastrophes, and pragmatic paradoxes—is theorizing its own evolution as discontinuous, catastrophic, nonrectifiable, and paradoxical. [...] It is producing not the known, but the unknown. And it suggests a model of legitimation that has nothing to do with maximized performance, but has as its basis difference understood as paralogy." (60, italics in the original)

One instance of such paralogy or search for new meanings in the old and established would be so-called "language games" (Lyotard 10), a concept borrowed from Ludwig Wittgenstein.[44] The plurality of these language games acts as a substitute for overarching legitimation discourses, so that contemporary science itself becomes a set of narratives. Lyotard thereby advocates the comprehensive dissolution of *grands récits*, but also maintains that "the little narrative remains the quintessential form of imaginative invention, most particularly in science" (60). Narrative knowledge, especially its local, heterogeneous, and small-scale dimension, is thus reaffirmed as a central mode of thinking "as legitimate as that of abstract logic"

---

43 Kuhn and Feyerabend are widely understood as the first philosophers of science to have called into question ordinary scientific research and conceptualized the act of 'doing science' in fundamentally different ways. Their historical theories thus not only "stand as crucial symptoms" (Jameson, "Foreword" viii) of a legitimation crisis of science, but also exemplify "the crisis of metaphysical philosophy" (Lyotard xxiv), which can no longer attribute legitimacy in the postmodern scientific landscape. Feyerabend's *Against Method* (1970) famously advocated the anarchic doctrine of 'anything goes' for the practice of science in order not to restrict potential progress. Science should ignore strict methodological rules and not preclude unorthodox ways of research. Feyerabend also opposed the understanding of science as an epistemologically exceptional endeavour that can be clearly delineated from other forms of human activity. For a closer exploration of Kuhn's theorem, see footnote 21 from the previous section of this chapter.

44 The definition of language games in Wittgenstein's philosophy is vague and ambiguous. Most commonly, the term is used to refer to semiotic practices, including the acts of giving order, telling jokes, or reporting an event (Wittgenstein 11-12). For a closer analysis of Wittgenstein's use of the concept, see Nicolas Xanthos (2006), who advances an understanding of language games as parameters for signification and representation.

(Jameson, "Foreword" xi) and a potential substitute for obsolete sources of scientific legitimation.

Although Lyotard's study was commissioned by the government of Québec as an independent *Report on Knowledge* (the subtitle of his book), it must be understood in its academic context, namely as a "thinly yeiled [sic] polemic" (Jameson, "Foreword" vii) directed at the Frankfurt School, especially Jürgen Habermas and his renowned *Legitimation Crisis* (1973, *Legitimationsprobleme im Spätkapitalismus*), which extends the problem of legitimation to the larger context of social and political institutions.[45] For Lyotard, Habermas's suggested remedy for a legitimation crisis, namely a return to the Enlightenment's moral principles of reason and rationality, seems hardly convincing: "[I]t seems neither possible, nor even prudent, to follow Habermas in orienting our treatment of the problem of legitimation in the direction of a search for universal consensus. [...] Consensus has become an outmoded and suspect value" (Lyotard 65-66). Consensus is viewed as detrimental in its totalizing tendency, and no longer feasible in a world that questions absolutes. Rather, Lyotard suggests an entirely novel form of scientific practice, a postmodern science, which will necessarily have to meet new standards of legitimation: "[T]he old humanist (emancipatory) narratives of legitimation are replaced by the new ideological legitimation of science promulgated by the state and corporations in terms of the value of efficiency, which has as its goal power rather than truth" (34), as Michael Peters interprets Lyotard's vision of scientific performativity. Consequently, contemporary sources of legitimation that live up to the challenges of the twentieth-century information society must bc sought elsewhere; mere agreement, let alone the belief in an absolute truth, no longer suffices to justify scientific principles, resources, and prestige.

In temporal proximity to Lyotard's foundational report, a myriad of studies emerged that testify to the momentous significance of questions of scientific legitimation for the budding information society. While *The Postmodern Condition* provides a philosophical framework for the exploration of the legitimation crisis, contemporary sociological and historical explorations attempted to trace its causes back to actual social, political, and economic transformations in American society.

45 While Lyotard postulates a caesura between the modern and the postmodern, Habermas (1976, 1981) views the present as a continuation of the as yet incomplete project of Enlightenment. In line with his philosophical forefathers Hegel and Lukács, he dismisses postmodernism as reductive and reactionary. His position must be considered in the context of his continuing dialogue with his former mentor Theodor Adorno, who critiqued the Enlightenment ideals of rationality and efficiency as having paved the way for the totalitarian regimes of the twentieth century. See Adorno's and Horkheimer's *Dialectic of Enlightenment* (1944) and Martin Jay's *The Dialectical Imagination: A History of the Frankfurt School and the Institute for Social Research 1923-1950* (1996).

Above all, the reason for why "legitimation becomes visible as a problem" (Jameson, "Foreword" viii) in postwar society is attributed to ethically problematic research projects and failed experiments after 1945, which severely challenged people's faith in the "endless frontier" (Vannevar Bush)[46] of science. Both the origin and the climax of this discourse are consistently located in the "1960s and 1970s, as reported abuses of [the public's] trust grew, and the image began to conflict dramatically with reality" (LaFollette 129). Prominent examples of such abuses—in addition to the already discussed Three Mile Island accident of 1979—are the 1959 core meltdown of the Sodium Reactor Experiment in California and the aborted lunar landing of Apollo 13 in 1970. In the wake of these incidents, a notorious *Science* editorial from 1971 stated that the "period of faith in science and technology as an engine of social progress has come to an end" (Schmandt 231), while the *Newsletter of the Program on Public Conceptions of Science* in its issue from April 1973 even noted a wholesale "rebellion against science" (Spencer 10). A decade later, Karin Knorr Cetina still detects the "symptoms of crisis of legitimation that disturb the relation between science and society, a crisis which seems as yet to be far from reaching a turning point" ("Reappropriation" 232).[47] These assessments further fuelled the debate on issues of the public validity and social legitimation of science.

The increase in reported abuses of national research projects can be linked to the aforementioned exponential growth of science in the postwar era. Due to its expansion and branching, it became acutely dependent on enormous Research & Development (R&D) budgets and other forms of state subvention. In the struggle for financial support from the state, science was one competitor among many; its "special political privileges" (LaFollette 16) resulting from earlier modernist legitimation discourses were eliminated after the Second World War. This particularly affected basic science projects, which had to face heightened public demands of applicability they simply could not fulfill: "A [...] result of [science's] expansion is the concomitant increasing need for financial resources and an accordingly growing dependency on public legitimation" (Weingart, *Stunde* 187-188).[48] The penetration

46 Vannevar Bush's 1945 report to the U.S. government suggestively entitled *Science—The Endless Frontier* insisted on the necessity of science funding for the advance of American society. By "arguing for patronage without external controls" (Rip 134), Bush's report established the independence of scientists as a vital prerequisite for progress. The report was also pivotal in shaping the contemporary "images of scientists as new national heroes" (LaFollette 107) of the period.

47 See also Weingart's "Science and Technology in a Legitimation Crisis" (1979), which Knorr Cetina's argumentation, among others, refers to.

48 Original wording: "Eine [...] Konsequenz ist der mit der Expansion einhergegangene enorme Zuwachs an erforderlichen finanziellen Ressourcen und damit eine erheblich gestiegene Abhängigkeit von öffentlicher Legitimation."

of science into all aspects of contemporary postmodern life thus also reflects back on science itself: The more public expenditures that science receives, the greater the political and public demands for its accountability. Furthermore, the already mentioned decline of traditional experimental culture as a result of the shift to collective, cost-intensive international research entails the loss of a formerly potent source of legitimation.

At the same time, the transformations of scientific research over the last sixty years have created heightened public awareness of our fundamental dependence on science. As a result, citizens may harbor disproportionally high and untenable expectations of technoscientific progress, which are habitually raised by science itself: Throughout the twentieth century, both academic and popular publications recorded "a litany of promises that science would cure every disease, fix every problem, and brush away every fear […]. This 'song of science' […] showed a public image of science that was not only inconsistent with reality but also politically unstable" (LaFollette 162). In the information society, technoscience is increasingly put under duress to provide rational solutions for all kinds of problems, ranging from health issues to ecological, financial, and, perhaps most profoundly, ethico-philosophical concerns. Ash thus notes a "pluralization and complication of the discourse of responsibility in the last thirty years" ("Verantwortung" 331)[49] affecting science and its utilization by policy makers. This consideration reflects that a crisis of scientific legitimation is not necessarily or exclusively linked to a greater skepticism toward scientific research, but may also evolve from a positive, even indeed utopian vision of science. Opposing discourses of science legitimation may hence coexist in postmodern society: Analogous to how a monolithic understanding and public image of science is both perilous and inaccurate, the reasons for a legitimation crisis can be manifold and contradictory.[50]

The same, of course, applies to reactions to counteract the crisis, both from within science and the outside. Generally speaking, commentators agree that the authority to attribute legitimacy has shifted from the scientific community to external factors, as Rip points out: "Instead of a struggle for facticity as within science, there must now be a struggle for 'legitimity' [sic] in the public domain" (146). Equally, Weingart observes "wide-reaching dependencies of science on public legitimation" (*Stunde* 28)[51] resulting from taxpayers' subvention of research projects. One of the

49 Original wording: "Vervielfältigung und Verkomplizierung des Verantwortungsdiskurses in den letzten dreißig Jahren."

50 For the plurality of the notions of 'science' and 'science legitimation,' see also Nikolow's and Schirrmacher's introduction to *Wissenschaft und Öffentlichkeit als Ressourcen füreinander* (2007) and its review by Barbara Orland (2009).

51 Original wording: "weitreichende Abhängigkeiten der Wissenschaft von öffentlicher Legitimation."

early remedies adopted by scientists, journalists, and policy makers to increase public acceptance was to encourage science education. These efforts were, however, often characterized by patronizing attitudes, denigration of lay audiences, and a lack of self-reflexivity, so that they have often resulted in science's self-delegitimation rather than affirmation (Wynne, "Understanding" 387-388). Today, mass and new media have been discovered as potential sources of legitimation and public validation, which has resulted, among other scenarios, in the creation of PR departments at major universities and research institutions. Sybilla Nikolow and Arne Schirrmacher (2007) refer to these forms of science communication as "legitimation and acceptance procurement practices" (17),[52] which may, however, backfire and undermine, rather than increase the authority of scientific knowledge.[53]

COOL SCIENCE: (DE-)LEGITIMATING SCIENCE IN POPULAR CULTURE

Notably more prolific in ameliorating the image of science than measures imposed from above are representations of science in popular culture. Gieryn (1987), for instance, shows how the use of science in advertisements not only serves to promote the advertised product, but, in return, reflects back on science itself by capitalizing on its epistemological authority. By displaying an elderly man in a white lab coat, who by virtue of modern science can provide unquestionable proof that Coca Cola is better than its competitors, the ad sells a soft drink *and* an image of science as "truthful, reliable knowledge, produced by disinterested experts, serving the good of us all" (Gieryn, "Coca-Cola" 13). A similar legitimatory effect can be achieved by the presentation of controversial research areas or theories in fiction films (see Kirby 2003, 2011). Science consultants for Hollywood productions may hence act as advocacies for their respective hypotheses, as the case of *Jurassic Park* (1993) or *King Kong* (1933) reflect: The image of an extinct species depicted in a movie fast becomes the most 'authentic' version in the public perception. Similarly, science-fiction and disaster movies as explored in chapter 7 of this study have been utilized to buttress the claims of astronomy, geology, and climatology as indispensable research branches for the survival of the human race and thus deserving of public resources.

52 Original wording: "Legitimations- und Akzeptanzbeschaffungspraxis."

53 Weingart (*Stunde* 232) reports how results are nowadays frequently already communicated to the media and, in consequence, the general public before their official approval by the scientific community. This can, in the long-run, severely damage the trust in science, especially when the supposed results are later revealed as less momentous than expected. See also Hagendijk and Meeus (1993).

In accordance with these preceding analyses, I argue that similar legitimatory effects can be observed in contemporary popular representations of cool science. The fact that the imprecise and non-factual notion of cool increasingly informs the exploitation of science in popular culture is hence understood as yet another response to the postmodern collapse of scientific legitimation. Cool may provide a viable source of legitimation for modern technoscience by responding to, complementing, and indeed substituting former (Enlightenment) discourses of scientific legitimation, which no longer exert actual cultural force in the modern information society. This line of argumentation is thus opposed to (sociological) explanations that exclusively attribute the growing visibility of science in popular cultural productions to the web 2.0 revolution and new channels of scientific popularization. While these technoscientific innovations guided by algorithms certainly play a considerable role in science's growing capitalization on cool in the popular domain, one must not forget that they are also only (by)products of the information revolution of the last decades. As the above survey has shown, the latter is, first and foremost, characterized by the substantial transformations that science had to undergo with regard to public esteem, institutional practice, interaction with related cultural domains, and legitimation of cultural and cognitive authority. I thus argue that the quantitative increase in what the analytical part of this study will define as cool representations of science is fundamentally connected to the postmodern crisis of scientific legitimation and must therefore be read as a response to the demise of former legitimatory sources. The semantically and ontologically evasive concept of cool paradoxically acts as both a challenge to and potential substitute for obsolete sociophilosophical discourses of scientific legitimation as explored by Lyotard's influential study.

This premise does not, of course, amount to the conclusion that the only reason why science is legitimized and publicly accepted today is because it is cool; quite to the contrary, science might be considered the epitome of the utterly uncool in more formalized and/or institutionalized contexts that noticeably differ from its popular cultural exploitations. The focus of analysis is accordingly placed on *representations* of science, specifically within film and television, which present an overall homogenous treatment of the subject matter and thus allow for informed conclusions. By locating new sources of legitimation within the realm of popular culture, the present study argues, on a more general level, that the rationale for science in the twenty-first-century information society is to be sought in the aesthetic domain, a proposition which corresponds to the characterization of postmodernity as an era that merges previously discrete realms of cultural activity: "[A]esthetic impulses have spilled out of the self-consciously defined sphere of art into the spheres of the cognitive and the scientific on the one hand and the practical or moral on the other" (3), as Patricia Waugh maintains by resorting to yet another spatial metaphor. In view of the above discussed permeation of technoscience into all aspects of society,

I would add that the cognitive-scientific can equally be regarded as extending its scope to the ethic and, above all, the aesthetic sphere. Before my analysis illustrates these dynamics in particular representations of cool science, the following chapter briefly discusses the methodological and disciplinary foundation of this study.

Nerd Alert

Science and the Popular

Natural science is not a natural kind.
RICHARD RORTY/
"IS NATURAL SCIENCE A NATURAL KIND?"

In 1926, two eminent physicists, most probably over a cup of tea and some biscuits, discussed the emergent and still controversial theories of relativity and quantum mechanics. They wrangled, as any true scientist would, over the question of objectivity and demonstrable truth. One problem with the revolutionary new paradigm introduced by relativity was that several vital magnitudes, such as electron orbits inside the atom, could in fact not be observed, which severely limited the possibilities of experimental control. The younger of the two researchers seemed deeply immersed in the doctrine of his time and advocated, just as he had been taught, the adherence to the scientific principle of empirical observation at all costs, maintaining that a good theory ought to be based on directly observable phenomena. All non-accessible parameters should not, consequently, be included in the paradigm, but rather ignored as beyond the scope of natural science, since a sound and proven theory can necessarily only make claims about actually observable quantities. The older and perhaps a tinge more experienced scholar dismissed the youngster's proposition as, while pervasive and compelling, nonetheless fictitious nonsense and enlightened his interlocutor about the contingency of scientific research: "In reality the very opposite happens. It is the theory which decides what we can observe."[1]

The two scientists engaged in this elementary debate on the very foundations of their discipline were Werner Heisenberg and Albert Einstein, and the latter's unorthodox view of the semantics of scientific enquiry not only anticipated several vital tenets of postmodern science studies as later initiated by Kuhn and Feyerabend, but also provides a suitable motto for this chapter. Before specific instances of cool legitimatory discourses will be explored in the analytical part of this study, the following pages are devoted to a discussion of its methodological foundations as an instance of the recently emerged discipline of cultural studies of science. Consequently, it will be imperative to delineate a number of key concepts for the ensuing analysis, including, similar to Heisenberg's thus inspired uncertainty principle, a problematization of the concept of representation within the framework of visual cultural studies. W. J. T. Mitchell's and Stuart Hall's pioneering poststructuralist studies on representation will help to prepare the ground for a practicable and elucidative analysis of selected popular cultural formations in the subsequent chapters of this study. Additionally, a brief survey of former and contemporary models of scientific popularization, science in public, and popular science in the historical and political context of the United Studies will outline the body of scholarship to which the methodological approach adopted by this study needs to respond. Finally, a closer inspection of the present study's own use of the term 'science' is in order, which, parallel to the lexical and ontological flexibility of cool, may be characterized by its performative mutability.

CULTURAL STUDIES OF SCIENCE

The theory that decides what shall be observed within the scope of this study is provided by the discipline of (American) cultural studies. More specifically, its main object of analysis, audiovisual representations of science, establishes the ensuing analysis as an instance of the still emergent field of cultural studies of science, a relatively new area of interdisciplinary and heterogeneous research combining philosophy, literary criticism, gender studies, sociology, and history. Similar to American studies, cultural studies of science creatively absorb insights gained by related and more established fields of enquiry like science studies or science and technology studies (STS), as exemplified by Latour's influential explorations of the construction of the laboratory and its objects (1983, 1987, and 1993) and Haraway's

1 See Heisenberg's *Physics and Beyond: Encounters and Conversations* (1969, engl. 1971), p. 63.

analysis of the metaphors informing the study of primatology (1989).[2] More recent scholarship such as, for instance, demonstrated by the seminal collection *Doing Science and Culture* (2000) edited by Roddey Reid and Sharon Traweek, equally acknowledges the influence of various academic backgrounds, research traditions, and institutional experience.[3] Nonetheless, the approach adopted by cultural studies of science differs substantially in several vital aspects from earlier (sociological) approaches to the study of science, as Joseph Rouse explicates in his paradigmatic essay "What Are Cultural Studies of Scientific Knowledge?" (1993):

"[C]ultural studies of scientific knowledge take as their object of investigation the traffic between the establishment of knowledge and those cultural practices and formations which philosophers of science have often regarded as 'external' to knowledge. […] Cultural Studies do not try to replace internalist accounts of knowledge by relying upon a privileged alternative explanatory framework (e.g. social factors), but neither do they grant epistemic autonomy to what is currently accepted as scientific knowledge." (4)

Rouse's polemic dismissal of the predominance of social factors for exploring contemporary constructions of science already suggests that cultural studies of science and their institutional establishment in the course of the 1990s must be understood in context, namely as a response to the until then dominant sociology of science and its social constructionist view of science.[4] The latter postulates that scientific facts

2 See chapter 3 for a discussion of Latour's and Haraway's work.

3 The international conference "Located Knowledges: Intersections between Cultural, Gender, and Science Studies" at UCLA in May 1993 is frequently cited as a founding moment in the history of cultural studies of science. Reid's and Traweek's collection largely derives from a research group established in the wake of this conference. In the introduction, Reid and Traweek relate that what emerged out of these efforts was an awareness "that it no longer makes sense to conduct the business of studying science, technology, and medicine as usual" (4).

4 Rooted in Karl Marx's and Émile Durkheim's groundbreaking work, social-constructivist studies were themselves a reaction to essentialist discourses of science, which postulated that science can be characterized by certain unique, inherent, and universal properties that set it apart from any other cultural domain and, consequently, account for its cognitive and cultural authority, financial resources, and public prestige. Famous examples are Karl Popper's concept of falsifiability (*The Logic of Scientific Discovery*, 1934) or the Mertonian norms of science as established by the American sociologist Robert Merton (1942). The latter proposed a set of four ideals which all scientists supposedly adhere to and which hence defines scientific endeavors. Consequently, epistemic practices that notably disregard one of these essential norms cannot count as true science, a verdict which

always emerge from already preexisting social relations and are thus invested with the preferences, motives, and epistemological preconditions of a particular social group (see Fisher 1990, H. Collins 1988). Sociological analyses thus strive to provide causal explanations for the particular outcomes of scientific practice by exploring "the moment-to-moment activities of scientists as they go about producing and reproducing scientific culture" (Restivo 107). The inescapable emphasis on the pivotal role of the scientist already suggests that such approaches, including the admittedly elucidative studies on boundary work, tend to center their analysis on what literary criticism might describe as the 'author' of a scientific fact, i.e. the respective research conglomerate and its laboratory. Not surprisingly, thus, cultural studies criticism of social constructivist approaches has always loomed large on the postmodern horizon.

In reaction to the strong focus of sociological scholarship on the social construction of scientific facts, cultural studies of science seek to problematize and eventually transcend the binary and only seemingly neat distinction between essentialist and social constructionist approaches to science. The indispensable 'outsider' position, which according to sociological approaches must unfailingly be adopted by the researcher to allow for a prolific analysis of the constitutive mechanisms of scientific enquiry, is dismissed as nothing more than a compelling myth. Quite to the contrary, cultural studies recognize their own epistemic and political engagement in processes of legitimation, scientific production, identity formation, and (dis)empowerment. 'Doing science and culture,' as Reid's and Traweek's seminal collection proposes, is understood as necessitating a basic awareness of how one's own production of (academic or other) knowledge will *nolens volens* constitute yet another, more or less authoritative definition of science, an example of boundary work. Remaining skeptical toward the proclaimed disinterestedness and objectivity of sociological accounts, cultural studies scholarship regularly sets its own, unavoidably subjective borders between science and related segments of professional activity, thus necessarily defining its object of analysis before exploring potential borderlands and zones of interaction. Gieryn's boisterous slogan—"[e]ssentialists *do* boundary work; constructivists *watch* it get done by people in society" ("Boundaries" 394, emphasis in the original)—might sound promising, but must eventually be exposed as inherently flawed. Indeed, by postulating what both essentialists and constructionists, scientists and non-scientists do or do not do, Gieryn is himself drawing an elaborate map of the scientific domain.

Any academic investigation into the cultural formation of scientific practice thus inevitably takes off from a biased position, which cultural studies of science are not only acutely aware of, but embrace as a necessary condition of cultural prac-

would, for instance, apply to genetic research in Nazi Germany (Gieryn, "Boundaries" 395-400).

tices. In contrast to sociological scholarship, the goal of such an analytic approach is anything but the causal explanation of the epistemological consequences of the cultural formations under scrutiny: Postcolonial theory and its cautioning against the non-reciprocal imposition of categories on an unwilling Other have demonstrated that this stance ultimately "reifies the boundaries between the interpretation and what it interprets" (Rouse, "Cultural Studies" 9). Retaining a basic "sensitivity to differences and contested meanings and identities" (Rouse, "Cultural Studies" 6) must thus constitute a central premise for cultural studies critique of science in order to counter universalizing trends and tendencies which propose a false unity of social categories in lieu of their historical and local contingency. Cultural studies of science must thus be understood as a radical form of science critique that "violates established notions of theory, practice, data, and policy" (9), as Reid and Traweek note at the turn of the century. Additionally, it is also the actual objects of enquiry which set cultural studies of science fundamentally apart from other explorations of scientific practices. The cultural productions thus analyzed might indeed by regarded as "'external' to knowledge" (Rouse, "Cultural Studies" 4) by the traditional philosophy of science, or as stemming from the periphery of established science.[5] Rather than 'explaining' the particular results of scientific practice, the focus of enquiry is put on processes of meaning making, articulation, and sustainment, refuting the idea of a stable and objective outside reality: "[T]he agencies and actors are *never* preformed, or prediscursive, just out there, substantial, concrete, neatly bounded before anything happens, only waiting for a veil to be lifted and 'land ho!' to be pronounced. Human and nonhuman, *all* entities take shape in encounters, in practices" ("Cat's Cradle" 65, emphasis in the original), as Haraway, a chief proponent of the discipline, stresses.

Cultural studies of science hence perpetuate several central premises of post-structuralist thought, which the present study is predicated on.[6] In accordance with

5 Famous examples are Haraway's analysis of the advertisements in *Science* (1992), Constance Penley's study of the cultural and symbolic co-option between *Star Trek* and NASA (1997), and Reid's study of the personal investedness of researchers exploring tobacco control (2000).

6 Both Hall's and W.J.T. Mitchell's understanding of representation harks back to Foucault's notion of discourse and his succinct formulation that "nothing has any meaning outside of discourse" (*Archeology* 32), a central tenet of postmodern cultural theory. Accordingly, representations must be regarded as contributing to the construction of shared attitudes, ideas, and possible statements about a given subject, which are, as Foucault meticulously pointed out, necessarily entwined with the (never singular or monolithic) negotiation of power and knowledge. Foucault' writings are generally understood as advancing a narrower concept of representation than employed in cultural studies scholarship today, which is also referenced by his focus on the production of power and knowledge,

the constructionist approach in cultural theory (Hall, *Representation* 25-26), the representations of science under scrutiny in this study are treated as 'texts' in the broadest sense, i.e. sites of meaning production and negotiation, rather than mimetic reflections of some objective, external reality. The present study thus largely follows Stuart Hall's seminal definition of representation as "the production of meaning through language" (*Representation* 28), wherein 'language' involves various representational systems, which can all function as "signifying—i.e. meaning-producing—practices" (*Representation* 28). Accordingly, the primary texts analyzed in the following analytical sections of this study were scrutinized, via close reading and sustained interpretation, for their potential to produce, rather than merely reflect knowledge through language, image, and discourse. In contrast to the more empirically oriented social sciences, the focus of the present study as an instance of a cultural studies critique is not the text's specific purpose as intended by its authors or producers, such as their objective of presenting science in a more or less favorable light, as these personal motivations are ultimately inaccessible. Ra-

rather than meaning, through discourse: "[O]ne's point of reference should not be to the great model of language (langue) and signs, but to that of war and battle. The history which bears and determines us has the form of war rather than that of a language: relations of power not relations of meaning" (*Power/Knowledge* 114-115). This tendency of Foucault's vast oeuvre must be seen in its context, as he specifically aimed to distance himself from structuralist approaches (Hall, *Representation* 42-43). Accordingly, his inclusive definition of discourse well exceeds the structuralist constraints of verbal language and has had a major impact on later theorizations, as both Hall and Mitchell acknowledge. Next to Foucault's notion of discourse, Jean Baudrillard's notion of simulation was equally influential for this strand of cultural critique. According to Baudrillard, visual and verbal signs (or 'signifiers,' in Saussurean terms) are understood as having no real content (or 'signified') they refer to. There is, in short, no underlying, relevant truth or reality which representations (in Hall's and Mitchell's sense) stand in for, a proposition which was subsequently popularized as the famous 'loss of the real' in postmodernist thought. While neither Hall nor Mitchell explicitly reference Baudrillard as a major source of influence, his radical notion of hyperreality, denoting the postmodern collapse of the distinction between the real and the simulated, vitally contributed to poststructuralist theorizations. The problematization of signification, luridly popularized as 'the crisis of representation' or 'the primacy of the signifier,' also emerges in the eminent works of Roland Barthes, Jacques Lacan, and Jacques Derrida, whose "axial proposition [...] that there is nothing outside the text" (*Grammatology* 163) epitomized the very crisis. See especially the following publications: for Baudrillard: *Simulacra and Simulation* (1981); for Derrida: "Structure, Sign and Play in the Discourse of the Human Sciences" (1966), *Of Grammatology* (1967); for Lacan: *Écrits* (1936-1966); for Barthes: "Death of the Author" (1968), *Mythologies* (1972), *S/Z: An essay* (1975).

ther, cultural expressions, images, and styles circulated in society (such as, in the present case, notions of cool) are treated as open to revision and active interpretation, since there is, bluntly put, no ultimate, external meaning to be explained for once and for all. Exceeding the realm of mere reflection, the cultural representations of science selected for close reading are necessarily treated as actively contributing to the specific construction of both dominant and emergent societal ideas.

Given its focus on film and television, i.e. cultural formations that foreground the visual dimension, the present study also resorts to the methodological framework provided by W.J.T. Mitchell, who announced the pictorial turn in cultural theory.[7] In contrast to structuralist analyses, which prioritize textual phenomena, Mitchell called for a greater scholarly attention to the role of pictures in shaping contemporary existence. However, he cautioned against a purified vision of representation as being either verbal or visual: "[T]he interaction of pictures and texts is constitutive of representation as such: all media are mixed media, and all representations are heterogeneous" (W.J.T. Mitchell, *Picture Theory* 5). This premise proves particularly viable for the purposes of this study, since the scrutinized primary texts always also involve a considerable audio-verbal element, which is equally taken into account in the analysis. Similarly, Mitchell's definition of the signifying work performed by linguistic and non-linguistic symbolic systems as a process which "not only 'mediates' our knowledge […], but obstructs, fragments, and negates that knowledge" (W.J.T. Mitchell, *Picture Theory* 188) also greatly informs the subsequent analysis of primary texts. The present study is thus founded on an understanding of representation, in the broadest sense, as transcending mere reflection by creating and circulating meaning, knowledge, and power. Neither pictures nor other forms of signifying practices, including spoken and written language, express some sort of pre-existing, single, and fixed meaning (possibly injected by the author of the particular text), but allow for a range of potential meanings constructed by the respective reader.

Additionally, the aesthetic representations found, for instance, in classic literature, visual arts, or, in the present case, popular culture "can never be completely divorced from political and ideological questions" ("Representation" 14), as W.J.T. Mitchell convincingly demonstrated. This concurs with the general methodological orientation of cultural studies of science, which reveal the "*semantic* realism"

7 In his proclamation of the pictorial turn, W.J.T. Mitchell is indebted to (as he acknowledges) Derrida's critique of logocentrism, as well as, somewhat paradoxically, Foucault's and Wittgenstein's alleged "iconophobia." Particularly with reference to Wittgenstein's philosophy, he writes that "[t]his anxiety, this need to defend 'our speech' against 'the visual' is, I want to suggest, a sure sign that a pictorial turn is taking place" (*Picture Theory* 11-12). Mitchell's pictorial turn must be understood as a response to Richard Rorty's *The Linguistic Turn* (1967), which set the paradigm for subsequent analyses.

(Rouse, "Cultural Studies" 17, emphasis in the original) adopted by many sociological studies as inherently flawed. There is no fixed content to be explained for once and for all nor can a given theory make determinate assumptions and observations about the world, a basic tenet of poststructuralist thought frequently theorized as the crisis of representation.[8] In addition to retaining a basic reflexivity about the present study's political and epistemic engagement in its objects of analysis, I thus also deem it vital to draw attention to the unavoidable contingency of any interpretation, which can never objectively explain cultural formations 'from the outside.' Every cultural reading is essentially also part of that culture. The present study is no exception to the rule, especially when considering its subject matter, the notion of science, which any cultural critique investigating its dynamics, legitimation, and representation will help to define and demarcate in the wake of it.

SCIENTIFIC POPULARIZATION, POPULAR SCIENCE, AND SCIENCE IN PUBLIC

Given the disciplinary emphasis on the fundamental situatedness of science within the broader societal fabric and the analytical focus on popular cultural formations, the present study necessarily enters into a close dialogue with previous (and, generally speaking, sociologically orientated) scholarship on the interaction between science and the public. This relationship has often been summarized by the buzz term 'popularization.' In the course of the immensely heterogeneous, transdisciplinary academic engagement with issues of popularization, popular science, and public understanding of science (PUS) since the 1950s, a range of models and methodological approaches have emerged, which the following section will briefly review

8 Owing to the poststructuralist problematization of cultural representation, the distinction between surface and depth (the 'underlying meaning'), which the traditional act of interpretation, be it in literary studies or the social sciences, fundamentally relies on, was exposed as artificial and methodologically flawed by poststructuralism. Paralleling the crisis of legitimation, the so-called crisis of representation challenges the idea that scholars can remain outside the limits of their object of enquiry and produce a stable and objective account of it. The question of how a theory can ever claim to be accurately and straightforwardly describing, reflecting, and corresponding to a given cultural condition is thus intrinsic to the postmodern academic landscape. Cultural studies of science react to this crisis by acknowledging that "the categories and practices of social explanation themselves belong to a scientific tradition" (Rouse, "Cultural Studies" 10), thus making it imperative for cultural theorists to "consider their own complex epistemic and political relations to the cultural practices and significations they study" (Rouse, "Cultural Studies" 10). See also Ebert (1986) and Lynch (1994).

and assess. Starting from this overview, I will consequently advance the specific approach the present study adopts with regard to its analysis of science-public interactions in the realm of popular culture. Considering the study's focus on popular cultural representations of science within film and television, it proves most feasible to combine and adapt different disciplinary outlooks on the complex relationship between science and the public. Additionally, it is the very concept of 'popularization' itself that this study necessarily needs to problematize in order to meet the demands of cultural studies critique.

The study of science and the public has long been dominated by the so-called linear (diffusion) model.[9] This model implies a hierarchy between the two poles, suggesting that scientific knowledge is first produced in a socially remote vacuum and, in a second step, imposed in highly simplified form on the largely passive and unquestioning public. Popularization was thus for a long time conceptualized as an epistemological one-way street, as the public was not found capable of reacting meaningfully to scientific findings. The linear model is based on electronic communications models (Gregory and Miller 86), which involve a sender ('science' as produced by trained experts), a transmitter (science journalists and, more generally, 'the media'), and a receiver (the humble, untutored public).[10] While the roles may (and in actual fact will) be habitually reversed in real communication, the linear model does not envisage such a possibility: The receiver is never able to react to claims by the sender. The terminology employed by studies predicated on the linear model already implies a strict hierarchical structure, which is rooted, as Stephen Hilgartner observes, "in the idealized notion of pure, genuine scientific knowledge

9 Jane Gregory and Steve Miller (1998) introduce a distinction between the linear model and the diffusion model, in which the latter is supposed to denote a "two-dimensional" (87) version of the former. In contemporary science studies critique, both terms are habitually used interchangeably to refer to the top-down, unidirectional view of scientific popularization (Wynne 1995, Cooter and Pumfrey 1994). Alternative terms include the 'educational model' (Weingart 2001, "Aufklärungsmodell"), the 'translation model' (Lewenstein 1992), or the diffusionist model.

10 Transmitters are given a particularly thankless task within this framework, as they will frequently encounter complaints by scientists for not having communicated their findings accurately. A case in point is the aforementioned Three Mile Island accident, whose disastrous effects on the public image of science were, among other factors, attributed to the flawed and contradictory news coverage by supposedly untrained science journalists, who were regarded as unable to understand and thus appropriately mediate the nuclear meltdown to their audience (Wilkes 2002). See also chapter 3 of this study and LaFollette (1990), in particular her chapter "Mass-Circulation Magazines and the Popularization of Science" (18-44), for closer information on science journalism in the United States.

against which popularized knowledge is contrasted" (519).[11] Derogatory synonyms like 'vulgarization' or 'debasement' reflect this attitude: Scientists are granted the ultimate authority to determine which simplified version of their pure results is appropriate and allowed to be disseminated, since the general public will not be competent enough to grasp the original formulations that were produced for the peer community. Accordingly, popularization primarily fulfills a political function, as it provides persuasive rhetorical and conceptual strategies to support the cultural authority of researchers against challenges by 'inferior' actors, including policy makers, journalists, and the general public.

Not surprisingly, the outdated linear model has been critiqued from a variety of disciplinary angles.[12] An early alternative was proposed by Hilgartner's study (1990) of the discourse surrounding the causality between cancer and diet, which suggested viewing popularization as a matter of degree: Simplified or readjusted versions of recent research are found not only in media accounts or school textbooks, but also in traditionally 'scientific' contexts, both within and beyond the respective disciplinary boundaries. The evident spectrum of popularization from sensationalist tabloid articles to recapitulating reviews in scientific journals, from highly professionalized jargon to everyday speech, exposes the binary distinction of epistemic production into either 'popularized' or 'genuine' knowledge as inherently problematic. This leads Hilgartner to conclude that "[t]he boundary between real science and popularized science can be drawn at various points depending on what criteria one adopts, and these ambiguities leave some flexibility about what to label 'popularization'" (528).[13] The dichotomy between 'authentic' and 'popularized' ac-

11 Historically, the linear model perpetuates the educational discourse of the Enlightenment and is grounded in the social and scientific landscape of the late eighteenth and nineteenth century (Schirrmacher 80). The period witnessed the rise of an educated, bourgeois public sphere (the Habermasian *Öffentlichkeit*), characterized by an increase in formal education for a greater percentage of the population and the development of a culture of thriving public dialogue owing to the establishment of newspapers, meeting places, and public lectures, like the Lyceum and Chataqua movements in the United States. See also Lewenstein (1992), Felt (1994), Ash (2002), and Myers (2003).

12 Additionally, the academic critique of the linear popularization model has also been criticized. David Edgerton (2004), for instance, claims that this view of popularization, in actual fact, never existed nor was it actively practiced and adhered to by scientists and policy makers. It is, as he maintains, a mere "invention of *academic* commentators" (Edgerton 31, italics in the original) for the purposes of historiography and scholarly propaganda, supporting what he calls a flawed "academic-research-centered model of science" (Edgerton 48).

13 Similarly, the dichotomy between expert and layperson should be exposed as a continuum: "Administrators, medical students, patent lawyers, post-docs, technicians, science

counts thus amounts to nothing more than a rhetorical distinction, frequently maintained by scientists themselves to lend weight to their concerns.

Another feasible metaphor is proposed by Ulrike Felt, who argues for an understanding of popularization as a process of "mutual influence and border re-adjustment" ("Wechselwirkungen" 47).[14] Felt specifically identifies four types of boundaries that are drawn, defended, and readjusted when scientific knowledge becomes popularized: borders between scientific disciplines, between science and non-science (or pseudo-science), between experts and laypeople, and between science and politics (Felt, "Wechselwirkungen" 50-51). Drawing on Felt, Mitchell Ash's notion of science and the public as mutual resources advances a similar understanding of popularization, which encourages us to envision their interaction as a tightly interwoven web with reciprocal effects. Ash and Felt thus advocate a "model of exchange" (Ash, "Ressourcen" 351),[15] in which science and the public make use of each other's resources to support, legitimate, or challenge their respective concerns. Accordingly, they both caution against a monolithic understanding of the public, which ought not to be conceived of as a preexisting, fixed, and coherent entity, but as only being constituted in the wake of scientific popularization endeavors, which tend to be directed at a very specific societal group (Ash, "Ressourcen" 349-350, Felt, "Wechselwirkungen" 49-50).

'The Web of Science Communication Contexts' suggested by Bruce Lewenstein (1995) takes a similar line and transcends bi-directional, but still linear conceptualizations of popularization by no longer conceiving of the relationship between science and the public in hierarchical terms, but in the shape of a network. Diverse scientific and presumably 'lay' discourses are interconnected, overlapping, and leading to each other. Thus, experiments in the lab may depend on prior meetings, which can in turn only be realized by a successful grant proposal. The grant proposal for a new research project may have been inspired by journal articles, informal talks with colleagues, and reports in the mass media. Popularization is accordingly no longer understood as external to the scientific production process, but constitutes an integral part of it. In correspondence with Felt's and Ash's model of exchange, the web model implies a mutual enrichment between science and the public, whereas both areas are indispensable for epistemic production and codification. This approach is particularly helpful in explaining how and why populariza-

journalists, and research scientists in commercial firms all take on, sometimes uncomfortably, an identity between expert and lay" (268), as Greg Myers, in discussing Hilgartner's study, points out.

14 Original wording: "Wechselwirkungen und Grenzverschiebungen."

15 Original wording: "Austauschmodel."

tion efforts may backfire and not achieve the desired effect.[16] Underlying the model is an understanding of the public as a dynamic, participatory collective and of science as a discursive, rather than naturally given and politically neutral concept.

A more recent model is offered by Arne Schirrmacher (2008), who introduces six stages of popularization that reference the plurality and social differentiation of science and the public, each of which can be grouped in three distinct categories. The range of audiences for scientific knowledge thus spans from 'broad public' to 'professional audience' to 'scientific community.' Studies on the specific dynamics guiding processes of popularization have also been advanced from the vantage point of linguistic discourse analysis, to which the journal *Discourse Studies* devoted a special issue in 2003. Its editor, Helena Calsamiglia (2003), for instance, argues for an understanding of popularization as the "*recontextualization of scientific knowledge*" (143, italics in the original), which is subject to conflicting interests just like any other social phenomenon. Another perceptive linguistic take on popularization is advanced by Sophie Moirand (2003), who proposes a circular model with a simultaneity and interchangeability of roles, so that the source and the consumer of media messages on scientific contents are conflated. The validity of scientific claims is thus not imposed from above, but subject to social negotiation and dialogue.[17]

16 While the linear model understood popularization as "an act of persuasion" (Gregory and Miller 85), based on the 'deficit model' of the public as a mass of ignorant laypeople, several astute studies demonstrate that the opposite may, in fact, take place (Wynne 1991, 1995). Increased understanding of science might lead to a critical public which is acutely aware of the threats posed by potentially detrimental research projects. The equation of popularization with greater public acceptance of and trust in science must thus be revealed as inherently flawed. Equally, attempts at procuring political patronage or resources by informing non-experts about new research directions do not necessarily yield the desired outcomes, a circumstance which Lewenstein's web model can help to account for.

17 Earlier models which also challenged the linear model of popularization, but were, in general, still committed to a view of science as exceptional and intrinsically authoritative were, for instance, developed by James Grunig (1980), Robert Logan (1991), and, to a lesser degree, the edited collection by Shinn and Whitley (1985). Grunig proposes a situational model of communication, which aims to correspond to the complexity of contexts and actors involved in popularization processes. Logan, on the other hand, argues for a view of popularization as secularization, since the sacral aura of science must first be rejected before it can be integrated into public discourse. Shinn and Whitley, finally, treat popularization as a rhetorical strategy that allows scientists to gather support and financial resources.

From a cultural studies perspective, the term popularization is to be revealed as inherently problematic, since it "belongs to a discourse of analysis that is ideologically and culturally loaded: as it assumes differences or boundaries between cultures and classes, so it assumes that the only object of analysis is the transfer of cultural items across such boundaries" (Cooter and Pumfrey 248). Indeed, popularization is far from an easily demarcated, transparent process, but should rather be thought of as a continuum. Just as the concepts of 'science' and 'public' must be revealed as flexible, relativistic, and locally as well as historically contingent, one must equally caution against a monolithic and essentialist understanding of 'popularization' and, accordingly, 'distortion' and 'accuracy.' Indeed, the public ought to be viewed as an active contributor to the process of knowledge production, rather than a passive receiver: "[T]he traffic across the boundaries erected between science and society is always two-way" (Rouse, "Cultural Studies" 13). It appears sensible to think of these resources as being structured on a spectrum, where boundaries are unavoidably blurred and fuzzy: The consumption of science always also involves its appropriation, adaption, and, in the broadest sense, (re)production.[18] Embarking from this basic problematization of popularization discourses, the present study aims to contribute to the debate by advancing its own approach for the analysis of interactions between science and popular culture.

SCIENCE AND/ IN/ AS POPULAR CULTURE: THE COOL APPROACH

In view of the above discussed critique of popular science models, it appears not only sensible, but indeed imperative for studies investigating the forms of exchange between science and popular culture to overcome the very notion of popularization, a culturally loaded concept which always already implies an inherent hierarchization and value judgment as well as an essentialist, homogenous, and esoteric view of science. In line with the above explored studies advancing continuum, network, or circular models to illustrate the relationship between science and the public, it might even prove more feasible to think of science *in* popular culture or science *as* popular culture (Cooter and Pumfrey 1994). As Rouse argues with reference to Latour's seminal laboratory studies, meaningful academic engagement with epis-

18 In that respect, so-called scientific illiteracy must equally be problematized: Wynne (1995, 1991), for instance, shows that "ignorance of science is actively constructed in tacit accordance with the contours of existing relationships, divisions of labor, dependency, and trust" ("Understanding" 380), which makes its conceptualization as a mere 'lack of understanding' seem premature and inaccurate. Ignorance can be discreetly structured, positively loaded, and strategically employed.

temic production needs to "destabilize[…] any distinction between what is inside and outside of science, or between what is scientific and what is social" ("Cultural Studies" 13). As an obvious example, the mere act of 'doing science' depends on material resources and social actors, which tacitly determine what kind of knowledge will be produced in what kind of ways and for what ends.[19] Rather than studying 'science and the public' in order to determine the degrees and processes of permeation of the one into the other, it seems more adequate to focus on the general appearance of science in popular culture beyond the limiting label of 'popularization.' Equally, the notion of 'popular science' as such appears problematic, since it is frequently not possible or analytically desirable, as the above account has shown, to draw a clear-cut line between original and mediated versions of scientific contents. 'Consumption' of scientific information, in whatever shape and by whichever means, is thus necessarily a creative process which always also involves an appropriation and reconstitution of the original resource. Additionally, it can lead to resistance just as frequently as it leads to acceptance of the obtained information.

The "ring-fenced methodological neutrality" (253) demanded by Roger Cooter and Stephen Pumfrey to approach 'science' as an object of analysis thus seems indispensable for any scholarly engagement with the concept.[20] Only when equipped with this strategic detachment will the researcher be able to circumvent, as Joseph Rouse argues, "the mistaken assumption […] that scientific knowledge belongs to a single kind, similar or distinguishable *in kind* in any interesting way from other kinds" ("Cultural Studies" 8, emphasis in the original). Regarding institutionalized, codified science as one sphere of discursive knowledge production among many others, on par with religion, superstition, common sense, astrology, and craft knowledge, does not, as PUS research frequently seems to imply, open the door to unbound relativity and a dangerous neglect of a rational, scientific worldview. Rather, such a view is crucial for discovering that science and its representations, just like any other cultural practice, are never neutral or aloof from profane matters, but

19 As an example, Rouse cites Sharon Traweek's study (1988) of the cultural and political dynamics accompanying the use of particle detectors in high energy physics. Detectors in the United States are short-lived, built by the scientists themselves, and replaced regularly, while Japanese detectors are built by industrial companies for a long period of time, usually surviving several generations of researchers working on them. These obvious differences in material circumstances necessarily affect the kind of questions asked, the knowledge produced, the style of articles submitted to academic journals, and the status attributed to the individual scientist in the research conglomerate.

20 This does not, of course, entail that studies investigating the various forms of entanglement between science and/ in/ as popular culture must bracket questions about the political uses of science or about their own epistemic investedness in the object of analysis. See above and Rouse (1993).

historically variant, locally and materially contingent, discursively constituted, and politically loaded. The same applies to related concepts like 'ignorance,' 'simplification,' 'understanding,' 'rationality,' or 'progress,' and, with reference to the analytical focus of the present study, 'popular culture.'

In that sense, the generally promising spatial conceptualization of science and its popularization as proposed by Thomas Gieryn's notion of boundary work (1983, 1995, 1999) and elaborated in subsequent studies needs to be revised and extended to meet the requirements of cultural studies critique: Rather than presuming coherent social identities and interests for the execution of scientific boundary work (the 'author' of a scientific fact or map), it might often prove more feasible to think of map-drawing in terms of the performative: It is through the "stylized repetition of acts" (Butler, "Performative Acts" 519) that specific maps gain a certain degree of stability and appear incontestable and intuitive, rather than heavily combated, hazy, and unstable. The performative, rather than essentialist nature of seemingly coherent spatial entities thus successfully camouflages the fact that all cultural spaces, be they real or fictional, are subject to potentially violent border negotiations, which regulate the attribution of meaning, truth, and power: "[S]pace," as Foucault maintains, "is fundamental in any exercise of power" ("Space, Knowledge, and Power" 252).[21] As with the production of gendered bodies through performative acts, the repetition and control of borders (and adequate punishment in case of noncompliance or transgression) aids to solidify them and thereby defines a certain cognitive area as the 'natural' territory of science. Eventually, this will create a profound "*appearance of substance*" ("Performative Acts" 520, italics in the original), as Judith Butler compellingly argued in quite a different context. The space we call science could hence be exposed as part of "prevalent and compelling social fictions [...] which, in reified form, appear as the natural configuration" ("Performative Acts" 524) of cultural spaces. As Butler's theory on gender performativity has, however, also demonstrated, this act remains quite independent of the individual ac-

21 Space and spatiality figure, of course, prominently in Foucault's thought. Long before contemporary demarcation theories of science by Gieryn and others emerged, Foucault asserted the fundamental spatiality of knowledge and argued for an adequate form of critique: "[T]he formation of discourses and the genealogy of knowledge need to be analyzed, not in terms of types of consciousness, modes of perception and forms of ideology, but in terms of tactics and strategies of power. Tactics and strategies deployed through implantations, distributions, demarcations, control of territories and organizations of domains which could well make up a sort of geopolitics [...]" (*Power/Knowledge* 77). The spatial conceptualization of science as one of the most regulative segments of American society thus appears particularly promising, if only, however, with the proposed shift of focus from social to cultural factors.

tor and is anything but a radically free choice, being both historically and culturally contingent.

In this respect, the notion of the performative can help to account for the great sedimentation, if not indeed naturalization, of vast areas of the scientific territory. Contemporary American science (and Western scientific practice in general) is, per definition, opposed to both religion and politics, and no elaborate argumentation, nor even deliberation, is needed to expose ufologists and phrenologists as attention-seeking pseudo-scientists. These examples vividly demonstrate that although its boundaries might be prone to blur, the vast majority of our common cultural map of science is presented to us as highly inflexible and hardly subject to revision at all. While these essentialist categorizations and hierarchizations are highly evident and, to a certain degree, expedient in everyday cultural encounters, academic critique is not immune to this pervasive "regulatory fiction" (Butler, "Performative Acts" 528) either: The historical, local, and material contingency of science may well be evoked as a fundamental premise in contemporary science studies scholarship, but the actual academic engagement with 'science and the public' continues to reveal a noticeable bias. Underlying a great number of studies, both dated and recent, is the tacit presupposition that a more scientific worldview, the integration of rational knowledge into mundane discourses, and a scientifically enlightened public will necessarily lead to a 'better' society, however vague its definition (Ash, "Ressourcen" 362). Contemporary controversial fields of research like eugenics or cloning demonstrate that the notion of scientific progress informing much of the academic discussion on popularization must to be treated with caution. Instead of unequivocally supporting the progress-orientated undercurrent of traditional popularization narratives grounded in the Enlightenment tradition, cultural studies of science must adopt a more ambivalent, postmodernist stance toward their object of analysis, showing that science (and its acceptance by the public) does not equal progress, nor does progress equal more livable lives for everyone.[22]

In addition to the necessary problematization of science as the key concept of analysis, there has been considerable debate over which sorts of texts should be analyzed as participating in popularization processes and hence taken as representative of popular science. While researchers like Hilgartner correctly establish a continuum between scientific and popular texts, the focus is, generally speaking, still placed on written material. Visual representations of science and their interaction with the public sphere are left largely unexplored, which Myers assesses as problematic: "The focus on the words in popular science texts tends to reinforce the as-

22 The notion of the 'livable life' is borrowed from Judith Butler as advanced in her *Undoing Gender* (2004) and *Precarious Life* (2004). For a critique of science as a value-free, apolitical arena of activity, see Mark Walker's *Science and Ideology* (2002) and Joseph Rouse's *Knowledge and Power* (1987).

sumption that popularization is just a matter of simplifying and perhaps distorting the original message provided by science" (272). Within the field of science and technology studies (STS), Derek de Solla Price already criticized this "papyrocentricity" ("Parallel Structures" 169) of the majority of science research in the early 1980s. A similar concern is expressed more recently by Cooter and Pumfrey (1994), who bemoan the lack of studies on the dynamics and effects of alternative, 'low brow' representations of science beyond the traditional, comparatively well explored realm of science journalism:

"[S]urprisingly little has been written on science generally in popular culture, past or present. [...] From coffee houses to comic books and chemistry sets, from pulpits to pubs and picture palaces, from amateur clubs to advertising companies, from Science Parks to Jurassic Park, our ignorance both of the low drama and the high art of science's diffusion and modes of popular production and reproduction is staggering." (237)

While there are already a number of (chiefly quantitative) studies on the depiction of scientific parameters in comics and popular literature (Haynes 1989, 1994, 2003), film (Pansegrau 2008, Weingart 2008, Shortland 1987 and 1989, Dubeck, Mosher, and Boss 1988), clipart cartoons (Schummer and Spector 2008), science fiction (Brandt and Casser 2007, Yergensen 2009), and television (Basalla 1976, Gerbner et al. 1981), popular renderings and their discursive construction of scientific characters, metaphors, and issues of authority and legitimation still leave much to explore from an American cultural studies perspective.

The present study is intended to counteract this perceived paucity of studies on fictional, popular cultural forms of scientific representation within the historical, political, and epistemological context of the United States. As Cooter's and Pumfrey's observation already suggests, much can be gained by shifting the focus of analysis to new and underexplored contexts and dissecting "the politics of the popular" (247). Especially the supposedly trivial and banal treatment of scientific matters in popular cultural texts, such as the nineteenth-century cabinet of wonders, twentieth-century advertisements (Gieryn 1987), or, in the present case, contemporary American film and television, can yield remarkable and otherwise inaccessible insights into dominant cultural images and discourses of science in society. An élite focus on dominant 'high brow' forms of scientific production, negotiation, and representation not only imposes a flawed cultural hierarchy, but neglects important discursive spaces of epistemic engagement.[23] Accordingly, the present study's focus

23 This view seems particularly grotesque when considering that historically, scientists did not necessarily enjoy a high social status nor was their activity regarded part of 'high' culture. Steven Shapin (1991), for instance, shows how the roles of gentleman and scientist were regarded as mutually exclusive in early modern England. This example reflects

on 'cool representations' is intended to instigate a fresh perspective in cultural studies scholarship of science by examining an *aesthetic* and *affective* category that would typically not be suspected to exist in close proximity to hard science. In that sense, the seemingly incontestable, coherent, and intuitive space of science may be thought of as expanding its repertoire via the repeated, strategic utilization and adaptation of cool strategies and encroaching into what are typically perceived as channels of popular culture—including the celluloid world of film and television. The scrutinized representations of science can thus be seen as inducing (and solidifying over time) a performative readjustment of boundaries or, conversely, an overlapping of the supposedly natural and distinct configurations of 'science' and 'popular culture,' which is understood as a characteristic feature of the growing scientization and technologization in postmodern information societies. Rather than exploring the popularizing effects of the examined representations or the accuracy of their scientific contents, the goal is to determine how traditional channels of American popular culture, informed by the all-pervasive notion of cool, are appropriated as resources for legitimatory purposes of science. The subsequent analysis of a number of selected examples from US-American film and television will demonstrate that cool as a popular and aesthetic, rather than cognitive or ethical form of scientific legitimation is highly viable and persistent.

how the study of science and/ in/ as popular culture is always also a study of power relations. See also Rouse (1991).

Cool Forensics and the Spectacle of Technoscience in *CSI: Crime Scene Investigation*, *CSI: Miami*, and *CSI: NY*

This is the show that has made science cool.
OPRAH WINFREY/THE OPRAH WINFREY SHOW

CSI: Crime Scene Investigation (2000-present) and its equally popular spin-offs *CSI: NY* (2004-2013) and *CSI: Miami* (2002-2012) are prime instances of how the rather timeworn genre of classic American whodunits has lately become invested with science and witnessed a glorious revival as a result. Interestingly enough, it seems to be their very emphasis on the promises and trappings of DNA processing, toxicology, and human dissection to which the shows owe their by now global success and inescapable coolness factor. *CSI*'s detective investigations hardly ever occur in the streets or the interrogation room, but are conducted in the subtly lit high-tech laboratory, topped off with catchy rock music and film-style camerawork. Far from stereotypical science nerds, the crime scene investigators are renowned for their indestructible cool pose, a direct effect of their professionally detached attitude. The show's cutting-edge splatter aesthetics, tabooing neither the graphic display of mutilated limbs nor necrophilic money shots seemingly penetrating the wounds of bloated corpses in fishnet stockings, further enhance its ubiquitous aura of coolness, as do frequent guest appearances of veritable MTV celebrities, such as Zac Efron, Tony Hawk, and LL Cool J. With its focus on forensic investigation,

CSI has thus introduced a new level of aesthetic stylization for the prime-time police procedural and successfully established its signature cool look.

This chapter[1] examines how cool is produced and circulated in the *CSI* universe, both as an aesthetic strategy and a form of individual performance that is intricately entwined with the show's particular constructions of science and technology. While *CSI* is frequently discussed with regard to its clichéd and morally reassuring plot formula and the unequivocal division of the American population into 'good' and 'bad' citizens, its representations of high-tech gadgetry, scientific procedures, and corresponding constructions of coolness have so far not been explored. Indeed, the escapist storylines noticeably clash with the program's progressive visual style, characterized by a sleek and glossy surface, hyperkinetic editing, and color-saturated imagery that make each episode look like a long-drawn, excessively retouched music video clip. Consequently, I will argue that underlying the *CSI* format is a profoundly ironic tension between a conservative narrative, on the one hand, and cutting-edge imagery, on the other. This juxtaposition, together with the proficient exploitation of the spectacular potential provided by high technology and autopsy, will be treated as a vital prerequisite for the show's constructions of cool forensic science.

In accordance with my initial definition of cool as both an aesthetic and a personal strategy (see chapter 2), the analysis in this chapter will be two-fold: First, I will explore how cool informs the implicit and explicit characterization of *CSI*'s protagonists, the omnipotent and omniscient scientist-detectives, who are frequently portrayed in iconic cool poses and endowed with the apposite props, costumes, and instruments. Their cool poise, as I will show, is directly linked to their scientific expertise, which calls for professional detachment and subdued emotionality. Secondly, my analysis will focus on the visual exploitation of high technology and forensic science for a cool visual effect through aesthetic juxtaposition. For this purpose, I will engage in a close reading of one of *CSI*'s signature scenes, namely an example of its notorious lab scenes, which heavily capitalize on the spectacular potential of technoscientific tools and gadgets. Science, as the close analysis of *CSI*'s depiction of lab work will demonstrate, is perceptibly stripped of its cognitive and ethical qualities in order to foreground the aesthetic dimension and provide genuine visual excitement to the lay audience. This visualization of scientific practice not only sets the program apart from classic police procedurals, but also from most other texts scrutinized in this study, which merely reference, rather than display scientific procedures.

---

1 A heavily abridged version of this chapter appeared in the edited collection *Is It 'Cause It's Cool? Affective Encounters with American Culture* (Fellner et al., 2014) under the title "C.ool S.exy I.ntelligent: Cool Forensics and the Spectacle of Technoscience in *CSI*." See Kohlenberger 2014.

The following pages shall thus serve to demonstrate that *CSI* is heavily informed by notions of coolness on both the visual-aesthetic level and the level of characters. Underlying the use of cool as both an aesthetic and an individual strategy is a juxtaposition of polarities, which mirrors the program's juxtaposition of conservative text with innovative image, sensational plotlines with rational law enforcement, visual quality with stereotypical character arcs and predictable narrative resolutions. What this chapter will thus reflect is how cool in *CSI* fundamentally depends on the visual dimension, which the program (in contrast to generically more conservative crime dramas) successfully exploits to an unprecedented degree. As I will argue, it is not despite but rather because of the show's pronounced focus on (techno)science, which effortlessly provides all the required visual props and accessories, that cool emerges as a dominant aesthetic code throughout the series.[2]

WELCOME TO LAS VEGAS/ MIAMI/ NEW YORK: THE *CSI* FORMULA

"Here comes the nerd squad," is how a regular police officer cynically warns us right at the opening of *CSI*'s pilot episode. This explicit introduction to the show's protagonists communicates straight from the very beginning that the viewer will not be witnessing the trials and tribulations of a hard-boiled private detective or a team of local cops, but a group of scientists, as this is how the investigators are repeatedly designated by themselves and by others.[3] While generally presenting a generic

2 My analysis in this chapter explores the construction of cool forensic science as shared by all three shows of the *CSI* franchise. Given the distinct location and visual composition of each version, an analysis of these particularities may, however, also yield interesting results with regard to specific contingencies of cool.

3 The *CSI* characters' real-life counterparts do not directly interact with suspects, conduct interrogations, or issue warrants. The strategic fusion of the roles of the scientist and the police officer is exclusive to *CSI* (and other forensic dramas it inspired) and again marks the significance of material, scientifically retrieved evidence for the show. With reference to the semantic expansion of the term 'detective' to include pathologists and CSI staff, Pierson perceptively notes that "[t]he *C.S.I* investigators both follow and extend the scientific, analytical tradition of the earliest crime fiction detectives, like Poe's Dupin and Doyle's Holmes, rather than the hard-boiled detectives of Chandler and Spillane" (198). See also Harrington 365-368 and Haynes, "Scientist in Literature" 394. In fact, it has been argued that the rise of crime fiction coincides with the birth of modern forensic science (see Thomas 2003). The works of Poe and Holmes, among others, reflect the invention of new identification technologies such as fingerprinting and photography. By extension, the decoding of the human genome (completed and publicized in February 2001)

crime drama in the great tradition of the American whodunit, *CSI* has innovatively shifted its focus to forensic investigation and exchanged standard detective work for cutting-edge science. As critics astutely note, the show has thereby "accomplished a rare feat for commercial television: bridging the divide between modern science (not science fiction) and entertainment" (Gever 450). Indeed, the crime in *CSI* is invariably solved thanks to the overwhelming weight of physical evidence, unraveled via the almost magical powers of science, especially (digital) DNA processing. Neither psychological expertise, applied in tedious interrogations followed by the suspect's tearful confession, nor wild car chases through the criminal underground may contribute to its solution: It is the lab that cracks the puzzle, not the interrogation room. As a consequence, the personal motivations behind the deed—why did the delinquent commit the offense?—are utterly irrelevant for the narrative. The show's procedural format, presenting one or two cases that will be solved in each forty-five-minute episode, does not allow for much attention to either the perpetrator's or the victim's emotional struggles.[4] As the text which most overtly relies upon the creation of suspense for audience involvement, *CSI* focuses less on psychological character arcs than any other cultural production analyzed in this study.

The fact that there is "little narrative or character continuity" (Allen 7) only abets the success of the show, as it allows the audience to watch episodes out of sequence without being forced to tune in week after week. With its straightforward plot formula, largely self-contained episodes, and scientifically founded division of the American population into the innocent and the guilty, *CSI* is immensely popular, not only in the United States, but worldwide. According to its network, the CBS International Studios, *CSI* is the most watched drama series in the world with more than seventy million viewers.[5] At the time of writing, the original *CSI* set in glitzy Las Vegas is already in its thirteenth season, has aired almost three hundred episodes, and inspired two equally successful spin-offs in equally hip and distinct urban centers, Miami and New York.[6] If one is to believe the show's creator, Anthony

---

can be seen as having sparked off a new wave of forensics-driven crime fiction, of which *CSI* stands as the most prominent example.

4 *CSI* thus essentially continues the tradition of the classic police procedural format, which was introduced in the 1950s by *Dragnet*, the United States' pioneering and most influential television crime drama to date. For an analysis of stylistic and thematic parallels between the two programs, see Turnbull 15-22.

5 See the recurrent CBS press releases at <http://www.cbspressexpress.com>, accessed 20 June 2012.

6 While *CSI: Miami*, the allegedly "most popular programme in the world" (Eden 11), has recently been canceled by CBS after its tenth season, the original *CSI: Crime Scene Investigation* was awarded the International Television Audience Award for a Drama TV Series at the Monte Carlo Television Festival for the fifth time. The award is based on

Zuiker, *CSI* airs in every country but six, including North Korea, Uzbekistan, and Iraq—by now a running joke that perfectly captures the show's global appeal and its investedness in processes of transnational cultural transfer.[7] In the UK, for instance, *CSI* was advertised as part of 'America's Finest,' thus representing the top league of what America had to offer crime-wise, at least on the small screen.[8] Guest appearances, cameo roles, and director slots for Hollywood icons like Quentin Tarantino have further solidified its cult status as one of the commercially as well as critically most successful American crime dramas to date, as have a number of novels, video games, toys, and traveling exhibits based on the series.

However, *CSI* only represents the tip of the proverbial iceberg: In the wake of the program's global popularity, American television has lately witnessed a myriad of new, science-invested crime dramas that focus on the reconstruction of the criminal deed via scientific methods. Thomas Doherty has referred to these shows, including Fox's *Bones* (2005-present), CBS's *Criminal Minds* (2005-present) and *Cold Case* (2003-2010), and NBC's *Crossing Jordan* (2001-2007), as "forensic noir" ("Cultural Studies" 15), a subgenre of the traditional detective story which is characterized by its focus on the almost magical powers of (forensic) science to solve any crime and a necrophilic interest in the dissection of the human body. With the exception of programs involving forensic pathologists (including the short-lived CBS program *Diagnosis: Unknown* and NBC's vastly popular *Quincy, M.E.*), the show's scientific storylines and their spotlight on the hybrid creature of the scientist-detective remain unprecedented. In fact, *CSI* and its various imitators seem to have been sparked off by nonfiction television programs and documentaries featuring criminalists and pathologists, such as Discovery Channel's *The New Detectives* (1996-2005) and *Cold Case Files* (1999-2006) on the Arts and Entertainment Network (Pierson 184). While the majority of these new science-driven shows do not, generally speaking, aspire to adopt a (semi-) authentic documentary style, they do borrow their plots from actual cases, which results in a curious "reality effect" that naturalizes crime and law enforcement as "a feature of everyday social commerce" (Gever 459). The straightforward narrative pattern of which *CSI* still stands as the first and most paradigmatic example—a tricky case is solved via the latest scientific

statistics from Eurodata TV Worldwide, in which *CSI* emerged as the show with the highest ratings worldwide. See the festival's official website <http://www.tvfestival.com/2012>, accessed 23 June 2012.

7 Zuiker in an interview for *CBS News* (T. Smith, "TV Crime Shows" n.p.).

8 In fact, *CSI* considerably helped to boost the British Channel Five's down-market brand image and soon emerged as its signature show. Together with several other high quality American drama series, such as *Law & Order: Criminal Intent* (2001-2011) and *The Shield* (2002-2008), *CSI* contributed vitally to the channel's new brand identity. See also Goode (2007) and Knox (2007).

methods and cutting-edge technology within less than an hour—has even infiltrated other genres: Medical shows like NBC's *Medical Investigation* (2004-2005) and Fox's immensely successful *House* (2004-2012) are structured, I would argue, along the same lines, with each episode bearing a much more striking resemblance to *CSI* than, for instance, *ER*, both in terms of thematic orientation and aesthetic stylization. In that sense, *CSI* "has effectively hybridized the crime drama television genre" (Dobson 76) and blurred the boundaries with adjacent genres.

While *CSI* is hence far from being the only program that provides forensics- and science-driven entertainment, it still ranks as the first and, perhaps, the most consistent one. The captivating "CSI shot" (Hamit 101), a largely computer-animated extreme close-up of a bullet or comparable weapon piercing the victim's flesh, has become notorious. Not coincidentally, Renaissance anatomical theaters and their public dissections of corpses for the titillation and entertainment (and alleged medical education) of the masses come to mind: Accompanied by classical music, medical professors performed aestheticized autopsies of human or animal bodies for students and general observers.[9] This cultural fascination with what lies inside the body, one of the last frontiers of modern science, seems to be rekindled through *CSI*'s digital images, which present us with an essentially anti-realistic because highly stylized and unfamiliar, yet visually appealing view of the interior. Similar to how public autopsies showcased innovations in the field of medicine and pathology, the program relies on state-of-the-art technology and television production to produce unprecedented visions of the human body. Not surprisingly, then, these images not only define the program's aesthetics, but also its market value, as Martha Gever argues: "What can be seen as the 'money shot' in *CSI* occurs when, during the autopsy, the camera appears to penetrate a wound or orifice and produce gushing blood, exploding organs or distressed viscera, simulating the damage inflicted by the fatal weapon or the disintegration of tissue resulting from some sort of toxic substance" (457). Gever's identification of this stock scene as the one moment that the audience pays to see and that is thus literally revenue-generating suggests its vital significance for the show's popularity, which is also reflected by the narrative space dedicated to such scenes in every episode.[10]

---

9 See Tait 49-52, as well as Jonathan Sawday's *The Body Emblazoned* (1995) and Elizabeth Klaver's *Images of the Corpse* (2004) for more background information on anatomical theatres and their civic functions in seventeenth-century society. Contemporary art installations, such as the highly controversial exhibition 'Body Worlds' by the German anatomist Gunther von Hagen, continue this tradition (Jermyn 84).

10 Furthermore, the classic 'CSI shot' is central to the whole show and each episode because of its expositional function, as it accompanies the dialogue during an examination and thus helps to "maintain the pace of the narrative" (126), as Ian Goode points out with reference to Hamit (2002). By visualizing the mundane scientific terminology through mag-

Labeling these penetrative trademark shots with a term originating in pornographic films is, of course, far from coincidental. Indeed, *CSI* has not only become infamous for its exploitation of the abject via the graphic display of intestines and disfigured body parts, but much more so for its conspicuous sexualization of the dead body. As Carlen Lavigne points out, especially a female corpse "is almost inevitably young, beautiful, and scantily clad" (389). Equally, Sue Tait observes that most victims are young, attractive people frequently murdered during or after sex, and thus, by analogy, punished for their (often deviant) sexual behavior, be it incest, sadomasochism, or diverse fetishes (52-53). In addition to the "forensic gaze" (Pierson 186), which turns the body into an object of knowledge and a piece of legal evidence, *CSI* thus also promotes a voyeuristic, necrophilic gaze for titillation by displaying the naked corpse on the examination table through panning shots and extreme close-ups, all of which are legitimized by their alleged scientific-educational purpose.[11] Although the autopsy room may, at first glance, be regarded as an essentially clinical, rather than erotic setting, the audience is invited, through slow camera movement, non-diegetic music, dim lighting, and distorting zooms, to let the gaze linger upon the naked bodies and make inferences about their sexually-connotated, violent deaths: "[O]n the autopsy table, we know, forensics and pornography meet and fuse" (25), as Mark Seltzer perceptively notes.

CONSERVATION VS. INNOVATION: COOL AS STRATEGIC JUXTAPOSITION

While *CSI* explicitly shows nudity, violence, and deviant sexual practices, it simultaneously condemns non-mainstream sexuality outside of the traditional two-person relationship. Lieutenant Horatio Caine's (David Caruso) cynical remark that "in a place like this, sex and murder might be indistinguishable" ("Killer Date") when arriving at the crime scene of a Miami swinger club can well be applied to the series at large: More often than not, promiscuity, even if engaged in by consenting adults, proves to be indicative of generally abhorrent, illegal behavior and a criminal character, or is, at the very least, curiously studied and subjected to ridicule. Equally, victims of crimes which involve deviant sexual relations are regarded as having been rightfully punished for their immorality: "When you go outside a marriage for

nified, provocative, yet aesthetically appealing images, the moment of the crime is reanimated and revisited, which "suspends the temporalities of classic editing" (Goode 126).

11 The notion of the forensic gaze is akin to Foucault's clinical and inspecting gaze, linking medical knowledge with power and visual observation with physical penetration and reassault of the body. See especially Foucault's *The Birth of the Clinic* (1961) and *Discipline and Punish* (1975) as well as Pierson 187-193.

passion, you're in trouble. And you're asking for trouble" ("Swap Meet"), as Sara Sidle (Jorja Fox), member of the Las Vegas CSI team, vigorously asserts. As one of the show's protagonists, her viewpoint firmly establishes moral guidelines for the audience, whose sympathy will accordingly be aligned with the detectives, rather than the blameworthy 'victim.' (Sexual) morality is hence depicted not as a matter of degree and cultural negotiation, but defined through clear-cut scenarios of unambiguously 'right' or 'wrong' social conduct.

This conservative undercurrent extends well beyond the realm of sexual relations and has, hardly surprisingly, fast become the focal point of academic critique. Lavigne, for instance, notes that the show may well be "considered in light of the current era of conservative backlash in the United States" (384), while Rachel Dean-Ruzicka argues that "after 9/11 the representation of familiar murders and their solution through good, first world, industrial science was a great comfort to viewers" (127). Ellen Burton Harrington detects a "reassuring ideological message" (366), which comforts its viewers "that identity, responsibility and truth are not relative concepts, even in the postmodern world" (366). This is especially true for the New York spin-off, which repeatedly references the World Trade Center attacks and their impact on public life. The detectives' roles as guardians of law and order are thereby directly linked to larger concerns, as Mac Taylor (Gary Sinise), head of the NY CSI unit, asserts: "There are three things that I'll protect at any cost: the honor of this country, the safety of this city, and the integrity of this lab" ("Grand Murder at Central Station"). Any of the three places most valued by Mac—his country, his city, and his lab—must be protected against threats from the outside, clearly establishing an unmistakable dichotomy between 'us' and 'them.' In contrast to the two other crime lab leaders in Las Vegas and Miami, Mac's identification as a fervent patriot is repeatedly emphasized through evoking his former career in the navy as well as the personal loss he suffered during 9/11. While *CSI* thereby incorporates, both tacitly and explicitly, the WTC attacks in its plotlines, it simultaneously strives to reassure its viewers that violence, terrorism, and crime can be kept in check through efficient, scientifically enhanced law enforcement, idealizing the role of the police for the preservation of national security and typically American moral values, including rationality and objectivity. Criminality is attributed to selfish and abnormal individuals, rather than the social circumstances which produce them. The United States, as Harrington perceptively notes, are hence glorified "as an enlightened and highly technological country that effectively manages its inevitable lawlessness at home" (379), as no crime will ever go undetected or unpunished. This moralizing and escapist undercurrent is greatly supported by highly formulaic story arcs and a stereotyped set of core characters, including a White Anglo-Saxon male as the leader of the respective CSI units, a White female as his second-in-command, and more conspicuously ethnic characters (an African-American scientist in the Las Vegas team or a Hispanic character in the Miami version) in lower positions. This "nuclear family arrangement" (Lavigne 385) extends to the show at large, where almost any deviation from the patriarchal structure leads to death.

CSI's conservative story lines are effectively mirrored by its fairly strict adherence to genre conventions. While the quite unusual focus on a team of forensic investigators, rather than classic police officers or private detectives, has led some critics to appraise the program as "generically innovative" (Goode 124),[12] it should nonetheless be noted that apart from its thematic orientation, the show closely follows the classic procedural format. As Nichola Dobson notes, the narrative set-ups of the three shows largely "conform to the generic verisimilitude of television crime drama" (75). Their emphasis on high gloss and glamour, particularly pronounced in the Miami version, is reminiscent of the well-known MTV aesthetic introduced by NBC's *Miami Vice*. Equally, the focus on crime solving, rather than any other element of the justice system, is in line with the tradition of the genre, as is the franchise's focus on what Dobson identifies as a "maverick 'lone' protector working outside the system" (84), which is realized through each of the crime lab leaders who as scientists rather than cops are thematically and spatially removed from classic police work. The opposing concerns of the forensic team, representing infallible science, and the standard police officers, representing manmade and hence fallible law, are frequently thematized in the show, thus establishing the CSIs as independent underdogs who do not shrink from transgressing social and legal conventions if it serves the forensic solution of the crime. This connects the CSIs to a long lineage of headstrong, unruly, and self-directed private investigators in American crime fiction and film. Additionally, each episode of the three versions works along a fairly predictable, repetitive narrative sequence, which effectively mirrors the highly ritualized scientific procedures it intends to display: The occurrence of a crime (or, in some cases, the detection of its aftermath) is followed by the arrival of the CSI team at the crime scene and the consequent collection and processing of evidence, which after a sufficient amount of dead ends and wrong suspects will help to identify the

12 Even this point is up for debate, as *CSI* does certainly not stand as the first cultural production which focuses on the scientific detection of crimes. Haynes, for instance, detects "the emergence of the detective-cum-forensic scientist in the character of Sherlock Holmes, who included chemistry amongst his other accomplishments, and Austin Freeman's Dr. Thorndyke, who also made frequent use of chemistry to track down evildoers" ("Scientist in Literature" 394). Especially *CSI: Crime Scene Investigation*'s team leader Gil Grissom, with his academic interests, odd hobbies, and celibate lifestyle, can be regarded as "one of the many heirs of Sherlock Holmes" (Gere 129). What may, however, be seen as a variation of the Sherlock theme is the focus on collective team effort rather than a lone detective and his subordinate assistant. See also Pierson 198, Harrington 365-368, and footnote 3 of this chapter.

true perpetrator (Dobson 81). Neither the trial nor the actual conviction of the suspect are usually part of the episode's narrative, but they may be referred to in an epilogue or final voice-over.[13] *CSI* could hence be classified as a typical formula story, which follows two main axioms: Everything that happens is important and there will always be a resolution (Dove 18).[14]

Interestingly, however, this clichéd, recurrent narrative pattern coincides with a progressive visual style. This expands well beyond the already introduced sexual and forensic graphicness. Produced by showbiz mastermind Jerry Bruckheimer, who signs responsible for box-office hits like *Pirates of the Caribbean* and *Armageddon*, the *CSI* franchise has adeptly appropriated Hollywood cinematography: Each series has its distinct color palette, ranging from neon in Las Vegas to exotic in Miami and noir in New York, and all three of them are characterized by a slick, high gloss surface, fast-paced editing, and expensive visualizations that aim to recreate the Blockbuster experience on the small screen. Whatever the merit of its fairly predictable plotlines, critics agree on the visual quality of the series and praise it as a particularly fine specimen of 'feature television': "The overexposed flashback images look like music videos, the lurid anatomical closeups [sic] like art film, the lab scenes like a lush photo shoot: cerulean blue trays, crystal glass, ruby chunks of human tissue laid out like a $300 sushi course" (Poniewozik n.p.), as *Time Magazine* enthused in a 2004 television special, whose hyperbolic tone, however, does not conceal that the show may actually be celebrating style over substance. Despite the fact that 'high quality' in *CSI* might be restricted to its mechanical proficiency, academic critique equally lauds the show's "*mise-en-scène* and the foregrounding

13 The omission of scenes which would present a (legal) evaluation and potential dismissal of the 'evidence' previously collected by the CSI team is in line with the show's presentation of forensic procedures as stable, unbiased, and infallible. Exceptions are rare episodes which thematize the aspect of subjectivity and potential human fallacy involved in the forensic process, such as portrayed in the episode "The Accused Is Entitled" in season 3 of *CSI: Crime Scene Investigation*: Here, Grissom and his team are accused of having compromised traces at a crime scene and must account for it in court.

14 Additionally, one may easily detect the typical suspense structure followed by each episode, which Dove separates into four central phases: cumulation, postponement, alternation, and potentiality, which may not necessarily be sequential and can also overlap (Dove 50-51). The basic structure guarantees that the viewer knows what to expect, which may only superficially appear like a paradox: As scholarship on plot formula has shown, reading (or viewing) pleasure is as much derived from the proficient use of familiar, hence reassuring structures as from invention and variation. This also helps to account for the success of a range of similar forensic crime dramas, which all work along the same narrative lines. See, for instance, Daiches (1956), Cawelti (1976), Eco (1976), and Dove (1989).

of visual style," which are appreciated as unmistakable "markers of quality in terms of production value" (Bignell 166).

It is not only the lavish design, however, but also the film-style lighting and framing, elaborate camerawork, depth and contrast of color, and complex postproduction that vitally contribute to the show's signature look. This includes the use of Hollywood-style special effects, especially the already mentioned extreme (computer generated) close-ups of body parts or lab equipment, through which the show "pushes the boundaries of realism on North American prime time television" (82), as Dobson argues. Additionally, the program engages in an intricate play with depths, shadows, and saturation, which counteracts its sensationalist necrophilia shots with a genuine artistic aesthetic. Susanne Turnbull has described the resulting imagery as a "neo-noir cinematographic effect" (27) that effectively transcends the format limitations of television. *CSI* could thereby be understood as a continuation of the 1980s televisual style as defined by John Caldwell (1995), which is characterized by a prominent emphasis on imagery and surface rather than sound and the creation of a distinct aesthetic that sets the program apart from its rivals.[15] Thus, its visual attributes endow the show with a high recognition value and make it an easily exportable production which has taken TV channels all over the world by storm. With its unprecedented representation of forensic investigation in a visually exhibitionist, highly stylized, and graphically embellished format, *CSI* has effectively redefined the parameters of its genre.[16]

As the above considerations have, however, shown, the high visual quality of the show does not pertain to its predictive narrative patterns, conventional dénoue-

15 See especially Gever (2005) and Goode (2007). With regard to the show's visual quality and its production of aesthetically appealing images, several critics have noted the "primacy of vision" (Gever 451) that informs almost every episode. Knowledge is primarily gained via visual observation, both at the crime scene and in the laboratory. The CSI investigators are frequently depicted with cameras or flashlights, which visually highlight a relevant detail of the scene. In the lab, these observations are complemented by the display of objects that are normally not accessible to the human vision, such as microscopic views or digital models of DNA sequences. Electronic instruments, however, not only enhance the limited human gaze, but also guarantee a depersonalized, objective perspective. Images, both actual and computer-generated, are hence vital for reanimating the moment of death, which reflects how *CSI* establishes "the notion of vision as an evidential-investigative-conclusive activity" (Bignell 163). In contrast to older crime dramas, which used to have characters relate the findings through narration, *CSI* chooses to show them directly and thereby invites audience participation. See also Seltzer 23-24, Tait 46-51, and Lury 46-51.

16 See also Dobson (81-85), who shows how *CSI* modified the generic boundaries of the classic American police procedural and thus heavily impacted the genre itself.

ments, and clichéd characterizations. In that respect, the world of *CSI* is an essentially paradoxical one, combining the old-fashioned whodunit formula with state-of-the-art cinematography, quality with trash, rational investigation with sensational images. Its episodes effectively join generic conventions in terms of narrative pattern and characterization with innovative elements such as visual effects, high production value, and thematic focus. Additionally, the show's thematic and stylistic orientation notably clashes: One may observe an evident "tension between spectacular, anti-realist excess on the one hand and (an apparent) investment in authenticity and forensic evidence on the other" (Jermyn 81). Rather than concurring with previous studies which tend to attribute the show's popularity to its morally conservative stance, I would thus argue that the show's appeal is primarily induced by this odd, yet successful marriage of the edgy and progressive with the safe and proven. Not surprisingly, it is exactly such a juxtaposition of mutually exclusive qualities that has been theorized as a vital prerequisite for a cool effect, both in terms of an aesthetic and a personal strategy. Following Marshall McLuhan's famous distinction between 'hot' and 'cool' media,[17] Jeff Rice argues that "cool function[s] by way of juxtaposition, [...] creating associations and emotional responses out of the combination of unlike words and images" ("Writing" 223). Similar to how a cool person will habitually be perceived as embodying conflicting qualities simultaneously, such as individuality and mass appeal, narcissism and modesty, a cool visual effect is equally produced in the interplay between polarities. In the case of *CSI*, the "ambiguous sensibility [...] that lies beneath Cool" (Pountain and Robins 43) is most feasibly played out in the ironic tension between image and narrative, which also informs the show's portrayal of its protagonists.

C.OOL S.EXY I.NTELLIGENT: *CSI*'S SCIENTIST-DETECTIVES

> For decades, scientists tried and failed to shed their fusty old egghead image, then along came *CSI*, and they became cool.
>
> LAURA SPINNEY/*THE GUARDIAN*, 13 APRIL 2010

17 In his landmark *Understanding Media* (1964), McLuhan establishes cartoons and television as examples of cool media, since in these cultural productions "so little is given and so much has to be filled in by the listener [or viewer]. On the other hand, hot media do not leave so much to be filled in or completed by the audience. Hot media are, therefore, low in participation, and cool media are high in participation" (36). Instances of hot media would be film, radio, and photography as well as printed matter.

My close analysis of how cool pervades the treatment of science in *CSI* shall thus, almost inevitably, begin with those who are repeatedly depicted as performing this 'cool science,' i.e. the respective CSI squad members. Indeed, one may discern a palpable focus on placid and emotionally composed protagonists, who are strategically juxtaposed with emotionally expressive and enraged criminals and suspects. In this sense, literally 'heated' characters serve as foils to the invariably cool protagonists, who will seldom lose control over their emotions. The following conversation between Lieutenant Taylor and his colleagues Lindsay Monroe (Anna Belknap) and Jo Danville (Sela Ward) establishes a clear-cut binary opposition between ridiculously sentimental suspects and rational detectives and, consequently, codes emotionality as inherently negative and potentially unlawful:

Mac Taylor: Michael had a brother, Tom Reynolds.
Lindsay Monroe: Mitochondrial DNA from the hair was a match to Michael.
Mac Taylor: Mito's shared by all maternal relatives, including brothers, so it would also be a match to Tom.
Lindsay Monroe: So he's our shooter. Why?
Jo Danville: Off to one side, unnoticed. Everyone else around him is celebrating Michael's return. Tom is just standing there.
Mac Taylor: Emotionless, neglected.
Lindsay Monroe: So you're telling me he starts shooting people because he didn't get enough hugs and kisses from mommy and daddy?

("Hide Sight")

As this short exchange vividly illustrates, excessive emotional responses are mostly attributed to perpetrators and accordingly represented as intrinsically problematic, if not indeed detrimental for themselves and others. The notion that emotional neglect may motivate and indirectly result in criminal deeds is regarded as utterly absurd by the detectives, who will invariably value cool rationality over (what the audience is instructed to perceive as) unrestrained demonstrations of sentimentality. This is also maintained in terms of physical appearance: While the CSIs will mostly don clothes in cool, neutral, and subdued colors, such as blue, gray, and black, the other (especially female) characters will be seen wearing dramatic evening gowns, colorful sequined party outfits, or expensive suits. Hence, one may argue that "the ultimate glamour is always reserved for the corpses" (König 106) and the suspects, whose literally 'hot' clothing style suggests emotionality, expressiveness, and eventually fatal passion.

The utmost affective control displayed by the protagonists by both their looks and demeanor is closely linked, as the show repeatedly reminds its audience, to their scientific professions. Indeed, they are frequently depicted as cool not despite, but because of science: In most instances, their coolness derives from their scientifically detached and (stereotypically) socially distanced attitude, which reflects how science and coolness are intricately interwoven in *CSI* on the level of character. The particular notions of coolness which inform the portrayal and characterization of the CSI team thus conspicuously deviate from that of the hard-boiled private detectives who dominated the golden age of American crime fiction in the 1920s and 30s. The investigators' dedication to the assumedly rational realm of Western science guarantees that they will hardly ever lose their tempers or indulge in unduly emotionality. Indeed, every new member to the team is explicitly advised to 'keep their cool': "Sometimes we deal with bugs, worms, waste, or worse. But as scientists, we look beyond the possibly offensive qualities of these things to what they might tell us about the puzzle we're trying to solve" ("Justice Is Served"), as Dr. Gil Grissom (William Petersen), leader of the Las Vegas CSI team, tells his underlings. As emotionally unruffled professionals and highly accomplished uber-detectives, the crime scene investigators must retain an indestructible cool pose in the face of the most gruesome atrocities, the horror of which they acknowledge, if at all, by a wry comment and a nonchalant lift of their designer sunglasses.[18] While the CSI officers do employ "the occasional self-reference as 'science geeks' or 'science nerds'" (Kruse 83), their actions repeatedly counteract and deconstruct this stereotypical reference and replace it with the image of the cool scientist-detective who works at the very core of twenty-first-century knowledge culture.

In fact, the program's very set-up calls for cool detachment rather than hot sentimentality. The core cast of the respective CSI team can hardly be described as round characters: While occasional glimpses into their domestic lives are granted in later episodes (Mac Taylor lost his wife in the World Trade Center attacks, whereas a *CSI: Miami* staff member escaped from an abusive marriage), it is, first and foremost, their jobs on which the storylines center. When the audience does get to know the personal world of a character, this insight usually contributes to the image of his or her professional persona: Grissom's preferred pastimes, for instance, such as his beloved documentaries and crossword puzzles or the fact that he prefers to live alone, only serve to solidify his "major trait of emotional disengagement/repression" (Pearson 42). These and similar 'hobbies' or eccentricities hence only mirror the protagonists' alleged priorities, as they are frequently depicted, in accordance with the stereotypical image of the scientist as a self-sacrificing idealist, as subordinating their personal lives to their work. Human drama is thus quite effectively replaced with "the drama of science" (Billen n.p.). The distinction between cops (i.e.

18 This cool scientific detachment may be exaggerated to a point where it appears almost ironic: Especially Horatio Caine's strategic use of sunglasses in *CSI: Miami*, usually in combination with a dramatic one-line pun, has given rise to ridicule by both critics and the general audience. On the other hand, the gesture soon became iconic among fans of the series and even inspired its own internet meme.

the protagonists of ordinary TV drama) and criminalists (who offer unique and fresh possibilities to the somewhat timeworn crime genre) is continuously emphasized, both through implicit and explicit characterizations of the team members: "I'm not a police officer, I'm a scientist" ("A Bullet Runs Through It") is, in variations, a recurrent theme in all of the three shows, aiming to accentuate the investigators' dedication to scientific knowledge, rather than the (potentially fallible) law. As scientists, they almost exclusively base their assessment on the samples from the crime scene rather than testimonies or witness reports, which cannot be scientifically verified: "I don't chase criminals," as Grissom asserts, "I just evaluate evidence" ("Pilot"). In contrast to cops or private investigators, the CSI officers hence "care more about the application of scientific technologies to generate and organize knowledge than crime and punishment" (447), as Gever persuasively argues. The respective crimes they investigate are thus effectively divorced from their social, cultural, and political contexts.

This decontextualization also entails that the detectives are seldom driven by intense emotions such as pity or rage, but by their pure and rational dedication to scientific principles and the resulting 'truth.' Undue emotional engagement is not encouraged, as Grissom clarifies: "There is no room for subjectivity in this department, you know that. We handle each case objectively, without presupposition, regardless of race, color, creed, or bubblegum flavor" ("Pilot"). This statement firmly establishes the CSI squad's work ethics and is not only directed at the team members, but also at the audience, who might, in all likelihood, be new to the world of forensic science and must thus, first of all, be introduced to its most important principles and guidelines. In view of that, the viewers are advised to refrain from unduly emotional attachment to any of the characters—a plea which is immensely abetted by the generally flat characterization and lack of personal background information, which hinders rather than encourages identification with the protagonists or, even less likely, the suspects. Within the laboratory or the autopsy room, human interaction and emotions would only counteract, or even indeed scramble the precise and clear-cut results yielded by technology. This denigration of excessive emotionality can already be detected in the show's camera work: *CSI* effectively substitutes the classic close-up of the face, zooming in on emotions, with close-ups of inanimate traces such as dust, hair, or body fluids. The show thereby shifts the focus from the embodied experience of feelings to the hyper-realistic display of evidence, which "promises access to knowledge" (166) rather than affect, as Silke Panse argues. The regular viewer will hence quickly gather from the confined world of *CSI* that only those who have trespassed will indulge in inappropriate emotionality: Cool detachment is portrayed as being the correct comportment of those who have nothing to hide.

In view of historical theories on the development of cool into a dominant Western sensibility, this appears to be ultimately paradoxical. As Daniel Harris, for instance, points out, cool is a "flamboyant charade of toughness and authority" (40) that conceals seemingly negative emotions, such as insecurity and anger. Several other studies underline the idea that cool may be understood as an emotional mask designed to suggest the very opposite of what one might actually be feeling. One particularly salient hypothesis of the origins of Western cool as suggested by anthropological research supports this perspective: In its original usage as an African tribal attitude, a cool attitude was valued as a viable survival strategy, translating emotional stress into apparent composure. As Robert Farris Thompson cites a West African definition of cool, the "'mask' of coolness" ("Aesthetic" 41) is regarded as the ability "to reveal no emotion in situations where excitement and sentimentality are acceptable […] to do difficult tasks with an air of ease and silent disdain" (41). Lying underneath cool is thus, somewhat paradoxically perhaps, an extreme emotionality:

> "What Cool actually refuses is the sentimental or incontinent display of negative emotions […] which are the very definitions of Uncool. However, Cool respects intense passion, and from some angles the look of Hollywood Cool is merely a mask covering an extreme form of romanticism […]. In short, Cool economizes on emotion rather than suppressing it altogether, preferring the bitter-sweet varieties it deems to be both intense and unconventional." (Pountain and Robins 119-120)

This paradoxical combination of intense emotionality lying beneath a cool surface evokes Rice's notion of cool juxtaposition and is noticeable in *CSI*'s depiction of several key characters. On the one hand, the CSI staff is portrayed as almost inhumanly objective and scientifically disinterested, both through recurrent dialogues which thematize their professional detachment as well as through implicit characterization via storylines that might implicate them personally (such as when friends, colleagues, or family members are involved in a crime). Through the same narrative strategies, however, we get to know that the CSI officers habitually act on strong emotions, above all the desire to do justice to the victims and their families and to avenge their deaths within the legal possibilities of a constitutional state. This is also observed by Harrington, who points out that the detectives frequently "temper their distancing irony and rueful comments over a brutalized corpse with a real sense of outrage at the human cost of crime" (372). Indeed, the generally subdued plot and (deliberately) minimalist acting is, time and again, interspersed with genuine outbreaks of passion, especially anger or hatred. Particularly the two spin-off series, *CSI: Miami* and *CSI: NY*, feature professional, yet vengeance-driven squad leaders who may repeatedly counteract their scientific pledge of objective disinterestedness. The head of Miami's CSI unit, Horatio Caine, frequently uses intimidation as an appropriate way for making suspects talk, which is portrayed, however, as a legitimate means for obtaining the truth and providing closure to the victims and their relatives. In a conversation with an FBI agent who refuses to collaborate, he ominously mutters: "I need answers for this little girl. Her father's just been killed, maybe her mother, too. And that's your fault. You understand that." When the FBI agent replies: "It sounds like you're making this personal," Horatio snaps back: "Dennis, you don't give me this link, I will make this very personal" ("Cross Jurisdictions"). This exchange quite overtly reflects how even outbursts of extreme emotions are verbalized (as well as visualized through the character's comportment and gestures) in terms of self-control and nonchalance, making Horatio paradoxically appear even cooler because he voices and acts on, rather than suppresses his emotions.

Furthermore, the emotional spectrum upon which Horatio and his colleagues are shown to be moving is typically restricted to the negative end of the spectrum, always seemingly on the verge of actual violence. Although in the end, Horatio's reasons for his unorthodox measures turn out to be, as the storylines repeatedly reaffirm, purely professional and ethically sound, they can hardly be described as legal. In addition to using threat as a means of approaching the truth, the CSIs "can lie in order to get the evidence that they need" (Dean-Ruzicka 124), especially when tricking suspects into unwittingly giving their DNA by letting them use tissues or lick envelopes.[19] In the same vein, the detectives are depicted as employing expressive emotionality to suggest danger, fearlessness, and capacity for violence if necessary. These qualities, as historical personalities acclaimed for their cool poise have amply demonstrated, suggest utter carelessness for social contracts. In *CSI*, such heightened affective responses are almost always linked to the investigators' work, rather than their private lives. Horatio's emotional expressiveness, for instance, seems to be reserved for his suspects: His relationship and later marriage with his colleague's sister Marisol, a minor plotline that offers rare glimpses into his domestic life, is marked by the utter absence of any displays of affection, such as kissing her on their wedding day or at her deathbed, which one critic described as "not just stoic, [but] frozen" (Bianculli 225). Hence, it is only his utter devotion to physical evidence and scientifically founded 'truth' that makes Horatio lose his cool, which seldom happens in private encounters. This also applies to the other team leaders, who are repeatedly portrayed as transgressing legal boundaries and

19 Through these and similar tricks that will eventually always be justified by their ends, *CSI* naturalizes the growing digitalization of our world and establishes surveillance techniques such as DNA profiling or CCTV as acceptable and "part of respectable citizenship" (Pierson 200). At the same time, its narrative pattern "mythologizes the investigators as servants of truth and protectors of society" (Harrington 372). Refusal to give DNA or fingerprints is thus seldom depicted as a civil right, but a sign of guilt. Consequently, deceit for the purpose of obtaining such data becomes an appropriate means for the detectives to chase and convict criminals.

exceeding their competences for the sake of scientific findings. Consequently, demonstrations of passion mostly prove to support and advance, rather than harm the (male) investigations. Science is thus constructed as the ultimate authority that may, if necessary, even trump the law: Legal justice, as *CSI* continuously reminds its viewers, is seldom as comprehensive and consoling as scientific fact, the production of which may, time and again, require the scientists to become bold outlaws. The authority required to make the investigators' cool pose work is thus largely gained and channeled through science, which not only legitimizes illegal or dangerous actions, but also attributes additional weight and urgency to their requests. In order to retrieve traces needed for scientific analysis and the subsequent conviction of the perpetrator, the detectives will thus repeatedly employ the concept of cool as an intimidating strategy that induces respect and awe. In the above exchange, for instance, Horatio can be read as being tacitly modeled on what Ralph Willett has identified as the character of the 'wisecrack,' an enduring technique employed by private investigators in hard-boiled crime fiction. Through his "irreverence towards authority and institutional power" (Willett 7), in the present case represented by the FBI agent, the wisecrack is understood as vigorously asserting his autonomy, toughness, and refusal to conform to other's rules, all of which are understood as central attributes for a cool poise. Accordingly, it is especially the stoic, detached, and pained Horatio who "evokes a much longer history of the lone troubled hero" and displays "traces of the noir detective" (McCabe 174), which may, to a lesser degree, also be detected in the other crime lab leaders. Their defiant comportment hence vividly harks back to the alleged cultural origins of Western coolness as a pose of utter self-confidence and self-reliance, even in times of great emotional turmoil and insecurity. As Richard Majors and Janet Mancini Billson observe, "coolness means poise under pressure and the ability to maintain detachment, even during tense encounters" (2). The exchange with the FBI agent, I would argue, qualifies as exactly such a tense encounter, in which stakes are high and emotions raw, but must be hid under a mask of cool control in order to tip the balance in Horatio's favor.

The said conversation with the FBI agent, which stands as one noteworthy example out of many, also reveals another noticeable dimension of cool which is reiterated throughout the series, namely the markedly gendered dimension of cool. As several critics have astutely noted, cool carries tenacious connotations of maleness and masculinity. Majors and Mancini Billson, for instance, establish cool as a tool originally employed by African-American men "for hammering masculinity out of the bronze of their daily lives" (2), while Pountain and Robins observe that "although Cool is a strategy that women understand and can wield as well as men, fewer of them actually choose to do so" (138). Especially with regard to African-American culture and folklore, cool has been described as a markedly masculine moral code signaling toughness, fearlessness, and readiness to assume risk, as Marlene Connor's (1995) and bell hooks's (2004) enticing studies on black manhood point out. Several case studies of cool historical personalities, such as George Clarke's analysis of Malcolm X and Miles Davis (1998), support this line of argument: While cool has been effectively appropriated by corporate culture and largely lost its ethnic connotations, it is still coded as a predominantly male phenomenon, drawing on the stereotypically masculine qualities of emotional detachment, mental control, and cocky egotism. Dominant constructions of female attractiveness and beauty, including modesty, womanly empathy, and literal 'hotness,' seem to preclude a cool stance from the outset.

In popular cultural texts, including mainstream television drama, the gendered vision of cool and its pertinent connotations of allegedly male virtues are highly palpable. While the patriarchal connotations of cool emerge even more conspicuously in some of the other texts analyzed in this study (see especially chapter 6 on *The Big Bang Theory* and chapter 7 on *The Day After Tomorrow*), *CSI* is certainly no exception to the rule. The notion of cool as "a hypermasculine folk religion" (Harris 40) is particularly noteworthy in the crime show because it is tightly linked to the characters' professions and the show's emphasis on technoscientific competence for successful crime detection. While each CSI team also includes female members who largely engage in the same activities as their male colleagues, there is a noticeable accent on men handling the equipment that can be most overtly classified as 'high tech,' such as cutting-edge voice or image detection software. Willett's optimistic prediction concerning the development of urban crime fiction from the vantage point of the mid-1990s, namely that "[t]he priest-like expertise of males in the management of technology (cars, guns, etc,) is increasingly challenged by women cops [...] as they become empowered by the resources of information technology" (134), has hence proven only partly true. As *CSI* and similar forensic crime dramas show, computer experts and engineers are almost exclusively male, while biological sciences, such as pathology, medicine, or chemistry (which are less dependent on heavy machinery and electronic devices), are performed by women.[20]

CSI and similar science-driven crime dramas can hence be described as being predicated on what Gray Cavender and Sarah Deutsch, with reference to James

---

20 This gendered division of labor is, generally speaking, not maintained with regard to other identity markers such as ethnicity or age. However, as Pearson notes, "[r]ace is [...] a key constructor and differentiator of identity" (49), as, for instance, in the character of Las Vegas team member Warrick, who is coded within (stereo)typical cultural conceptions of the African-American community, such as having been raised by strong matriarchs and struggling with gambling problems. Not coincidentally, thus, Warrick figures as the criminalist who is most overtly associated with cool as a marker of (black) masculinity, both through his demeanor and clothing style. See Majors and Mancini Billson (1992), Connor (1995), Clarke (1998), and chapter 2 of this study.

Messerschmidt (1993), term "techno-masculinity" (69). As Messerschmidt compellingly argued, the image of the cool cop has shifted from depictions of physical strength and aggressively male behavioral patterns to displays of technical expertise for the construction of patriarchal authority.[21] The same, I would maintain, can be applied to current representations of coolness within contemporary autopsy-focused crime drama: In contrast to hard-boiled private investigators, who embodied the very ideal of the aggressive and unsympathetic macho figure, the CSI detectives' coolness is predicated on their technical competence and scientific ingenuity. Accordingly, it appears sensible that the absence of passion and emotionality is no longer required as the prime marker of cool masculinity. Rather, the crime scene investigators are repeatedly shown as both harboring and voicing (negative) emotions, such as Horatio in the above exchange with the police officer. His coolness is constructed through other character traits, such as professionalism, absolute dedication to science and objectivity, and the courage to violate the law if it is rationally justifiable and serves the establishment of scientific knowledge.

Female emotionality, which in the show is mostly clad as empathy, is primarily coded as unduly compassion that will ultimately prove a hindrance to objective scientific analysis. As Mac has to remind his colleague and subordinate Stella (Melina Kanakaredes): "Use your head, not your heart!" ("Creatures of the Night") to which she retorts, in a later episode: "I have no choice but to be emotionally involved in this case" ("Open and Shut"). These exchanges establish a firm dichotomy between the rational male supervisor and the emotional female subordinate or assistant, who continuously struggles, by virtue of both her gender and her lack of scientific work experience, with not letting her emotions impede her intellect. One of the most persistent pairings of the New York version, Sara Sidle and her boss Grissom, who will later engage in a love affair and are eventually revealed as married in season ten, displays similar dynamics. As Sara, a physicist and material analyst, tells her supervisor (and later love interest): "I wish I was like you, Grissom. I wish I didn't feel anything" ("Too Tough to Die"). While her emotional involvement with the case makes Sara the more attractive character at this point of the story and constructs Grissom as the heartless scientist manically obsessed with facts, it also markedly foregrounds her femininity and alleged softness of character. As Sara's supervisor and her elder by several years, Grissom represents the role model every young investigator should aspire to emulate. This hierarchical set-up ultimately exposes Sara's excessive emotionality as dangerous for both her personal well-being and her scientific work:

21 See also Cavender's and Deutsch's compelling analysis of how *CSI*'s focus on techno-masculinity reflects larger cultural changes regarding dominant notions of masculinity and heroism (69).

Gil Grissom: You have empathy for her [i.e. the victim], Sara. You want someone to pay for what was done to her. That's normal.
Sara Sidle: You want to sleep with me?
Gil Grissom: Did you just say what I think you did?
Sara Sidle: That way, when I wake up in cold sweat under the blanket, hearing [her] screams, you can tell me it's nothing. It's just empathy.

("Sex, Lies, and Larvae")

This exchange overtly illustrates that Sara is not capable of deploying "Cool as a body armour" (Pountain and Robins 41) against disproportionate and eventually detrimental affective involvement in the case, which results in largely negative consequences: She becomes vulnerable and suffers immensely. Her emotions, as the last line evoking nightmares and 'cold sweat' particularly suggests, take on an almost pathological quality: Her empathy is not 'normal,' as Grissom courteously suggests, but excessive, leaving her personally damaged and, as one would expect, unable to work as meticulously and diligently as her responsible position requires her to. In contrast to Horatio's outbursts of anger, Sara's emotions are not primarily nor solely attached to her actual investigations and may hence prove to be harmful, rather than conducive to her scientific work. Indeed, her excessive empathy makes her transgress professional boundaries and voice inappropriate suggestions to her boss, highlighting her need for male protection. Sara is thus portrayed as being unable to mobilize coolness as an effective mask to control her emotionality both for own sanity and her job. The lack of cool ultimately damages her professionalism and scientific reputation.

In fact, Sara will later be compromised in court for her lack of professional detachment. Being accused of having forced her boyfriend Hank, a paramedic involved in the case, to strategically place traces at the crime scene, Sara asserts: "I collect evidence without emotion," to which the attorney replies: "You do get emotionally involved, though with the men on your cases. Hank Peddigrew isn't the first time" ("The Accused Is Entitled"). Sara then has to go through the shame of having her past romantic involvement with Grissom publicly exposed in the courtroom. What is more, the attorney suggests that the love affair heavily impinged on Sara's professional work and even made her manipulate traces at the crime scene: "Just how far will Ms. Sidle go on the evidence to please her boss, Gil Grissom, whether he returns her attentions or not?" ("The Accused Is Entitled"). At no point in the series is the affair regarded as equally compromising Grissom, who as team leader and most experienced scientist of the CSI unit is never required to prove his scientific disinterestedness and professionalism. While cool is hence constructed as working naturally for Grissom (and most other male scientists in the series), Sara as the youngest and hierarchically lowest female team member is depicted as continually struggling for cool detachment, and frequently failing in her efforts.

Hence, whereas *CSI* does acknowledge the importance of emotions for efficient police work, it is the (male) team leaders such as Las Vegas's Gil Grissom or Miami's Horatio Caine who are able to successfully temper their scientific detachment with just the right amount of empathy to guarantee that they are not perceived as heartless robots. Female CSIs such as Sara Sidle are, however, frequently depicted as dangerously exaggerating their compassion for victims or aggression toward perpetrators, which is explicitly linked to their roles as mothers or lovers who (wrongfully, as the show implies) prioritize the well-being of their loved ones over their scientific work. Consequently, much more background information is usually supplied on their private lives, which serves to explain and motivate their emotional engagement. Not only do family and dating play a much larger role for their characterization, but it is also noteworthy that, as Roberta Pearson observes, the "boundary between the female characters' public and private lives is blurred to a far greater extent than with the male characters" (49). Hence, through her romantic involvement with co-workers or her boss, Sara is continuously depicted as losing her cool and, even more consequentially, her scientific credibility: Indeed, she will eventually be suspended from work due to a heated argument with her supervisors after having lost her temper with a suspect accused of domestic violence. Within the overall story arc of the whole series, this can be seen as the logical climax of her character's development, since she is from the outset portrayed as acting aggressively in times of pressure or stress.

The notion of professional detachment hence perceptibly harks back to visions of cool as part of an emotional survival strategy, a form of "*transcendental balance*" (Thompson, "Aesthetic" 41, italics in the original) that helps to disguise extreme affective states such as anger, compassion, or fear for the sake of others and oneself. In *CSI*, this is explicitly connected to scientific credibility: Coolness, as the program repeatedly demonstrates, is the only appropriate mindset for the devoted and meticulous scientist, who must by all means guard against unduly emotional involvement in each case. Subjectivity and human interpretation are directly linked to affective engagement, which the investigators ought to avoid at all costs. Practicing true science, as *CSI* constantly reminds us, requires a cool set of mind. Cool as an affective personal strategy is, however, only one dimension that remains highly visible in *CSI*'s treatment of science. Equally, one can also perceive a pronounced focus on visual-aesthetic coolness that again noticeably thrives on dominant representations of science within the show. The following section will thus more closely examine what 'doing science' entails in the fictional world of law enforcement and how the latter mobilizes cool as an aesthetic strategy for entertaining and visually appealing displays of scientific practice.

The Spectacle of Science

Evidence only knows one thing: the truth. It is what it is.
Gil Grissom/"Friends and Lovers"

As the preceding analysis of *CSI*'s cool scientist-detectives has already hinted at, the practice of cool juxtaposition, which evolves from the interplay of oppositional qualities, emerges as a structuring principle and aesthetic strategy throughout the series and becomes particularly pronounced with regard to the show's treatment of science. On the one hand, the conservative and reassuring undercurrent informing large parts of the show is immensely abetted by the work of science, which almost always manages to provide comforting closure by differentiating unambiguously between the deserving good and the punishable bad. In that respect, the program's underlying philosophy is heavily based on 'Locard's exchange principle,' which is named after the founder of modern forensic science Edmond Locard and postulates that every contact between two people will inevitably leave traces of evidence on both of them.[22] *CSI* takes this basic theory to its extremes, showcasing a veritable "crime lab utopia" (Dean-Ruzicka 124) and an almost fanatical belief in the powers of DNA processing to solve any case. This attitude inevitably results in a glorification of science and of the (largely Enlightenment) ideals it ostensibly stands for, such as objectivity, universality, social progress, and control. As this chapter's epigraph already suggests, science is constructed as infallible, unequivocal, and definite, leading to absolute truth, moral authority, and justice. This idealized image of science is already asserted in the pilot episode, in which Grissom advises his team to "concentrate on what cannot lie—the evidence" ("Pilot"), by which he means the material traces that can be touched, weighed, measured, calculated, and visually presented to the viewer. Grissom's equation of physical evidence with unquestionable truth renders a trial literally unnecessary: The fact that the program hardly ever shows courtroom scenes in which the evidence is assessed and potentially disputed by attorneys substantiates *CSI*'s underlying credo that "there is no better judge than science" (Cavender and Deutsch 78).

The following encounter between Grissom and a perpetrator's family, tellingly taking place at a community meeting in a local church, further establishes science as the source of absolute truth:

22 It is thus, quite paradoxically, the (body of the) victim, rather than the perpetrator who confesses the crime. See Weissmann and Boyle (2007), Chisum and Turvey (2000), as well as the *Encyclopedia of Forensic Science* (2002) edited by Barbara Gardner Conklin et al.

Gil Grissom: Hello, my name is Doctor Gil Grissom. I'm the night shift supervisor of the Las Vegas Police Department's crime lab. I'm not a police officer, I'm a scientist.
Suspect's brother: You work for the cops, that makes you a cop. You're not on our side.
Gil Grissom: Actually, I'm a forensics expert. My job is to collect physical evidence from a crime scene to determine who did what to whom and how did they do it. I've been asked to come here today by the Mayor and Sheriff Berdic to present our analysis of the evidence in this case to your community.
Suspect's mother: Why here? Why should we believe your evidence?
Gil Grissom: Physical evidence cannot be wrong; it doesn't lie. It's not influenced by emotion or prejudice, it's not confused by the excitement of the moment. I'm here in God's house to explain to you the truth about exactly what happened the other day.

("A Bullet Runs Through It")

This exchange not only promotes science as the epitome of safe, proven and, most importantly, unbiased knowledge, but also sets the CSIs categorically apart from regular police officers, who due to their lack of scientific disinterestedness might be much more inclined to prejudice and partiality. Science, once again, is to be kept free of emotions or the 'excitement of the moment' to produce unambiguous facts about the delinquent. Although the series piques itself on employing real-life criminalists as writers and technical consultants,[23] the validity and reliability of scientific methods like DNA processing is hardly ever questioned, nor does *CSI* distinguish between data and evidence or acknowledge the vital role of human interpretation and subjectivity for the production of scientific knowledge (Pierson 188, Kruse 84-88).

In the post-9/11 American TV landscape, science hence becomes celebrated as a normative, markedly Western force and the last resort of stability in an increasingly unstable world. The CSI team's reliance on genetic information and physical properties for identification affirms the notion of individual, racial, national, and sexual identity as unambiguous, coherent, and stable, rather than fluctuating, uncertain, and culturally constructed (Pierson 199). Throughout the series, science is constructed as an integral part of American national identity (Dean-Ruzicka 129), which citizens should take pride in and support to the best of their abilities: "Science," as Mac Taylor confidently asserts, "is our integrity" ("Murder Sings the Blues"). It hence also presents a viable weapon against all sorts of threats from outside, juxtaposing industrialized scientific progress and Western rationality with primitiveness, brutality, and foreignness. The underlying reason why criminals will, eventually, always be caught is either their lack of cunning or their hubris or, most often, a combination of both:

23 For instance, one of the show's regular writers, Elizabeth Devine, is a former LA County Sheriff's Department criminalist (Lury 45).

Mac Taylor: What kind of killer photographs his crime and then emails it to the cops?
Don Flack: A show-off. Someone who thinks he's smarter than us.
Mac Taylor: We're about to prove him wrong.

("Prey")

Through this and similar encounters, the audience quickly learns that the detectives will invariably outsmart the perpetrators because of their superior scientific knowledge. Hardly surprisingly, Mac's prophecy will soon become true, proving once again that science works in favor of justice and the law and will invariably succeed: Although the material evidence is repeatedly constructed as unbiased and unequivocal, science will ultimately turn out to be on the detectives' side, supporting the good in their fight against evil. The evident popularity of this heavily idealized image of science is reflected by the drama show's real-life impact on discourses of science, crime, and justice via the so-called 'CSI effect': Both crime victims and jury members have been shown to increasingly expect a wealth of physical evidence to help decide a case, which puts prosecutors under pressure to deliver more extensive forensic proof.[24] The appeal of the 'hard scientific fact' seems to reach well beyond the narrative confines of the *CSI* universe.

Providing comfort, stability, and closure in a world shaken by terrorism, financial crises, and climate change is, however, only one function that science fulfills in the program. While I do concur with previous studies which establish science and its normative powers as one of the show's central topoi, I would nonetheless propose to consider its exploitation in a more differentiated light. Despite the program's undisputable celebration of rationality and materiality, science is, in fact, paradoxically re-enchanted through *CSI*'s imagery and narrative: Scientific methods appear like pure magic when a tiny skin particle on the bottom of a wine glass not only reveals the murderer's identity, but, first of all, their credit card details and Facebook profile. This absurd, yet common scenario also demonstrates how DNA profiling, like most other procedures visualized in the show, necessarily depends on digitalization and technologization to become efficient. The dominant image of science within the narrative world of *CSI* is one that heavily relies on an amalgamation with high-end technology in order to yield the desired result. Even more importantly for the audience, it has to be "represented as a visual spectacle" (Pierson 190).

24 The actual existence of the 'CSI effect' is subject to vigorous academic debate. While the results of the empirical study conducted by Shelton, Kim, and Barak (2006) "show that specifically watching *CSI* or a similar show did not have a causative impact on juror demands for scientific evidence as a condition of a guilty verdict in most criminal case scenarios" (367), the data obtained by Schweitzer and Saks (2007) "support the claims of those who have argued that the *CSI* effect increases the prosecution's burden" (363). See also Hooper (2005), Podlas (2006), and Mopas (2007).

Rather than being lulled with theoretical science, abstract graphs, and ancient test tubes, the viewer is immediately immersed into the blinking, beeping, and awe-inspiring world of high-tech gadgetry, be it via the media-saturated laboratory, the state-of-the-art voice recognition software, or computer programs that magnify CCTV images without pixelating them in the least.[25] When exploring the promises and trappings of science in *CSI* and similar forensic shows, it might hence prove feasible to shift the focus of analysis to the already introduced notion of 'technoscience.'[26]

Despite the pervasiveness of technoscientific acquisitions in the real world of crime detection that *CSI* purports to reflect, the show's pronounced focus on technology when displaying scientific procedures primarily serves the aesthetic stylization of the prime-time police procedural. Sleek techno gadgets, flashing multimedia appliances, and shiny electronic devices are seldom shown in use (Kruse 84-85), but seem to be staged for ornamental reasons, guaranteeing visual entertainment and providing compelling illustrations when tedious procedures or obscure theories are explained. As David Pierson argues with reference to the re-enchantment of modern science through the invention of photography (Slater 1995), "the technical ability to produce realistic simulations, such as CGI representations of the inner organs of the body, is both a tool of science and its magical outcome" (191). Following this line of argumentation, I maintain that the exploitation and the appeal of science in *CSI* cannot solely be attributed to its normative power to produce safe and secure knowledge for solving the case and restoring order in an increasingly unstable world. Much more so, the specific technoscientific imagery that *CSI* heavily capitalizes on, from the chrome-and-steel laboratory to larger-than-life computer simulations, is exploited to produce visually gripping scenarios. It is only due to its combination with gleaming machinery that science can be meaningfully commodified and transformed into a visual spectacle, as the iconography provided by technology allows the show to focus on the aesthetic, rather than cognitive or ethical dimension of science. This is also enhanced by *CSI*'s signature camerawork, which employs "techniques that enlarge, magnify, manipulate and expose the depths and details of the surface" (Goode 127), resulting in the aforementioned slick, glossy, and distinct cinematic visual style. Through intricate lighting and postproduction, the metallic texture of electronic instruments is highlighted and re-

25 Gever even notes, with reference to a panel discussion with the show's producers, that the *CSI* laboratory is better equipped than any real crime investigation lab in the United States (447-448). The fact that the science represented in *CSI* is nonetheless far from accurate has been commented on extensively. See, for instance, Harrington 374-376, and Cavender and Deutsch 78-79.

26 See chapter 3 for a detailed overview of the origins and the use of 'technoscience' in STS scholarship and beyond.

peatedly scrutinized by the camera's gaze, drawing attention to the preponderance of the technological, rather than human element in the modern crime solving enterprise. The omnipresence of machines and gadgets thus supports the show's pronounced emphasis on surfaces, (artificial) images, and appearances, as it allows a high frequency of shot compositions that prominently foreground props, costumes, and objects like sunglasses, computers, or the investigators' massive flashlights (Bignell 167). This is already established in the show's credit sequences, which can be taken as "indicative of the genre's iconography" (Dobson 83): The montage shows the characters at work, both in the lab and at the crime scene, zooming in on traces, bullets, and technological equipment. These shots are intercut with short scenes showcasing the respective locations, mostly through signature sights and landmarks such as Brooklyn Bridge in the NYC version, and overlaid with obscure 'scientific' imagery, including long rows of neon-colored digits, test tubes, colorful fluids, microscopic views of hair or skin, bar codes, and virtual maps, all of which are evidently computer-generated. In the Miami version, the actors' names even evolve from mathematical equations. The opening sequence hence clarifies from the outset that this show is not just about science: It is also about the spectacular, high-quality visualization and aestheticization of science.

This overt emphasis on the visual and spectacular potential of technoscience, I would argue, serves as a vital prerequisite for instigating moments of 'cool science,' since coolness in the program very much emerges as a visual effect that relies on exactly such forms of excessive stylization and aesthetization. In apparent accordance with Rice's succinct formulation that in order "to be cool in the twentieth [and, one may safely add, the twenty-first] century, one must be connected to electronic culture" ("Writing" 222), *CSI* strategically marries the moral security guaranteed by good old science with the flashy, exciting, hyper-detached world of high-tech. In contrast to the generally more colorful outdoor shots, the settings in which actual science happens (i.e. *CSI*'s laboratories, autopsy rooms, and media-saturated offices) are dominated by a highly restricted color scheme, ranging from glacial whiteness to a variety of bluish tones. The sleek surfaces of the technical equipment seem literally cool to touch and appear to have a distancing effect, since the scientific production of knowledge is rarely portrayed as team work. Yet, the solitary toil is never frustrating or mundane, but fast-paced, varied, and, above all, visually appealing. Even the professional detachment expected from the stereotypical, emotionally dysfunctional scientist is successfully translated into an enviable nonchalance that makes the lab coat appear cooler than any leather jacket.

This seemingly intrinsic coolness of twenty-first-century information culture is overtly acknowledged through the show's protagonists. Warrick (Gary Dourdan), the lab's audio-video analyst in Las Vegas, longs for a new high-tech machine, which he intended to purchase, as he eventually confesses, for purely ornamental reasons rather than utility:

Catherine Willows: So how's your new toy working out?
Warrick Brown: It's been downsized.
Catherine Willows: Bummer. I know how you wanted to see that thing work.
Warrick Brown: Well, it's the same difference, really. Air is drawn into the last tube, the chalk absorbs the chemicals from the air and mass spec will break it down at the lab.
Catherine Willows: So why did you need the expensive one in the first place?
Warrick Brown: 'Cause it was cool.

("Bully for You")

Warrick's deadpan one-liner "'cause it was cool" (delivered in the appropriate nonchalant manner) summarizes the appeal of science in *CSI* in a nutshell: It is not so much the functions that technical instruments fulfill, such as yielding intricate DNA analyses or mass spectrometry, but their aesthetic qualities, epitomized by the elusive attribute of coolness, that make them ideal additions to the show. Within the hypermodern slated glass walls of the *CSI* lab, cool thus emerges as what Alan Liu has astutely termed "the nascent, everyday aesthetics of knowledge work" (8), which thrive on connotations of technological expertise, social distance, and an emphasis on visuality and surface. *CSI* strategically capitalizes on exactly these properties and thereby manages to fully exploit the spectacular potential of technoscience for a cool visual effect.

LAB WORK: COOL AND THE APORIA OF INFORMATION

In order to illustrate the above considerations, the following section will focus on one example of *CSI*'s signature scenes, namely the notorious lab scenes in which one or more investigators conduct scientific analyses, primarily DNA processing, of the samples they have previously collected at the crime scene. Each episode contains at least one of these scenes, which highlights the centrality of the lab for the *CSI* universe. The franchise hence effectively contrasts the urban jungle of Las Vegas, Miami, or New York, dominated by violence, danger, and lawlessness, with the safe interior of the laboratory, which always succeeds in re-establishing law and order through the unfailing powers of science: "The three cities, together, present the best and worst of America and, in doing so, offer their *CSI* teams an inexhaustible (for inexhaustible investigators) supply of dead bodies and cases to solve" (8), as Michael Allen observes in his introduction to *Reading CSI* (2007). As such, the lab also fulfills a vital psychological function: It provides shelter and restores stability in an increasingly complex, overwhelming, and fragmented world, which the metropolitan centers in which the shows are set exemplarily reflect. The lab thus effectively counters the moral and material decline witnessed in the city (including, but not limited to the decline of family life and neighborhoods) with the promises of progress and security provided by Western industrialized science. While the urban space is frequently coded in terms of a riddle or jigsaw puzzle that runs counter to transparency and intelligibility as well as "a body of costumes and traditions" (Willett 8) that can, if at all, only be read by the knowledgeable native, the *CSI* laboratory, by contrast, is the place where the city's (literal) traces are deciphered and endowed with meaning through science. The challenges of modern American cities, above all their size and ethnic diversity, can adeptly be met by the technical equipment of the forensic lab, whose microscopic images put urban vastness into perspective and whose DNA analysis stabilizes (racial and gender) identity. These technologies also represent an effective antidote to the cities' emphasis on surface, superficiality, and simulation, be it the deceptive neon world of Las Vegas, populated by magicians and showgirls, or the cheerful Art Deco hotels of Miami that belie the city's enormous social inequality. The lab effectively inverts this cult of the surface by burrowing past the exuberant exterior into the gritty interior by virtue of scientific dissection, rendering visible what would usually remain hidden from the public eye.[27] Hence, the intimate setting of the lab, which for reasons of production efficiency is employed far more frequently than outdoor shots, also mirrors the show's "preoccupation with the body's interior" (Panse 165) and the secrets it harbors.

The particular scene I want to explore in greater depth here is taken from the episode "And They're Offed" from Season 7 of *CSI: Miami*, whose urban backdrop lends itself particularly well to the recurrent dichotomy between safe lab and glamorous, yet dangerous city. The show exploits the ironic contrast between the tropical postcard image of Florida as the state of sunshine, beauty, and affluence, epitomized by gorgeous young models who will repeatedly feature as crime victims, with Miami's darker sides. As the so-called 'murder capital' of the USA in the 1980s and the country's hyped 'new frontier,' the city's defining images are those of (illegal) immigration, drug trade, prostitution, and corruption among the rich and sophisticated (Willett 73-76). The examined episode indeed dips into these thematic concerns, as it revolves around the manipulation of a high-end horse race. Its lab scene shows crime scene investigator Eric Delko (Adam Rodriguez), a young and handsome Latino in fancy designer garb and lab coat, who analyzes a range of samples to convict (or exonerate) the horse keeper Terrance Chase, who is accused of having doped his boss's horse. We get to understand that his presence not only haunts the whole lab procedure, but serves as the main motivation for Detective Delko's diligent work—despite repeated avowals of objectivity and professional disinterestedness, the investigators frequently already entertain a strong suspicion

27 See Rahilly (2007), who elaborates on this dichotomy between surface and interiority with regard to *CSI: Crime Scene Investigation* and its setting in Las Vegas.

and merely look for material evidence to scientifically bolster their claims and ensure that the delinquent will be rightfully sentenced in court. The lab's technology hence also has a vital healing function: It allows the dead crime victims, as the show continuously stresses, "to haunt the living and [...] continue to speak" (Gere 133) to the forensic expert in order to assist the detection of the murderer. In this case, Horatio and his team are convinced, by a somewhat esoteric intuition, that Russian Mafia boss Ivan Sarnoff is involved in the horse race fraud. Sarnoff has already figured prominently in previous episodes and is held responsible for a series of murders, for which the CSI staff, however, had not yet been able to convict him due to lack of evidence.

Before the camera zooms in on the Miami-Dade Police Department, the audience is presented a glimpse of the city's breathtaking sky line and its deep blue ocean through a high-speed establishing aerial shot, which emphasizes the contrast between the two settings and solidifies the lab's status as a refuge from the dazzling metropolitan life outside. The scene itself is set in the department's glass-walled laboratory, with catchy rock music in the background. Close-ups of the impressive high-tech equipment interchange with medium shots of the scientist preparing his samples, blending them with a transparent fluid, and subjecting the results to a variety of scientific procedures. The piece of evidence thus analyzed is presented from several angles, while Miami's ocean again provides a colorful and contrastive background. What the viewers are thus presented with is a dazzling and highly entertaining "spectacle of microscopic and magnified visual effects" (Goode 136) that is overtly intended to leave them in utter awe of the gizmos on display. The editing between the takes is fast, abrupt, and fragmented, interspersing the bright and color-saturated shots of the laboratory with coarse black-and-white flashbacks to the crime scene and the suspect. This abrupt cross-cutting adds to the generally agitated impression, which is further enhanced by the incessant blinking of lights and the rapid movement of blurred figures in the background. Toward the end of the forty-five seconds scene, Detective Delko seems to have reached a major break-through: In perfect synchrony with the last beat of the obtrusive music, he nonchalantly flings away his gloves and hurriedly leaves the lab, assumedly off to confront the true culprit.

What is most striking about this scene, first of all, is that it heavily resembles a music clip or an advertising video, complete with popular rock tunes, hyperkinetic editing, and jerky flashes of overexposed, color-saturated imagery. The chrome-and-steel high-tech gadgetry, impressively augmented by the elaborate lighting in bluish tones, makes the background look like a photo shoot setting rather than a sterile and bleak laboratory. While Detective Delko might seem to be looking for it, the overall mise-en-scène of the lab analysis rather suggests that he has already long found the formula for cool. Indeed, the sleek state-of-the-art look of *CSI*'s laboratory "mirrors the advanced technology with which the series is produced" (Panse

162), not only in terms of set-up, but also with regard to its inherent purpose: *CSI*'s lab scenes not only aim to make the rather mundane scientific procedure appealing to the (largely lay) audience by providing overt visual entertainment, but also strive to reinforce this visual appeal by a distinctly 'cool look.' Similar to the show at large, this is achieved through Hollywood cinematography, an elaborate color scheme, and a noticeable emphasis on visual appearances such as the lab's various tools and machinery. Crime solving, as *CSI*'s lab scenes amply demonstrate, is primarily a "visual enterprise" (Gever 460), composed of aesthetic flourishes and an emphasis on literally cool, metallic surfaces that underline the overall high-tech feel of the show.

The foregrounding of the visual dimension, which pervades *CSI* as a whole, is perceptibly enhanced in this scene via the complete absence of plot-related information. It thus stands as a prime example of how in *CSI*, "dramatic developments frequently pause for dazzling displays of computer graphic virtuosity" (Gever 446). Owing to its music video style set-up, the scene is dominated by non-diegetic music at high volume and does not feature any form of verbal utterance by the character. The lack of dialogue is highly characteristic of *CSI*'s lab scenes: Even if there are several scientists in the laboratory conducting an analysis side-by-side, one will hardly ever witness verbal communication or extensive plot development *during* the procedure.[28] The rapidly alternating cut-ins and close-ups show impressive, yet murky images in the microscope or the translucent computer screen, but the actual results of the analysis will only be revealed at a more advanced stage in the narrative for strategic effect. In that respect, the lab scene is, generally speaking, not necessary for plot development and could easily be omitted without impairing or disarraying the overall storyline of the episode. If all that the viewers took an interest in were merely the actual results of the analysis convicting the criminal, it would certainly not be relevant to incorporate relatively long-drawn and incomprehensible lab procedures. Again, a comparison to the pornographic money shot suggests itself, as the lab scene, just like the even more visually expressive 'CSI shot,' is "divorced from the fictional narrative" (Weissmann and Boyle 93) and rather invests in the visual and technical aspects of the scientific analysis.

CSI's generic lab scene thus proves to be a viable counter example to what a range of notable studies, such as the quantitative analysis of 222 Hollywood fiction films featuring scientists conducted by Weingart, Muhl, and Pansegrau (2003), have repeatedly confirmed, namely that "the representation of the scientists' actual methods of work will be of interest only if they reveal the problematic, even criminal, nature of these methods. Where these methods are not foreign to the everyday practices of the lay public, the representation would tend to focus on the results on-

28 This does not pertain, however, to moments after the lab procedure, when observations or findings must necessarily be discussed with colleagues to convey them to the audience.

ly" (284). Indeed, not only classic science-fiction and adventure movies, as the other texts analyzed for this study will demonstrate, but also traditional crime dramas tend to skip such lab scenes completely. While the German *Tatort*, for instance, has lately made concessions to *CSI* and similar autopsy-driven shows by equally highlighting the relevance of forensic science for crime detection, the audience will usually be spared the minutiae of a DNA analysis or an autopsy.[29] Rather, scientific results are summarized and verbally communicated, instead of visually presented to the detectives (and thus the viewer) at strategic points of the story. Classic American police procedurals, such as the paradigmatic *Dragnet* or the immensely influential *Miami Vice*, equally refrain from displaying the actual collection and processing of evidence. The standard lab scene exposing scientific processes is thus very specific to *CSI* and has become one of the show's trademarks, on par with its signature color scheme and film-style cinematography.[30]

Indeed, the above described lab scene perfectly fits the highly paradoxical world of *CSI*, which combines formulaic story arcs with progressive, high-quality imagery. The aforementioned ironic tension thus created is also maintained during the execution of the scientific analysis. While there is, strictly speaking, no information forthcoming to the viewer throughout the procedure, the whole scene is, in fact, exclusively concerned with the retrieval, manipulation, and organization of relevant data: The analysis conducted by Detective Delko is meant to distil facts about the criminal from the samples he collected at the crime scene, so what the viewer is witnessing here is the process of the scientific production of knowledge. Interestingly enough, even the obtrusive rock song in the background thematizes knowledge and the access to it, denied or granted: "Here's a story that you ought to know / I had a little fight, I was dancing at a go-go / I'll tell you anything, baby, except the truth" (Eagles of Death Metal, "Anything 'Cept the Truth"). According to the rationale of *CSI*, the story we ought to know must be told by the evidence, because any other testimony would constitute 'anything except the truth.' At the same time, however, the audience is denied access to this truth. While Detective Delko apparently experiences an epiphany at the end of the scene, suggesting that he has hit upon a crucial piece of evidence, this discovery is not explained to the viewer.

29 *Tatort Münster*, for instance, features a forensic pathologist as part of the detective duo.

30 Several critics (Gever 2005, Pierson 2010) observe that *CSI*'s focus on autopsy and the human body as legal evidence resorts to the expository techniques of NBC's *Quincy, M.E.*, with the significant modification that the latter does not visually represent the actual performance of science. Equally, the graphic explicitness of the British program *Prime Suspect* can be seen as a vital source of inspiration (Harrington 373). None of these prominent forerunners, however, exploited the visual-aesthetic dimension of forensic science and technology as meticulously as *CSI*, nor did any of them feature extensive scenes set in the laboratory or the autopsy room.

In this respect, lab scenes in *CSI* habitually present us with a reversal of classic dramatic irony, as it is not the characters who are left in the dark (usually for a comic or tragic effect), but the audience, who can, at this point, only wildly guess at what the investigator has detected in the microscope. This strategy subverts traditional film theory, which suggests that by showing, rather than narrating a process of (criminal) cognition, the audience "is placed in a position of greater knowledge" (Weissmann and Boyle 94) and induced to active narrative involvement: In the lab scene, however, the highly magnified images and detailed experimental set-ups on display are certainly no help to a viewer not versed in forensic science, since they are not mediated through dialogue or voice-over.[31] This is particularly unusual for the classic police procedural format, which *CSI* essentially continues, since balancing information distribution in favor of the audience is a key variable for creating suspense: "One thing that keeps us interested in a story is our access to privileged information, whereby we know something that has been withheld from the characters themselves and hence enjoy a sense of superiority" (5), as George Dove argues with respect to the classic detective formula story.[32] In *CSI*, however, this typical suspense situation is complicated through the use of high-end technology and science practice. Paralleling its exclusionary and intimidating mechanisms in the contemporary information society, technology in *CSI* does not merely fulfill a vital aesthetic function; it is also employed to establish a strict hierarchy between those who are granted access to electronic instruments and the knowledge they produce (i.e. the investigator) and those who are denied access (i.e. initially both the audience and the suspect). Since the lab scene focuses on visual properties and does not mediate the scientific process via dialogue or voice-over, it serves to illustrate how the use of technical expertise becomes a "primary exercise of power" (Ross, "Challenge" 304) in a culture that fundamentally relies on knowledge work.[33] While sci-

31 This resistance to information transmission by the visual runs counter to the program's overall structure, which consistently prioritizes image over text. According to director Danny Canon, "our philosophy was that if the sound on your TV set went out, you should still be able to know what the story's about" (qtd. in Panse 155). While the lab scene indeed does not require any sound (apart from its background music), the viewer will hardly be able to decipher, from this scene alone, 'what the story's about.'

32 *CSI* thus effectively violates what Dove identifies as "the principle of *confidentiality*" (32, italics in the original), which states that readers (or viewers) have to be made privy to information not shared by the characters in order to guarantee their attention to and participation in the story. In the lab scene, this is replaced by visual excitement, which equally manages to capture the audience's attention.

33 Additionally, technology in *CSI* not only serves to empower some characters while excluding others, but also establishes a "hierarchical human-machine dyad" (453), as Geyer argues. It is, in actual fact, the machines rather than the scientists that are portrayed as do-

ence and technology are constructed as pivotal sources of authority in the series, a lack of verbal explanation (or, conversely, highly specialized terminology) ensures that the audience will primarily appreciate the gleaming tools and instruments for their decorativeness.

Hence, the lab scene stands as a prime instance of how *CSI* consistently privileges the visual over the verbal dimension. The function of the DNA analysis or, more precisely, its representation in the lab scene is primarily an aesthetic one; it provides narrative space within each episode to parade and flaunt all the exciting props that technoscience has to offer: the media-saturated laboratory, its high-tech equipment, the young and casual scientist, and, above all, the coolness that is apparently inherent in such performances of science. Consequently, the lab scenes manage to exude coolness not despite, but indeed because of science, which provides all the right costumes, attitudes, and paraphernalia for a cool visual effect. Via such signature scenes, *CSI* thus proficiently exploits (and simultaneously promotes) the spectacular potential of forensics, autopsy, and digital DNA processing to create provoking and aesthetically appealing images. It is in such scenes that science most conspicuously induces the "stylish thrill" (Seltzer 24) that lies at the heart of any cool affective encounter. The transformation of science into a veritable spectacle is greatly enhanced by the aforementioned complete absence of verbal communication and complicated plot development, which allows the viewer to fully concentrate on and savor the image.

Indeed, the paucity of information in this scene may be understood as yet another vital prerequisite for the production of cool, as cool relies on exactly such a lack of content and factual knowledge. Again, this is exemplarily reflected in the lab scene, which remains on the level of surface by refusing to leak relevant plot information and thereby add depth to the story; its purpose is primarily an aesthetic one. Through this evident recourse to form rather than content, the scene emerges as a perfect breeding ground for cool, which equally thrives on surfaces and superficiality. This line of argumentation is indebted to Alan Liu's classification of cool as a "cultural dominant" (76) of the techno-scientific age, part of intricate new information aesthetics that shape contemporary existence. Liu argues that cultural productions, such as websites or blogs, are at their coolest when no information is forthcoming, or, as he phrases it: "Cool is the aporia of information" (179), denoting a state of contradiction, perplexity, doubt, or even indeed loss of information. The examples which Liu analyzes in his study are, for instance, websites that have

ing all the work. Since subjectivity is regarded as adverse, rather than conducive to processing evidence, the human element is relegated to the very margins of the scientific analysis. While the *CSI* scientists are as ingenious and objective as can be, they must nonetheless rely on technology to augment their abilities and produce unequivocal results that will withstand a court challenge. See also Kruse 84-85 and Tait 47-48.

been voted 'cool' by internet users. As his analysis compellingly shows, the top-ranking websites are those that are either completely devoid of content or that obviously contain information, but refuse to disclose it to the user.

This circumstance, as Liu argues, is grounded in cool's resistance to definition and knowability. Cool thrives on the very fact that it can never be fully identified, measured, and clearly demarcated; its cultural force and timeless popularity derive from its ontological elusiveness, changeability, and semantic flexibility. In a perceptive article on "The Coolhunt" (1997), Malcolm Gladwell notes that cool appears to exist in a hermeneutic circle, as it "cannot accurately be observed at all, because the act of discovering cool causes cool to take flight" (n.p.). This dazzling assertion not only explains why those asking to define cool are almost certainly not cool (of which the recent academic treatment of cool might serve as a compelling example), but also clarifies why we may tend to attribute an air of mystery and inscrutability to those individuals we perceive as cool. This is, in essence, also insinuated by Liu when he argues (in unsuspecting unison with Rice's theory of cool juxtaposition) that cool is produced in an oscillation between two polarities, namely the evident existence of information, on the one hand, and a reserve of that information, on the other. This paradoxical limbo state is also present in *CSI*, whose narrative structure, as Charlie Gere suggests, "is predicated on the tension between the known and the unknown" (138). This inherent tension is particularly feasibly carried out in the lab scene: Detective Delko seems to have discovered a vital piece of evidence in the microscope, but he does not bother to communicate his findings to the viewer. While the audience thus gets to understand that the scene contains vital information for the plot, they are not granted access to this information. On the visual level, this juxtaposition of contradictory ideas is played out via the cross-cutting of conflicting shots: High-resolution images in lurid colors alternate with short sequences in grainy black and white; blurred flashbacks to the suspect and the crime scene, filmed in a deliberately amateur fashion with hand-held cameras and low saturation, are interspersed with highly stylized computer-generated close-ups of the analytic process, presented on the deep blue background of Miami's ocean.

The standard lab scene hence reflects one particular way that science, the very epitome of codified and consolidated knowledge, can be represented as cool in popular cultural contexts such as prime-time police procedurals: The focus is noticeably shifted from the knowledge that science produces to a situation which communicates no relevant information at all. Rather than underpinning the cognitive or ethical value of science (i.e. its ability to ensure justice and moral stability through the production of hard and testable facts), it is the aesthetic dimension that such scenes exploit. Granted, *CSI* and similar forensics-driven shows do employ science for the convincing retrieval of intra-diegetic knowledge, which is essentially needed to advance the plot and solve the case. Moments of cool science, however, such as the lab scene, primarily capitalize on the spectacular, visual potential of science and technology through extreme stylization and aestheticization. While the legitimation for science within the narrative world and rationale of *CSI* is, first and foremost, its ability to provide closure by miraculously revealing the murderer's identity, the audience is, in fact, greatly encouraged to appreciate, or even indeed tolerate the heightened narrative attention to science because of its (seemingly intrinsic) visual coolness.

RECAPITULATING THE LAB

The lab scene as one of *CSI*'s core ingredients reflects how forensic crime shows succeed in endowing the hardly glamorous practice of science with coolness. The specific visual strategies employed, including Hollywood cinematography, a glossy surface, and distinct camerawork, enhance the production of cool as a central aesthetic theme. This results in a notable tension between a normative, morally reassuring narrative, on the one hand, and cutting-edge, cool imagery, on the other. As the above analysis has shown, the images of science presented in *CSI* are noticeably informed by dominant notions of coolness, which are largely induced by the show's prominent focus on the technological, visually exciting aspects of scientific practice and lab work. Hence, it is exactly the overt emphasis on science as a key narrative and visual element that contributes vitally to the program's signature cool aesthetics. Coolness in *CSI* can thus be understood as a direct consequence of, rather than antonym to, its scientific storylines, rational and detached worldview, and sleek high-tech surfaces. The same pertains to the protagonists of the show, who convey the "ambiguous sensibility" (Pountain and Robins 43) that produces cool through their simultaneous display of an extreme, negative affective range (including, above all, anger and vengeance) and a cool exterior that is portrayed as camouflaging their actual state of mind. This tendency, as the preceding sections suggest, perceptibly mobilizes the conceptual origins of modern-day notions of cool as a survival strategy for conveying composure and inducing respect in times of peril and distress. Furthermore, the analysis has shown that *CSI* reiterates highly gendered visions of cool as a moral code that thrives on masculine stereotypes of toughness, fearlessness, and emotional control. Especially when wielded by male characters, the particular displays of cool are perceptibly linked to technical expertise and portrayed as enhancing, rather than impeding the scientific ingenuity of the respective characters. Similar to the above described lab scene, their cool pose strategically draws on distinct gestures, habits, and props such as gleaming machinery and electronic instruments.

CSI as the perhaps most paradigmatic instance of contemporary science-driven crime dramas hence reflects how the heightened (and, considering contemporary

American society's hostility to science in formal rather than popular cultural contexts, audacious) use of scientific storylines and highly specialized terminology is legitimized and made not only tolerable, but indeed entertaining to the largely lay audience. While the representation of science in *CSI* is informed by a highly moralizing undercurrent that promotes the normative powers of scientific knowledge, scenes dedicated to the display of pure scientific practice capitalize on its aesthetic potential to produce gripping images. It is in these moments that 'cool science' most conspicuously emerges as one of the show's dominant topoi. Additionally, the above analysis also attests to the viability of the apparent shift of focus from science to the nowadays much more pervasive concept of technoscience. *CSI*'s media-saturated homes, offices, autopsy rooms, and laboratories not only reflect how technoscientific tools and gadgets have thoroughly pervaded our workspaces and private lives, but also demonstrate that they can serve as effective props for a cool televisual effect that is hinged on surfaces, cold textures, and spectacular visual qualities. The combination with cutting-edge technology helps to spectacularize science and relocate the focus from the knowledge that scientific practice produces to its visual-aesthetic properties in a situation that communicates no plot-related information at all. The legitimation of science within the narrative confines of the prime time police procedural must thus, first and foremost, be sought in the aesthetic sphere, rather than the cognitive or ethical. The extreme stylization of the show, its pronounced televisual style and foregrounding of surface, mise-en-scène, and cinematic imagery construct forensic science as the ultimate harbor of twenty-first-century information cool. While my analysis in this chapter was deliberately focused on laboratory science and explored the uses of autopsy only in passing, it should nonetheless be noted that the splatter aesthetics achieved through techniques like the notorious 'CSI shot' and an overt emphasis on the assaulted human body additionally contribute to this effect, aligning forensic science and pathology with toughness, visual excitement, and danger, rather than mundane lab work. This is augmented by controlled and detached scientist-detectives, who will retain their enviable nonchalance in the face of the most shocking homicides. On both the personal and the aesthetic level, the body of evidence will hence allow for only one plausible conclusion: *CSI*'s alleged nerd squad is, in fact, a pretty cool gang.

Geek Cool and the Comedy of Science in *The Big Bang Theory*

> [T]here's something about the nerd's principled disdain for (or inability to abide by, same difference) ordinary social conventions that strikes Americans—a nation of nonconformists—as noble.
> *TIME MAGAZINE*, 25 SEPTEMBER 2005[1]

"Smart Is The New Sexy" is the fitting tagline that the CBS hit sitcom *The Big Bang Theory* (2007-present, henceforth in this chapter *TBBT*) immediately popularized with media outlets all over the world.[2] Indeed, the provocative slogan was soon to be proven accurate by the rapid and enduring commercial success of the show: With its celebration of science and geek fan culture, *TBBT* has fast become immensely popular, not only with viewers in the United States, but almost worldwide.[3] Scientifically educated characters obsessing over *Star Wars* and action figures no longer serve as slightly pathetic sidekicks, such as the legendary Steve Urkel of ABC's *Family Matters* or the obligatory nerd gang in every American high

1 See Grossman (2005).

2 See, for example, Strauss (2007) for *USA Today* and Workneh (2011) for *CNN*.

3 In fact, the show was even plagiarized on Belarusian television under the title *The Theorists* (Lyons 2010), which, despite ensuing legal disputes, attests to the global appeal of its central theme.

school drama from *Beverley Hills, 90210* to *American Pie*. In *TBBT*, they have taken center stage and proven to be more accessible and likeable than the markedly non-nerdy characters with whom they are repeatedly contrasted. The appeal of the geeky scientist seems to have reached well beyond television: Critics have come to speak of a veritable 'geek-chic' inspired by the sitcom, making it fashionable to wear horn-rimmed glasses, backpacks, and electronic accessories, and intersperse one's dictum with scientific references. The former polarities of 'Dr. Nerd' and 'Mr. Cool' no longer seem to hold.[4]

This chapter will examine how science in *TBBT* is presented as a source of comedy as well as a strategy for allowing particular constructions of (geek) coolness to emerge. As the only (intentionally) comic text analyzed in this study, the show holds a special place in the contemporary, popular cultural representation of scientists and scientific storylines, since the portrayal of its subject matter is necessarily complicated by the potentially subversive effects of humor. Indeed, it may serve as a prominent counter-argument against the particularly pervasive assumption in science studies scholarship that "there are hardly any comedies about science. Evidently our society does not find much to laugh about in science" (Weingart, Muhl, and Pansegrau 286).[5] While this claim certainly seems to prove true for the majority of popular cultural productions thematizing science, both generally and within the constraints of the present study, the international success of *TBBT* demonstrates that there are exceptions to the rule, exceptions which transcend a niche market and gain widespread popularity.

Hence, one central question which this chapter attempts to explore is how and why science can figure as a major source of laughter in the traditional domestic sitcom format as witnessed on mainstream American television and beyond. I will argue that in the case of *TBBT*, the comedy of science is notably enhanced by the protagonists' representation as 'geek' characters, a recently pervasive term which has been used to describe "any unsociable person obsessively devoted to a particular pursuit" and is specifically directed at individuals who, just like the show's scientists, are "extremely devoted to and knowledgeable about computers or related technology" (*OED*, s.v. geek). The misfortunes and failures arising from the characters' various geeky obsessions propel a myriad of comic situations, which have traditionally formed the very basis of the domestic sitcom. Additionally, as I will argue, it is through this alliance with geek culture that the scientists emerge as bearers of an alternative, geeky cool. Referencing Daniel Harris's elucidative thoughts on

4 This pun draws on the *Family Matters* episode "Dr. Urkel and Mr. Cool," in which Steve Urkel invents a 'cool juice' to transform himself into the good-looking Stefan and woo his beloved Laura.

5 In their quantitative analysis of 222 movies featuring scientists, only 5% of the scientists portrayed could be classified as comic (Weingart, Muhl, and Pansegrau 286).

cool as a clandestine aesthetic code as well Ilan Dar-Nimrod's differentiation between 'contrarian' and 'cachet' coolness, the present chapter will propose that Leonard and his friends can be read as wielding a particular form of cool that thrives on notions of rebelliousness, unconventionality, and emotional restraint, while at the same time disregarding more socially desirable traits such as trendiness, peer approval, and physical attractiveness. Similar to the previous one, this chapter will conclude with a close reading of one key episode of the show, which serves to connect and apply my considerations on science as a source of comedy and coolness. The final episode analysis shall thereby help to demonstrate that rather than engendering the much proclaimed death of cool (cf. Gioia 2009, Geiger 2010), the revenge of the nerds as portrayed in *TBBT* and a myriad of other contemporary texts effectively updates cool to the demands of twenty-first-century information culture.

"IT ALL STARTED WITH A BIG BANG": THE COMEDY OF SCIENCE IN *THE BIG BANG THEORY*

TBBT is generically conservative and follows the traditional (domestic) sitcom format, with each episode amounting to about half an hour, including the obligatory commercial break. At the time of writing, the program is currently in its sixth season and has aired more than one hundred episodes. With a peak of over twelve million viewers, it repeatedly ranks among the highest-rated shows of the channel and has inspired a lively fan community.[6] The story is centered around a stable set of characters in recurrent environments in and around Pasadena, California, including Sheldon (Jim Parsons) and Leonard's (Johnny Galecki) shared flat, the university cafeteria, Penny's neighboring flat, her workplace *The Cheesecake Factory*, and the local comic book store. In line with dominant sitcom conventions, the comedy of the show is augmented by an artificial laugh track, which is partly generated through the live studio audience. The series revolves around a straightforward, yet highly successful formula: Four scientists (three physicists and one engineer) in their twenties are juxtaposed with an aspiring actress, who, due to a lack of acting jobs, has to earn her living as a waitress. The four male characters—experimental physicist Dr. Leonard Hofstadter, theoretical physicist Dr. Sheldon Cooper, astro-

6 At the time of writing, *TBBT* draws a greater audience than the last run of *American Idol* and has repeatedly been named the most popular sitcom in the United States, second only to *Two and a Half Men.* The fact that its theme song, describing the formation of the universe according to the big bang model after which the show is named, was written and recorded by the Canadian alternative rock band *Barenaked Ladies* only adds to its cult status among viewers. See Moraes (2012), Carter (2012), and Prince (2012).

physicist Dr. Rajesh Koothrappali (Kunal Nayyar), and Howard Horowitz (Simon Helberg), doctoral candidate in engineering—are portrayed as highly intelligent and brilliant academics. Additionally, they regularly indulge in what are depicted as 'geeky' pastimes, such as their frequent visits to the local comic book store or laser tag games. Especially Leonard's romantic pining for Penny (Kaley Cuoco), their new neighbor living across the hall, allows for a myriad of humorous situations mobilizing the show's prime comic trope, i.e. the contrasting of two highly disparate subcultural groups. *TBBT* is thus a typical hybrid between family and workplace sitcom, as it blends "family comportment (living together, couch-centric) and workplace (sexual exploration, flirt-centric)" (Hartley 67). Its comedy is chiefly created through a repeated juxtaposition of the highly divergent worlds that the male protagonists and Penny inhabit and the misunderstandings that inevitably result. While the male characters are from the outset established as uber-nerds with superior intellect, they are portrayed as socially dysfunctional, shy, and utterly unaware of basic social conventions. Penny, on the other hand, is repeatedly dazzled when the guys discuss physics, but displays common sense and superior social skills. The core set of characters has in later seasons been expanded to include two female scientists, Bernadette (Melissa Rauch) and Amy (Mayim Bialik), as love interests for Howard and Sheldon.

While the primary concern of this study is certainly directed at the textual level and the thematic aspects of the show, I largely concur with Brett Mills's seminal work on the contemporary sitcom, in which he vehemently argues against sidelining the comic aspect when analyzing the genre. While much of the scholarship on television comedy formats has traditionally marginalized the examination of humor, Mills argues that "the sitcom is only meaningful—and explicable as a genre—if its comic intent is understood; it is this which drives and defines it" (*Sitcom* 5). The sitcom's affective engagement of its viewer in terms of humor and empathy is crucial. Before one can meaningfully assess the social, political, economic or other 'situation' of a situation comedy, it is first of all imperative to examine its comedy. As one of the oldest televised forms and a quintessentially Anglophone genre,[7] the sitcom is indeed, first and foremost, defined by its comic rather than serious intent, which distinguishes it from the narratively similar formats of the drama series or the soap opera. As with television comedy in general, any textual analysis of the sitcom hence faces the decision of how to evaluate the humor of the respective show:

7 The origins of the sitcom are generally traced back to the British music hall and American vaudeville, burlesque, and minstrel shows (Neale, "Sketch" 62). With the advent of the radio, sitcoms began to venture into broadcasting, but it was only in the late 1940s that they were established as a staple of American popular culture. To this day, American sitcoms continue to be successfully exported around the globe and dominate network ratings at home and abroad. See Mills, *Sitcom* 106-113, Hamamoto 1-3, and Hartley 65.

Treating comedy as either undermining or confirming the presented situation must necessarily have significant repercussions on the analysis of the show's thematic aspects. For the purposes of the present study, this essentially amounts to the question of whether the science nerds in *TBBT* generate humor through exposing *themselves* or *other characters* as laughing matter.

Seeking to meaningfully address this central question, academic analyses of (television) comedy such as those offered by Palmer (1987), Greeber (2001), Mills (2001, 2009), and Morreale (2003) traditionally draw on humor theory for exploring the role of laughter in the portrayal of specific social scenarios, groups of people, or political ideologies.[8] Broadly speaking, the audience's (real as well as augmented) reaction to a specific scene can be interpreted as laughter *at* or laughter *with* the character that utters the joke or is caught in a comic situation. In the case of *TBBT*, this means that the portrayal of science in the sitcom is not as unambiguous as it might be in more 'serious' genres, such as the classic drama series or the action movie. Rather, jokes, punch lines, innuendo, and instances of slapstick repeatedly disrupt the straightforward story arc of a given episode, as does the obtrusive laugh track. Depending on which perspective one wishes to adopt, the humor evoked by the sitcom can hence be understood as either affirmative or subversive of the specific social situation it displays. One frequent source of laughter in *TBBT*, for example, is Sheldon's inability to understand social conventions, rhetoric flourishes, or metaphorical statements; in accordance with the stereotype of the pragmatic, no-nonsense scientist who values facts over fancy, he is represented as taking each utterance literally. Equally, he regards social etiquette as unnecessary and arbitrary. This results in repeated scenes in which Sheldon upsets and deliberately disrupts the conventional social protocol, such as when Penny in her function as a waitress at the local restaurant wants to take his order:

Penny: Hey, you guys ready to order?
Sheldon: Since we come in every Tuesday night at 6:00 and order the same exact thing, and it's now [checks his watch] 6:08, I believe your question not only answers itself, but also stands alongside such other nonsensical queries as "Who Let the Dogs Out?" and, uh… "How are they hanging?"
Penny: Okay, so the usual with extra spit on Sheldon's hamburger.

("The Love Car Displacement")

In this scene, the artificial laugh track is sounded after both Sheldon's and Penny's final repartee, thus diligently marking the funny parts for the audience at home. The question that hence suggests itself at this point is why Sheldon's and Penny's ex-

8 For classic scholarship on humor theory, see, for instance, Morreall (1983, 1987), Latta (1999), Boyd (2004), and Billig (2005).

change is (intended to be) comic. Depending on which humor theory one wishes to support, the scene could alternatively be read as poking fun at either Penny's or Sheldon's expense or, conversely, at both of them. The conclusion one reaches in this question is far from inconsequential, as it also impinges on any reading of the larger ideological and cultural concerns treated by the sitcom.

LAUGHING AT SCIENCE OR LAUGHING WITH SCIENCE? SOME PRELIMINARY REMARKS ON THE SUBVERSIVE POTENTIAL OF (SITUATION) COMEDY

Following Mills's seminal studies on the (British) sitcom (2001, 2004, 2009), there are three main possibilities of meaningfully dissecting the humor in the above scene. According to the so-called superiority theory, which can be traced back to as far as Greek comedy in the fourth century B.C., laughter is primarily understood as a negative emotional response that confirms the existing social order and supports hegemonic, rather than alternative interpretations.[9] Underlying this theory is the belief that "people laugh when they feel a kind of superiority, particularly over other people" (Mills, *Sitcom* 77). In modern television comedy, this frequently affects a specific social group and their supposedly generic behaviorisms, such as ethnic, religious, or sexual minorities, or, in the present case, a group of professionals. Hence, the audience's laughter at Sheldon's lengthy explanation would be read as laughter of derision at the scientist's ridiculous insistence on linguistic accuracy, his slave-like devotion to facts, and his inability to grasp basic social conventions. Consequently, the joke allows the audience to feel 'superior' as they will effortlessly decode Penny's query as a polite conversation starter typical for a standard service encounter. On the intra-diegetic level, it is hence Penny who emerges as the superior conversation partner, making Sheldon the emblematic 'butt of the joke.' Laughter after Penny's witty retort would accordingly be understood as laughter of assent, indicating agreement with her public shaming of Sheldon in front of his friends and the triumphant countering of his scientific arrogance. Thus, if one chooses to under-

9 Both Plato and Aristotle's writings offer perspectives on the reasons and sources for humor, which the superiority theory heavily draws on. Greek philosophy understood humor as an essentially immoral act, suggesting that people primarily laugh at someone's misfortunes and failures. Mills (2009) observes that this theory was elaborated by Thomas Hobbes in the seventeenth century, who saw humor as a tactic of the lower social classes for mocking the powerful and thus potentially asserting their own dominance. Other famous takes on comedy and laughter which essentially support the superiority theory include René Descartes, George Meredith, and Mikhail Bakhtin. See Mills, *Sitcom* 77-82.

stand laughter as indicating assent, the power structure suggested by the exchange is not only condoned, but indeed naturalized by the artificial laugh track, suggesting that amusement is a justified reaction to Penny's mocking depreciation of Sheldon.

Conversely, humor can also be read as subverting and questioning social norms, as suggested by both the incongruity and the relief theory. While Mills and other theorists discuss these theories separately from each other, I want to suggest that vis-à-vis the question of whether humor is to be decoded as subverting or preserving social structures, they both allow for largely similar conclusions. Essentially, the relief theory harks back to Freud's theory on the formation of psychoses, which were seen as arising out of internal suppression of subjects tabooed by 'civilized' society, such as death, sexuality, violence, or the abject.[10] Humor is accordingly understood as an inherently positive and highly valuable form of interpersonal communication, allowing the expression of normally unacceptable or highly controversial ideas, thoughts, and topics. Similarly, the incongruity theory relies on the reversal of what is expected and condoned, hence demonstrating "that humour rests on diversions from social norms" (Mills, *Sitcom* 87). Having its origins in Kant's work on humor in his eminent *Critique of Judgment* (1790), the incongruity theory suggests that humor arises out of the clash of opposing and highly disparate situations, discourses, or social groups. Like many other classic sitcoms,[11] *TBBT* indeed places a basic incongruity at its core, namely the clash of two highly disparate professional

10 See especially Freud's treatise *Jokes and their Relation to the Unconscious* (1907). Freud locates the source of comedy and humor in the id, which is in a constant struggle with the regulatory forces of the super-ego. Jokes must accordingly be valued as a means of reconciliation between the demands of the id and the super-ego, since they offer a socially acceptable outlet for the discussion of otherwise tabooed topics.

11 See, for instance, sitcoms based on the clash of different social classes (*The Fresh Prince of Bel-Air*, *The Nanny*), genders (*Maude*, *Butterflies*), lifestyles (*Dharma and Greg*, *Family Ties*, *Two and a Half Men*), or generations (almost all classic domestic sitcoms centered around a nuclear family). *TBBT*, I would argue, is essentially based on two highly divergent professions, i.e. white versus blue collar work and the according world view associated with it. Furthermore, this distinction notably aligns with gender. However, Jane Feuer notes that "the basic dualism of such shows tends to thin out as time goes by and more and more situations are required" ("Situation Comedy" 70). This is clearly the case in *TBBT*, in which especially Leonard, but also the other male protagonists increasingly manage to approach 'normal' social behavior, indicated by the circumstance that three of them have managed to find a girlfriend by season five. Furthermore, the humor in later episodes frequently stems from a reversal of the polarities established in the first season, such as when Penny demonstrates scientific or geeky knowledge ("The Dead Hooker Juxtaposition," "The Gorilla Experiment") or Sheldon has learned to grasp and employ sarcasm more readily than the other characters ("The Bus Pants Utilization").

occupations, the scientist and the waitress. More profoundly, it is based upon the ideological opposition between a (ridiculously) rational, highly pragmatic world view, probably best exemplified by Sheldon but, to varying degrees, adhered to by all scientific characters, and the 'normal,' conventional world of common sense and emotionality, represented by Penny and other non-scientific characters. This "binary, dualistic structure [is] typical of the domestic sitcom" (Feuer, "Situation Comedy" 70) and serves as its main source of humor by inspiring a wealth of situations which simply reiterate this basic opposition, such as the above scene in the restaurant.

Drawing on both the relief and the incongruity theory, Sheldon's mocking comment would accordingly be read as exposing and simultaneously critiquing arbitrary social norms and Western society's tendency to polite, yet meaningless small talk. Conventionally, one would expect a typical service encounter between waitress and guest, as suggested by the setting, Penny's initial question, and the other protagonists' adherence to the conventional routine. Sheldon's reaction, however, undermines these expectations and hence succeeds in drawing laughter from the audience, which is delightfully dazzled by his unexpected fury. The audience's laughter can hence be read as an appreciation of the fresh and unusual, perhaps even audacious and politically incorrect perspective offered by Sheldon as representative of a highly abstract and pragmatic world view. Accordingly, the viewer does not laugh to deride Sheldon, but to support the unusual take on standard social situations offered by the show's scientific perspective. Regardless of whom exactly laughter is directed at, however, humor according to the incongruity (and, to a lesser degree, the relief) theory is a proficient vehicle for voicing concerns that might otherwise be highly offensive or inappropriate to articulate and for thereby indicating possible alternative viewpoints. Hence, Sheldon's deviance from the norm is not only pleasurable (rather than irritable) to the audience, but also reveals a deeper truth, namely the "tenuous and artificial nature of social norms" (Mills, *Sitcom* 87), which the viewers themselves may already have experienced as suffocating, limiting, or inflexible. In that sense, Sheldon's scientific abstraction is offered as a powerful alternative to the repetitive small talk routine at a typical American diner, as his reply indeed only reveals its absurdity and futility. Such a perspective grants humor a highly subversive potential and appreciates comedy as a valuable vehicle for social critique, questioning established norms and dominant ideologies. In its most radical sense, it can be seen as presenting potential alternative lifestyles and propelling social change.[12]

12 This argument becomes particularly pronounced in respect to the sitcom's portrayal of gender relations, as is the case in shows which revolve around an 'unruly woman,' a term employed by Kathleen Rowe (1995) with reference to *Roseanne*. Female characters who notably diverge from the stock character of the domestic, physically attractive, and de-

The examination of a sitcom's comic intent as either affirmative or critical of the presented situation thus offers a viable point of departure for the analysis of its thematic aspects. Not surprisingly, it is the latter two humor theories that have most readily been utilized for the analysis of individual shows and are frequently cited in support of a vision of the sitcom as a powerful critique of contemporary American society and its hegemonic discourses. While the superiority theory would suggest, for the show at large, that the humor is primarily derived from ridiculing and debasing the scientific world view presented by the protagonists, one may also, conversely, read *TBBT* as a powerful advocate for science, as it repeatedly exposes the idiosyncrasy of irrational, non-scientific behavior steeped in illogical social conventions, such as in the first encounter between Leonard, Penny, and Sheldon:

Penny: I'm a Sagittarius, which probably tells you way more than you need to know...
Sheldon: Yes, it tells us that you participate in the mass cultural delusion that the sun's apparent position relative to arbitrarily defined constellations at the time of your birth somehow affects your personality.

("Pilot")

While Sheldon's reaction is, once again, a highly inappropriate and incongruous reply to Penny's attempts at small talk, it can nonetheless be easily read as effectively exposing Penny's belief in zodiac signs as superstitious and absurd. Penny's apparent modeling on a stock character of the traditional American sitcom, the 'dizzy blonde,' would support this reading: The punch line (additionally marked by canned laughter) would accordingly indicate that the comic situation is built around Penny as the 'fool' and source of humor, while the scientists emerge as superior and provide the necessary contrast to her naïve notions. This is reiterated in the show at large, where science can be understood as the subject that disrupts and challenges societal expectations; it provides, in other words, the necessary opposition for incongruity and, consequently, humor to emerge. Like many other classic (domestic) sitcoms, *TBBT* can be read as being predicated on a stable, traditional set of social norms, from which its protagonists, due to their unusual scientific professions and geeky interests, repeatedly deviate. Science hence acts as a force that disrupts the

mure housewife counter Western expectations of femininity and may thus not only offer a vehicle for comic failure, but also a powerful critique of contemporary gender relations (see, for instance, Mellencamp 1986 and Feuer 2001). While the female scientists appearing in later seasons of *TBBT* may be read as contemporary versions of the unruly woman (in particular Sheldon's love interest Amy), the main character Penny fulfills all expectations of classic female beauty and behavior. For the representation of gender and the scientific profession, see also feminist science studies work, e.g. Keller (1985, 1992), Schiebinger (1989), and Harding (1992, 1994).

safe and conventional world of the sitcom, to which the audience, according to the above cited humor theories, grants approval through their laughter.

Despite these convincing preliminary considerations, I will nonetheless strive to refrain from adopting a one-sided stance toward the general ideological outlook of the show, especially where its representation of science is concerned. In many instances, laughter clearly seems to be directed *at* the respective character's misfortunes, which in the case of *TBBT* typically involves situations that non-scientists would (supposedly) easily manage, but which the protagonists find trickier than their graphs and calculations. Thus, science is constructed as the 'reason' why the characters emerge as comic failures. Examples typically include their attempts at flirting (e.g. Howard in "The Dead Hooker Juxtaposition"), dating (e.g. Leonard in "The Fuzzy Boots Corollary"), sports (Sheldon and his arch enemy Kripke in "The Rothman Disintegration"), or mere small talk (e.g. Sheldon in "Pilot," Raj in "The Maternal Capacitance"). It would not be difficult to argue that in these cases, laughing is the acoustic signal of a sense of superiority (perhaps even relief) harbored by the average viewer, who will most likely find small talk with their colleagues and friends easier to master than a mathematical formula. Accordingly, the protagonists would be mocked for their lack of social skills, which the show represents as a pitiable, yet comic deviation from the norm. The use of typical sitcom conventions, such as the stereotyping and evident de-masculinization of the protagonists, would support this reading: Through their clothes and their appreciation of video games, toys, and comic books, they are physically and metaphorically likened to children rather than grown men, which can be seen as an additional source of humor through devaluation.[13] On the other hand, laughter can also be interpreted as a form of assent, suggesting that many viewers share the experience of finding their professional work easier to manage and more rewarding than personal encounters or physical activity. In that sense, the violation of a social norm or custom is valued by the viewer, suggesting that the protagonists' perspective is supported, rather than mocked or derided. Just like the above quoted scene in the restaurant, this reflects

13 Butsch observes that this is a frequent tactic applied to the portrayal of working-class men in domestic sitcoms, where "humor was built around some variant of working-class man's stereotypic ineptitude, immaturity, stupidity, lack of good sense or emotional outbursts, traits that have been culturally defined as feminine or child-like" (21). Alternatively, these portrayals could, of course, also be understood as critiquing dominant notions of masculinity and the cliché of the working-class man as the emotionally uptight, brutal, and insensitive head of the household. Within the scope of the present study, one should also note that the big icons of Western cool, such as James Dean or Marlon Brando, have traditionally been described as "adopting a look synonymous with failed manhood: perpetual boyhood" (Noble 156), which is more generally connected to the circumstance that contemporary cool thrives on connotations of youth and youthfulness.

how comic texts are highly ambiguous, as the inherent humor complicates any traditional, straightforward interpretation. It thus remains difficult, as Mills also remarks, to decide "what a comic representation is 'about'" (*Sitcom* 10). Similarly, Joanne Morreale agrees that "it is no simple task to read a particular sitcom, or to determine whether a particular sitcom serves an ideologically conservative or subversive function" (xix). *TBBT* in general and its treatment of science and scientific characters in particular are certainly no exception to the rule.[14]

While the sitcom's representation of family, gender, profession, race, or class is far from definite and straightforward, but allows for a multiplicity of different, or even indeed diametrically opposed readings, it seems safe to argue that the genre has always offered a viable space for the negotiation of changing cultural attitudes toward these and similar identity markers.[15] As one of the perhaps most ideological-

14 The contradictory position of the sitcom in general and *TBBT* in particular is further complicated by its theatrical performance and camera work as well as the artificial laugh track: All of these generic features point to the artificiality of the sitcom and obstruct the easy and willing suspension of disbelief. In contrast to the verisimilitude opted for by drama series or, to a lesser extent, soap operas, a sitcom's audience is constantly reminded that they are watching a play being performed, not least because of the rather static position of the cameras and the reactions from the live studio audience. At the same time, however, the sitcom invites its viewers to become emotionally invested in the show through laughter, empathy, or suspense. According to Mills, this contradictory position oscillating between emotional distance and engagement "is one vital to comedy generally, in which laughter requires an involvement with, and a detachment from, that which is funny" ("Comedy Verité" 68). This is particularly true for *TBBT*, which in contrast to many other recent sitcoms (*The Office*, *Scrubs*, *Kath and Kim*) has neither abandoned the artificial laugh track nor the typical shooting and editing style. *TBBT* is hence also a viable and prominent example of countering repeated calls announcing the 'death of the sitcom' in terms of its creativity (Mills, *Sitcom* 125). While it remains conservative in its strict adherence to genre conventions, the show's focus on a yet unrepresented set of characters can certainly be understood as proof of new creative pathways taken by the industry.

15 This point has been heatedly discussed since the sitcom's inception in the late 1940s. The genre has repeatedly had to meet severe criticism or, perhaps even more consequential, utter disregard from academic commentators. One prominent perspective is offered by David Marc, who proposed a dichotomy between the supposedly radical, critical nature of stand-up comedy and the formulaic, conservative world of the domestic sitcom, which has proven highly influential for subsequent academic analyses: "Whereas sitcoms depend on familiarity, identification, and redemption of popular beliefs, stand-up comedy often depends on the shocking violation of normative taboos" (24). Indeed, it is the former's low cultural status which is frequently evoked to account for the paucity of aca-

ly invested genres of American popular culture, the sitcom with its specific social situations offers a subtle commentary on contemporary conflicts, struggles, and discourses in American society: "The television situation comedy—the most popular American art form—is a virtual textbook that can be 'read' to help lay bare the mores, images, ideals, prejudices, and ideologies shared—whether by fiat or default—by the majority of the American public" (10), as Darrell Hamamoto persuasively argues. Hence, whatever the comic value of its representations, the mere fact that *TBBT* emerges as the first American sitcom revolving around a group of scientists and their geeky pastimes ought to be taken note of as a highly remarkable circumstance, not only in the history of the genre, but also with reference to popular culture's treatment of science. Accordingly, one may concur with Michael Wiatrowski, who notes that *TBBT* "appears to be an examination of the American geek's growing relevance" (n.p.). More profoundly, it can be understood as a reaction to (or, even more radically, a promotion of) the growing pervasiveness and significance of science and technology in our everyday lives. In that respect, the show is in accordance with the general development of the American prime-time sitcom from domestic shows centered on a nuclear or alternative family to a group of friends, as well as a shift from the suburban setting to the city (Hamamoto 1-2). While in most other sitcoms revolving around friends or roommates, the characters' jobs are merely peripheral to the story (e.g. *Friends*, *How I Met Your Mother*, *Will and Grace*) and seldom depicted as the central source of laughter, *TBBT* is principally based on the tensions arising from the encounter between scientists and 'normal' people. Indeed, the show frequently not only contrasts the scientific aloofness of its protagonists with neutral standpoints, such as Penny's down-to-earth common sense and trivial knowledge of celebrity culture and fashion, but also with more overtly science-opposed perspectives, such as the Christian fundamentalism presented by Sheldon's Texan mother (e.g. "The Electric Can Opener Fluctuation") or the blatant scientific illiteracy of Penny's dim-witted boyfriends (e.g. "The Justice League Recombination"). The following section shall thus more closely explore this central subject matter, which notably distinguishes *TBBT* from its generic forerunners.

---

demic studies on the television sitcom (Mills, "Studying" 61). Similarly, the genre is regarded has having undergone "few fundamental changes" (Hartley 65) since its beginnings, which suggests an inherent unwillingness to adapt to changing social and political circumstances. Additionally, its generic characteristics, such as the use of stereotypes and the focus on a simplistic, limited social milieu, can be seen as reasons for the sustained academic dismissal of the sitcom.

"...A WORKING KNOWLEDGE OF THE UNIVERSE AND EVERYTHING IT CONTAINS": SCIENCE, GEEK CULTURE, AND THE OTHER

TBBT explicitly foregrounds science (more specifically physics) as one of its central themes and seldom tempers it with another, potentially more mass-appealing topic. This is already firmly established by the sitcom's title, referencing a dominant theory of how the universe came into being, and by the titles of the individual episodes, which mock the structure of typical physics formulas, such as "The Middle Earth Paradigm" or "The Panty Piñata Polarization." The show has enlisted one of UCLA's leading physicists, David Saltzberg, as its official science consultant to guarantee maximum scientific accuracy.[16] The science, as *TBBT* franchise and fans never tire of boasting, is accurate to the tiniest detail, from the protagonists' obscure monologues to the figures and calculations on the whiteboards in Sheldon's and Leonard's shared apartment. Cameos of real-life scientists, rather than movie stars or pop singers, further solidify its status as a show that strives to take its science seriously. Indeed, the guest appearances of Nobel laureate George Smoot ("The Terminator Decoupling") and scientific media celebrity Stephen Hawking ("The Hawking Excitation") drew a particularly large number of viewers. Similarly to how *CSI* has impacted popular discourses of forensic science and pathology via the notorious 'CSI effect,' *TBBT* is reported to have significantly ameliorated the public perception of physics: In November 2011, *The Observer*, for instance, headlined that "interest in A-level and university courses rises as US comedy [meaning *TBBT*] makes the subject 'cool'" (Townsend n.p.).

Within the narrative of a single episode, science generally assumes the role of an insider joke, as the general viewer can hardly be expected to fully understand each and every scientific reference, especially since the academic disciplines evoked not only comprise physics, but also the humanities, social sciences, and engineering. This is especially true with regard to the extensive scientific explanations and the highly specialized vocabulary, primarily employed by Sheldon, who is constructed as the most stereotypical and socially unskilled of the four. The following conversation between him and Penny takes place immediately after Sheldon has cooked scrambled eggs to "test my hypothesis about the separation of the water molecules from the proteins and its impact vis-à-vis taste" ("The Luminous Fish Effect"). Leonard and Sheldon have just agreed on pursuing the experiment when Penny enters the apartment:

16 Nonetheless, viewers have been complaining about inaccuracy both in terms of the scientific knowledge and the details of geek culture represented by the show. See, for instance, Overbye (2010) and Jensen (2012).

Penny: Hi! Hey, I'm running out to the market. Do you guys need anything?
Sheldon: Oh, well this would be one of those circumstances that people unfamiliar with the law of large numbers call a coincidence.
Penny: I'm sorry?
Sheldon: I need eggs. Four dozen should suffice.
Penny: Four dozen?
Sheldon: Yes, and evenly distributed between brown, white, free range, large, extra large, and jumbo.
Penny: Okay, one more time?

("The Luminous Fish Effect")

In such situations, Penny serves as a stand-in for the audience, who will, most likely, be equally dazzled by Sheldon's statements, typically delivered at high speed with flat intonation and no breaks. According to Wiatrowski, "Penny clearly represents the unmarked category of dominant American culture and thus a surrogate for 'us' in the audience" (n.p.), which the above scene exemplarily reflects: Much more so than in their discussion of zodiac signs (in which the norm may, depending on one's stance on astrology, be represented by either Sheldon or Penny), science here signifies a definite, comic deviation from expected behavior, which is endorsed by Penny and, one may assume, the audience. As the main and even indeed only source of humor in the above conversation, science is constructed as special, unique, and culturally estranged, rather than being naturalized for the viewer, as is the case in cultural productions that strive to smoothly incorporate scientific storylines into the narrative.[17]

While science is markedly foregrounded as one of the show's central topoi, its construction as comic, different, and abnormal concurs with the circumstance that the audience hardly ever gets to see actual scientific practice, which clearly differentiates *TBBT* from the text analyzed in the previous chapter. Certainly, this circumstance must also be attributed to the fact that "the depiction of scientific activity does not lend itself well to story telling because abstract thought or the pursuit of knowledge as such is difficult to present in images" (Weingart, "Chemists" 32).[18] This reluctance to visualize scientific work coincides with the representation of professional life in the classic sitcom. With reference to domestic sitcoms, Richard Butsch argues that "[o]ccupation and thus the class portrayed is not an artifact of the genre as it is of police, lawyer or medical drama series. [...] [T]he major portion of action is among family members usually in the home" (18) or, in the case of

17 See, for instance, autopsy-driven science drama and disaster movies as discussed in chapters 5 and 7 of this study.

18 With the exception of *CSI* (see chapter 5), this premise has been verified by almost all texts analyzed in this study.

TBBT, among friends at home. While one or more scenes per episode do take place at the university, it is mostly the cafeteria, the hallway, or the protagonists' offices, rather than the actual laboratories. In stark contrast to *CSI*, there are hardly any lab scenes in which the audience is presented with the actual production of scientific knowledge that would subsequently provoke an otherwise improbable plot twist or prepare the dénouement. In reaction to the low visual appeal of most scientific practices, a frequent cinematic device, as Weingart notes, is to shift the focus to alternative, visually exciting concerns and "show scientists in adventure or love" ("Chemists" 32). The latter certainly applies to *TBBT*, as its plot lines primarily revolve around the scientists and their fruitless endeavors at dating.[19] Again, this concurs with standard sitcom conventions as concerns shows that incorporate the characters' workplace, which seem "generically driven to be about sexual chemistry rather than occupational specificity" (Hartley 67). Indeed, the university is the setting where Leonard dates his co-worker Leslie, Sheldon meets his girlfriend Amy, or the guys discuss their love lives over lunch. Science (or, more precisely, the results it yields) is hence not as directly integral to plot development as is the case, for instance, in *CSI* or *The Day After Tomorrow*. Rather, it is the accessories of science, such as specific props, jargon, and, above all, the (largely detrimental) effects of their scientific professions on the characters' private lives that are foregrounded. In narratological terms, science in *TBBT* is a means to provoke narrative conflict (such as when Leonard fails to communicate efficiently with Penny due to his highly professionalized vocabulary), whereas most other texts analyzed in this study employ science to solve a narrative conflict already generated through other elements (e.g. crime, disaster) and to thereby provide closure.

When we do get to see science in *TBBT*, it is, as the above discussed scenes have demonstrated, for a comic effect as well as, to a lesser degree, visual entertainment. Both purposes are primarily fulfilled by applying scientific methodology to everyday routines, such as using a laser to heat up a cup of instant noodles ("The Fuzzy Boots Corollary") or basic extrapolation to calculate Penny's number of sexual partners ("The Robotic Manipulation"). Experiments are thus mostly meant for the scientists' diversion, illustrating their eccentric predilections, rather than their everyday work. In accordance with the dualistic structure of the show, they are typically employed to facilitate mundane routines or presented in non-academic contexts, hence further emphasizing the comic contrast between two highly disparate worlds. Indeed, science in its natural habitat and applied to typical scenarios would hardly strike the average sitcom viewer as comic or absurd, or even indeed moderately interesting. The binary structure of the typical sitcom thus also serves to ex-

19 In correspondence with the changed premises of the information age, however, the aspect of 'adventure' is reduced to a virtual video quest or a game of paintball, which the protagonists regularly engage in.

plain how a subject as mundane and austere as natural science can effectively be employed as a source of humor in mainstream television comedy; after all, science in formal contexts such as higher education will seldom be appreciated as a laughing matter. Only the sustained juxtaposition of scientific storylines, characters, and dialogues with the world beyond the ivory tower (which, as the above discussed scenes have amply demonstrated, is typically represented as the norm) allows the protagonists' profession to be ridiculed and mocked, as well as employed for spectacular effects in a slap-stick tradition. In that way, science in *TBBT* serves the show's comic intent and provides for its comedy.

Additionally, the comic effect contrived by science is enhanced through a powerful auxiliary. While science in *CSI* is intricately entwined with crime-detection and corresponding technology (see chapter 5), *TBBT* teams it up with geek culture and fandom. In the framework of this study, such connections have emerged as a frequent topos surrounding the representation of science in popular cultural formations: Not only in sitcoms, but also in classic television drama, action movies, and biopics is science joined and contrasted with other thematic concerns, such as crime (*CSI*), dating and geek culture (*TBBT*), adventure (*The Day After Tomorrow*), or business (*Jobs*, *The Social Network*). Pure science, it appears, can hardly be entertaining and enjoyable on its own; rather, repeated references to the protagonists' various obsessions and pastimes, including video games, comic books, costume play, robots, fantasy and sci-fi, are frequently drawn on for additional comic potential. The following conversation between Leonard and Sheldon illustrates how the comedy of an absurd and inappropriate scientific explanation offered by Sheldon is augmented by his identity as a science-fiction fan.

Leonard: You'll never guess what just happened!
Sheldon: You went out in the hallway, stumbled into an interdimensional portal which brought you 5,000 years into the future where you took advantage of the advanced technology to build a time machine, and now you're back to bring us all with you to the year 7010 where we are transported to work at the thinkatorium by telepathically controlled flying dolphins?
Leonard: No.
Sheldon: Aw.
Leonard: Penny kissed me.
Sheldon: Well, who would ever guess that?

("The Boyfriend Complexity")

On the one hand, the scene evokes laughter because of Sheldon's inability to correctly decode and apply the social conventions of informal conversation, which the audience is led to directly attribute to his scientific occupation associated with abstract thinking and a pragmatic mind-set. Hence, the comic deviation from the social norm consists in Sheldon taking literally a rhetorical exclamation typically employed as a conversation starter and elaborating on it disproportionately. Additionally, however, the situation he imagines to have happened to Leonard is evidently inspired by his exuberant appreciation of *Star Trek*, time travel, and science fiction. Indeed, from a purely scientific standpoint, the scenario he conjures up appears absurd and highly improbable and thus, in fact, runs counter to his otherwise pragmatic attitude. For the purposes of the sitcom, however, the combination of his inability to conform to social convention (caused by his identity as a scientist) with his proclivity for obscure references from science fiction (motivated by his identity as a comic book fan) establishes him as foreign, different, and, ultimately, comic. Furthermore, the fact that Sheldon is portrayed as finding time travel more likely than Leonard's successful romantic encounter with a girl not only makes his viewpoint appear absurd to the general viewer, but also underlines the protagonists' inexperience with the other sex. Again, this adds to their narrative construction as geeks or nerds, who stereotypically fail at or display utter disinterest in successful human interaction.

It is in scenes capitalizing on these prototypical geeky qualities, I would argue, that science most conspicuously comes to assume the role of what postcolonial theory has classified as the Other.[20] I am here following and elaborating on

20 As a key concept of postcolonial theory, the other is, first of all, defined as "anyone who is separate from one's self" (Ashcroft, Griffiths, and Tiffin 169). Thus, the notion strongly relies on binary oppositions, as the othered, colonized subject serves as "a means of establishing a binary separation of the colonizer and colonized and asserting the naturalness and primacy of the colonizing culture and world view" (Ashcroft, Griffiths, and Tiffin 169). Postcolonial theory distinguishes between the 'other' on the one hand and the *grande-autre*, the capital 'Other,' on the other hand, which harks back to (post-)Freudian psychoanalysis, specifically to Jacques Lacan (1936-1966, 1964) and his notions of subjectivity and the self. In most simple terms, Lacan established that the 'other' resembles the self and provides the basis for the self to define its own identity and sense of itself. Conversely, the so-called *grande-autre* describes the symbolic interlocutor via whose gaze the subject gains identity. Lacan's distinction is translated into the imperial, colonizing Other, which is constructed in the very same process that produces the colonial, inferior other (Ashcroft, Griffiths, and Tiffin 171). This process is known as 'othering,' which allows the self to draw definite boundaries between itself and the other, thus literally de-fining its own identity as well as that of those outside the boundaries (Spivak 134-135). While critics like Gayatri Spivak maintain the basic distinction between the 'other' versus the 'Other,' the terms are frequently used interchangeably, as Said's seminal *Orientalism* (1978) exemplifies. Thus, the colonizer's construction of the colonial other is often defined as the construction of the Other, whose spelling aims at accentuating the abstract and symbolic meaning (Ashcroft, Griffiths, and Tiffin 171-172). My own use and spelling of the term 'Other' in this chapter concurs with the latter practice.

Wiatrowski's notion of the "Comedy of the 'Other'" (n.p.), which he identifies as one of the main comic strategies in *TBBT*. While Wiatrowski argues that the show's humor is, among others, derived from a sustained othering of some of its main characters, above all Sheldon and Raj, I would, more generally, attribute this position to the notion of science itself, which, as the above analyzed sequences have demonstrated, repeatedly serves to provoke and justify comic deviations from the norm. Otherness is thus not only achieved, as Wiatrowski argues, through the characters' extreme social disorders bordering on the pathological, but indeed through their mere adherence to the world of science and research, which allows them to continually showcase its internal codes, aesthetics, and terminology. This is made even more effective, as the above paragraph has shown, through its strategic combination with geek subculture. Similar to how the process of othering in a colonial context serves as "a means of establishing a binary separation [...] and asserting the naturalness and primacy of the [...] [dominant] culture and world view" (Ashcroft, Griffiths, and Tiffin 169), the peculiarities and absurdities generated by the protagonists' identities as scientists and geeks equally establish them as less 'natural' than the unmarked characters, above all Penny. In literary texts, the otherness of characters is habitually emphasized by their origin and illustrated via their looks, dress, and behavior, which are portrayed as deviant, abnormal, or strange in contrast to the norms represented by dominant characters (see Templeton 2007). This can certainly be observed in *TBBT*, in which their scientific profession as well as their participation in various geek subcultures inclines the adult characters to wear hoodies, colorful t-shirts, old-fashioned hair-cuts, and horn-rimmed glasses, which provide a notable contrast to Penny's more contemporary clothing and hairstyle.

As Wiatrowski rightly argues, however, othering literary characters traditionally also involves explicitly constructing them as mad, diseased, or criminal.[21] In *TBBT*, it is the whole world of science and sci-fi fandom that is repeatedly presented as a form of mad obsession, as the following encounter between Penny and the boys, proudly presenting their newly purchased prop from the movie *The Time Machine*, illustrates:

Leonard: The lights flash, and the dish spins. You wanna try it?
Penny: No, I don't wanna try it! My God, you are grown men! How could you waste your lives with these stupid toys and costumes, and comic books, and, and now that, that...
Sheldon: Again—time machine.
Penny: Oh, please, that's not a time machine. If anything, it looks like something Elton John would drive through the Everglades.

---

21 The latter is represented, for instance, in Charles Dickens's *Great Expectations* by the Australian convict Magwitch (Templeton 2007).

Sheldon: It only moves in time. It would be less than worthless in a swamp.
Penny: Pathetic! All of you, completely pathetic!

("The Nerdvana Annihilation")

By verbally denigrating the guys' gadgets (the terms 'toys' and 'costumes' evoking the childishness of the whole endeavor), Penny exposes their hobbies as unnatural and alien. Derogatory terms, such as 'stupid' and 'pathetic,' serve to emphasize their narrative construction as deviant and mad. The humor in this scene is undoubtedly directed at the scientists, although the studio audience's reaction after Penny's final repartee suggests laughter mixed with (feigned) shock at her harsh words. While geeky hobbies thus certainly contribute to overt denigration, it is the scientists' unexpected and deviant mannerisms, frequently exceeding mere social unease, which most conspicuously establish them as Other: "The performance of extreme social ineptness, caused by social disorders, builds a comedic formula of Otherness that completes the fundamental theory of comedy within *TBBT*" (n.p.), as Wiatrowski argues. The "social disorders" which Wiatrowski refers to include a range of compulsive behaviorisms that, in his opinion, verge on the pathological, such as Raj's selective mutism (he cannot speak around or with women) and Sheldon's restricted use of gestures and facial expressions, monotonous and highly formal speech (even in informal contexts), compulsive adherence to repetitive routines, and inability to feel empathy and to understand implied meanings beyond the literal (such as irony or sarcasm). While it has so far not been explicitly thematized within the show, critics have noted that these characteristics may suggest that Sheldon is suffering from Asperger's Syndrome (see Soraya 2009, P. Collins 2009), which would transform his stereotypical social awkwardness into a genuine pathology and, consequently, make his manifest deviance from mainstream culture even more pronounced.

In literary theory, such extreme (pathological) deviations from the intradiegetically established norm are usually understood as establishing a clear hierarchy among the characters, since the Other typically allows the rest of the characters to emerge as superior, as they adhere to as well as perpetuate the dominant normative discourse of the fictional world and the reader (see Templeton 2007). In line with the above explored theories on the ambiguous and potentially subversive effects of humor and comedy, however, I would argue that the specific effects of othering in *TBBT* are notably more complex. Within the narrative world, Sheldon's quirks and peculiarities are not attributed to a pathological condition, but to his extreme scientific inclination, which he transfers directly and unabashedly to private encounters with friends or family. Accordingly, what makes Sheldon (and, to a lesser degree, Raj and the other male protagonists) different from the unmarked, 'normal' characters is, as the show's overall narrative seems to imply, science. Hence, it is their scientific profession in combination with typical geek stereotypes that serves as the main attribute through which Leonard and his friends are habitually constructed as different, "weirdo" ("The Roommate Transmogrification"), "creepy" ("The Panty Piñata Polarization"), "bat-crap crazy" ("The Cooper-Hofstadter Polarization"), and downright "pathetic" ("The Killer Robot Instability"). It seems that science not only needs to be toned down by joining it with geek culture, but also effectively denigrated and othered by contrasting it with allegedly more worthy, 'normal' pursuits in order to guarantee its crucial role for the comedy of the show. As a practical instance of the Other, science then lends itself perfectly as a subject matter for the sitcom format.

Despite the guys' repeated ridicule and comic devaluation through their association with science, however, one must not forget that they still serve as protagonists of the show and can hence hardly be described as inferiorized or mere foils. Additionally, the preceding section has made us aware that humor may undermine any such straightforward interpretation of the role that science allegedly comes to play and that comic representations ought to be treated as ambiguous, indefinite, and potentially subversive. It is for this reason that their superior subject position as comic heroes also helps to explain how the scientists, despite serving as a pivotal source of the show's comedy by being repeatedly forced to assume the position of the Other, will eventually still emerge as cool role models inspiring an enormous fan culture and merchandise. What is more, this coolness paradoxically thrives on their very devotion to and cultivation of geek chic and technoscientific expertise.

GEEK COOL: NONCONFORMITY AND THE REVENGE OF THE NERDS

And The Geek Shall Inherit The Earth.
A POPULAR T-SHIRT SLOGAN

It seems that any exploration of the alleged 'coolness' of *TBBT* must unavoidably begin with the concession that there are, in fact, very few instances in which the show's protagonists are, by any objective standards, depicted as (mainstream) cooler than and thus superior to other characters. Most frequently, these instances occur when the protagonists use their advanced scientific understanding to underline their superiority, such as when conversing with one of Penny's boyfriends, who usually represent the stereotypical handsome, yet dim-witted high school jock. In these rare situations, the science characters are portrayed as adroitly employing strategies more overtly associated with mainstream cool at the expense of the less science-savvy antagonist. The first of these situations occurs in season one during Penny's

Halloween party, in which her ex-boyfriend Kurt, scantily dressed as a sort of modern caveman, repeatedly ridicules the nerds until they decide to strike back with their own weapons. This involves, among other hardly comprehensible insults, Leonard calling his rival an ape—and proving it en suite:

Leonard: A homo habilis just discovering his opposable thumbs says what?
Kurt: What?

("The Middle Earth Paradigm")

The second example, taken from season four, employs a similar set-up and also involves one of Penny's boyfriends, this time called Zack, who remains completely unaware of the scientists' prank. Again, it is the boyfriend who actually initiates the encounter by making an offensive or, in this case, downrightly absurd statement, to which the scientists feel obliged to reply:

Zack: You know, I saw this great thing on the Discovery Channel: turns out that if you kill a starfish, it'll just come back to life.
Sheldon: Was the starfish wearing boxer shorts? Because you might have been watching Nickelodeon.
Zack: No. I'm almost sure that it was the Discovery Channel. It was a great show. They also said dolphins might be smarter than people.
Leonard: They might be smarter than *some* people.
Zack: Well, maybe we can do an experiment to find out.
Sheldon: That's easy enough. We'd need a large tank of water, a hoop to jump through, and a bucket of whatever bite-size treats you find tasty.

("The Justice League Recombination")

Sheldon and Leonard are in utmost control of the situation, while Zack remains completely oblivious to the insider knowledge shared by them. Both instances draw upon dominant notions of coolness as a form of secret code shared exclusively by an elite set of insiders, who, consequently, appear mysterious and obscure to the admiring and slightly bedazzled onlooker. The coolness they display in this scene could well be classified as "an aesthetic [...] of the passwords and Masonic handshakes through which card-carrying initiates gain entrance to the clubhouse" (Harris 48), which suggests that the attitude only becomes effective (and cool only becomes cool) by conveying an impression of elitism, secrecy, and exclusion. As a personal strategy inspiring control and emotional composure, cool has repeatedly been understood as sharply distinguishing between those who are familiar with and those who will forever remain ignorant of the required gestures, props, and poses. Opaqueness, impenetrability, and resistance to the transmission of inside information are hence vital ingredients of a cool poise. In the above conversation, Leonard and Sheldon convey exactly these qualities, sharing a common code and language which expertly excludes their conversational partner. This dimension of cool as an elite aesthetic code that relies on a mechanism of exclusion is highly present in the encounters between the scientists and Penny's boyfriends, as the unequal distribution of information also gives rise to the comedy of the scenes: It would not be far-fetched to identify Kurt and Zack respectively as the infamous 'butt of the joke' due to their lack of participation in the common code shared by Leonard and Sheldon. The above scenes hence also reflect how coolness frequently amounts to a fluency in relevant styles and demeanors, whose adoption equals "a process of becoming literate, the acquisition of a cultural technique" (Geiger, Schröder, and Söll 16).[22] In the case of *TBBT*, the cultural techniques shared by the protagonists (including, most overtly, highly formal language which is alien and confusing to the other characters) are supplied by their scientific profession, which, ultimately, boils down to the fact that they can only meaningfully wield and employ (their own version of) cool *because of*, rather than despite their formal education.

Essentially, this understanding of cool harks back to Alan Liu's definition of cool, which I already mobilized in the previous chapter's close reading of *CSI* and its lab scene, namely as an aesthetic code that conceals, obscures, and displaces information, but does not share it with the onlooker. "[C]ool," as Liu compellingly argues, "achieves what may be called an 'ethos of the unknown'" (9), suggesting that a cool attitude relies, among other perceived qualities, on displays of seeming impenetrability and incomprehensibility. The situation in the above example from *TBBT*, however, noticeably differs from the particular form of cool unknowability that emerges in *CSI*. In the case of the sitcom, the audience is privy to the joke that is being played on Zack, as the references employed by Sheldon and Leonard hardly involve any fundamental scientific facts that would require superior knowledge to understand.[23] While *CSI*'s lab scene excludes the audience as well as any non-scientific character of the show (such as the suspect), *TBBT* creates its comedy by ensuring that the audience will easily comprehend Leonard's and Sheldon's dialogue so that they may laugh at Zack being mocked. Within the narrative world of the sitcom, however, they nonetheless emerge as those possessing superior and literally exclusive knowledge, as their adoption of a common code only works by excluding others. As delineated further above, this strategy can be understood as a vi-

22 Original wording: "ein[…] Prozess der Alphabetisierung, […][das] Einüben einer Kulturtechnik." See also Danesi (particularly 38-46), who supports this argumentation by demonstrating that North American teenagers' adoption of cool depends on a variety of shared semiotic codes, including body image, social cognition, and language.

23 The only overtly scientific idiom employed is, in fact, the reference to the 'homo habilis,' which as an anthropological term does not necessarily rely on the protagonists' intimate knowledge of physics.

tal source of cool: Not coincidentally will those individuals who are acclaimed as particularly 'cool' be attributed an air of mystery and inscrutability that makes them stand out from the clueless masses.

More frequent than such scenes which display dominant notions of coolness suggesting nonchalance and mental control, however, are situations which exploit and develop alternative forms of cool.[24] In contrast to Ted Gioia's referencing of *TBBT*, among other shows, as pivotal evidence for the alleged "death of cool" ("Death of the Cool", 8), I will argue that cultural productions appraised for "celebrating the joys of nerdiness" (Gioia, "Death of the Cool" 221) advance and successfully promote their very own version of cool that draws on exactly these 'nerdy' properties. Far from subscribing to a vision of cool as obsolete and outdated, I thus propose that the former nerds and geeks as portrayed by *TBBT*'s protagonists have clearly and overtly shifted their roles from former antagonists and ridiculed side-kicks to the new gurus of cool in twenty-first-century information culture. It is this notable shift in cultural visibility, rather than Gioia's somewhat bleak outlook on the demise of coolness as a Western sensibility, that I would identify as the "revenge of the nerds" (Gioia, "Death of the Cool" 15). This is, time and again, explicitly thematized in the show, such as when Leonard, moments before his encounter with Kurt the caveman, confidently asserts: "Our society has undergone a paradigm shift. In the information age, you and I are the alpha males" ("The Middle Earth Paradigm"). While the uncomfortable exchange with Penny's muscular, rather than brainy ex-boyfriend may ultimately suggest the opposite (Kurt effortlessly lifts Leonard off his feet to intimidate him), the overall story arc of the episode supports this premise: Not only have the scientists taken center stage and will, consequently, be those who the audience sympathizes with, they also display a range of common (and unique) insignia for creating a special form of 'in-group cool,' which has been picked up by its fans and viewers.

In analyzing the show's production of a subcultural, alternative cool that capitalizes on distinct 'geek' attire, behaviorisms, and storylines, I have found Daniel Harris's afore-quoted article on "Coolness" (1999) particularly valuable. Harris primarily defines cool in terms of a countercultural stance intended to decidedly set its bearers apart from whatever they may perceive as ordinary, normal, and mainstream. Cool hence varies from peer group to peer group and is always an insider phenomenon, which only those who have accordingly been initiated may understand and meaningfully appropriate. Indeed, the specific sort of coolness displayed

24 These may, however, also inform more dominant, established definitions of cool, as the ready transfer of 'geek chic' insignia into the mainstream reflects. The appropriation and commercial co-option of subcultural aesthetic codes by mass culture has, of course, always played a vital role in the cultural history of cool. See Frank (1998), Lasn (1999), Pountain and Robins (2000), McGuigan (2009), and chapter 2 of this study.

by the geeks in *TBBT* plays upon a particular "look of coolness, which serves a specific sociological function for tightly knit groups [in this case geeks and/or scientists] whose very survival depends on their ability to create a unified aesthetic identity" (Harris 48). Similarly to how "[m]embers of gangs use a complex set of visual symbols to express their fierce allegiance to one another" (Harris 48), one may also discern a distinct aesthetic code, popularized and marketed as the alleged 'geek chic,' shared by all four of the protagonists, such as backpacks, t-shirts with scientific jokes or slogans meaningless to the outsider, black-rimmed glasses, hooded sweaters, and training shoes. What is, of course, highly ironic in *TBBT* (and the source of its recurrent comedy) is that the particular forms of insider insignia chosen to mark the protagonists as members of the same socio-cultural group are defined by their very 'un-coolness' by mainstream standards, i.e. their divergence from other, dominant definitions of cool, such as maintaining control in social encounters, leading a dangerous and adventurous life, being a suave and smooth talker, and wearing fashionable clothes. These traits would, at first glance, rather be attributed to Penny's boyfriends, who might be less intelligent than Leonard, but, contrary to his confident assertion, still display a pervasive dimension of coolness as tall, handsome, and tanned athletes who will effortlessly impress the girls the geeks pine after.

The authority of such hegemonic, though perhaps slightly worn notions of cool is not, however, supported by the overall narrative of the show, in which those character types and physiques who would traditionally serve as protagonists (the basketball player, the aspiring movie star, the blonde, yet naive waitress) are relegated to the margins and subjected, as the above examples reflect, to ridicule, while it is the scientists whose trials and tribulations dominate the storylines and who will eventually get the girl. Within the narrative of *TBBT*, cool is thus more notably assigned to the geeks, rather than the unmarked characters such as Penny, the boys' varying love interests, their family, or their neighbors. This is strongly connected to the fact that since its cultivation as a rebellious stance adopted by rebels, underdogs, blues singers, and various youth cultures, cool has gained strong connotations of individuality and individualism. In the formative years of modern-day Western cool after 1945, it became closely associated with countercultural avant-garde movements that propagated the unconditional cultivation of the self regardless of (bourgeois) society's norms and expectations. It is for this reason that cool today is still described as a markedly "individualistic attitude" (Pountain and Robins 23), cherished for its "healthy celebration of the self" (Pountain and Robins 26) and "free expression of […] individuality" (Pountain and Robins 120). Equally, Geiger, Schröder, and Söll note "a strong tendency for individualization" (16),[25] which characterizes the development of Western cool after the Second World War. Self-

25 Original wording: "mit einer starken Tendenz der Individualisierung."

stylization and self-fashioning are thus key to a convincingly cool pose, as the recent empirical study by Ilan Dar-Nimrod et al. (2012) also confirms: The attributes 'individualist,' 'unique,' and 'unconventional' rank particularly high in the overall frequency of coolness-related adjectives (177). In a nutshell, cool may be understood as the attitude that most unabashedly treasures non-conformity and difference while simultaneously despising conventionality and mass culture.

In that respect, the scientifically neutral characters (above all Penny's boyfriends, who accomplish the dominant ideal of Western masculinity in every aspect) are, bluntly put, too normal, conventional, and unmarked to be cool, a perspective which the show repeatedly asserts. While the scientists are, in accordance with the conventions of a sitcom, also valued by the audience for their comic failures, this is, I would argue, not the only characteristic that makes them more attractive than any of the unmarked characters. Indeed, Harris's definition of cool as a "democratic" (46) and "physically forgiving" (46) moral stance seems to discard as utterly uncool the very stereotypes which the *TBBT* protagonists are repeatedly juxtaposed with, namely those displayed by Penny and her athletic boyfriends:

> "Part of the success of coolness among self-conscious teenagers stems from the fact that it is [...] so inclusive, rejecting as it does the cookie-cutter aesthetic of 'normal' people, of Barbie and Ken, of the voluptuous blonde cheerleader and the all-American quarterback boyfriend. Cool people actively deride conventional notions of physical beauty, associating wholesome, rosy-cheeked good looks with conformity and devoting themselves instead to the cult of the grotesque." (Harris 46)

In particular, Harris's evocation of cool as a countercultural attitude that rejects 'normality' as an unappealing category and aims, on the contrary, for the special, unique, and different seems to be played out perfectly by the geeks of *TBBT*. The individual episodes' story arcs continually stress that the geeks' prime asset is not their suavity, their steeled bodies, or their courage, but their scientific knowledge: When Leonard first meets their new neighbor Penny, he almost instantly inquires whether she wants to see his white board, suggesting that the enormous calculations on it may well impress her—a notion which Penny's astounded reaction confirms. The appeal of Leonard, both as a love interest for Penny and as a key character for audience identification and empathy, can indeed hardly be described as being induced by "conventional notions of physical beauty" (Harris 46) or masculine strength. Rather, it is his repeated deviance from social, physical, and professional norms that notably contributes to his construction as an icon of subcultural, geeky cool. Hence, deviance from the norm not only serves the comic intent of the sitcom, but also helps to construct its protagonists as cool by accentuating their unconventionality, individuality, and detachedness from the mainstream, which, in almost all of the cases, relies on their identity as scientists and geeks.

Indeed, there are several instances in the show in which the alternative, geeky cool displayed by the protagonists effectively aligns them with more traditionally contrarian values, such as rebelliousness, hedonism, roughness, and irony—a circumstance which is, once again, heavily dependent on their status as stereotypically rational, emotionally insensitive scientists. Frequently, supposed 'failures' to communicate are revealed as being based on utter indifference, rather than ignorance; in particular, the uber-rational character of Sheldon is frequently portrayed as consciously and deliberately choosing not to conform to the required social protocol, whose purpose he refuses to understand. When his girlfriend Amy asks him to meet her mother, he seeks his flat mate's advice:

Sheldon: What am I supposed to do?
Leonard: Well, have you considered telling her how you feel?
Sheldon: Leonard, I'm a physicist, not a hippie.

("The Desperation Emanation")

In this case, Sheldon not only violates social regulations (a romantic relationship being based on mutual trust and sentiment), but also advocates and displays utmost emotional restraint. Both qualities can be classified as demonstrating "the anti-establishment attitude associated with cool" (Rice, "What is Cool?" par. 2) by thriving on "potentially antisocial content" (Pountain and Robins 13), which Sheldon seems to eagerly indulge in. Other instances of his deliberately anti-social, disaffected behavior include him publicly denouncing Bill Gates during his speech at the university ("The Toast Derivation"), interrupting or completely ruining academic events ("The Cooper-Hofstadter Polarization," "The Cooper-Nowitzki Theorem"), being issued with restraining orders due to repeated domestic disturbances ("The Excelsior Acquisition," "The Vacation Solution"), going to jail for insulting a judge ("The Excelsior Acquisition"), or, most commonly, repeatedly offending his friends and relatives by his typically direct, unrestrained answers. Quite notably, all of the above listed situations align Sheldon, despite his lack of suavity and poise, with the eternal icons of Western cool, be it James Dean, Marlene Dietrich, or Kurt Cobain, whose veneration by popular culture is still based on the pervasive notion that they acted on their personal beliefs, rather than in accordance with society's expectations. It is in this respect that Pountain and Robins define cool as a "*permanent* state of *private* rebellion" (19, emphasis in the original) that runs counter to generally accepted norms and rules of conduct. The scientists' continuous (conscious or accidental, but always comic) violation of such rules can thus be read as a particularly pronounced stance of individual defiance.

Indeed, it may be their very uncoolness by conventional standards (represented by Penny and other neutral characters) that stimulates such alternative notions of cool. In that respect, the particular cool displayed by the scientists could also be understood as originating in traditional conceptualizations of coolness as an emotional, cognitive, and physical mask against harassment, discrimination, and abuse. Historically, cool has been synonymous with "the ability to maintain detachment, even during tense encounters" (Majors and Mancini Billson 2), endowing the cool person with (seeming) composure and invulnerability. In that sense, the function of cool has been that of a veritable survival mechanism that allows its bearers to preserve their moral and mental integrity in times of great distress.[26] Similarly to how African warriors, plantation slaves, and African-American jazz and blues musicians are understood as having employed cool as a form of self-defense, the geeks may equally be seen as adopting indifference and disaffection as well as irony and self-mockery to maintain their self-esteem despite their obvious outsider position. The resulting stance may again be read as a less engaged, opaque, and independent form of coolness that is more notably aligned with the historical roots of the term than with present-day popular conceptualizations. Indeed, Dar-Nimrod views defense and protection against adverse social forces as a vital function of particular personality orientations of coolness which "may shield a self-perceived outsider's sense of self-worth by creating cognitive (rebelliousness), emotional (detachment), and behavioral (roughness) defenses against the judgment of mainstream culture" (183). In *TBBT*, this is, again, particularly noticeable in the character of Sheldon, but also exists, to varying degrees, in the portrayal of the other male protagonists, as the following 'experiment' conducted by Raj, an astrophysicist, and Howard, an engineer, reflects:

Raj: We now have the address of the top model house.
Howard: God bless you, Google Street View—registered trademark.
Leonard: Okay, for the record, what you guys are doing is really creepy.
Howard: You know what? If it's creepy to use the Internet, military satellites, and robot aircraft to find a house full of gorgeous young models so that I can drop in on them unexpectedly, then fine, I'm creepy.

("The Panty Piñata Polarization")

Howard effectively employs ironic detachment, one of the three "core personality traits" (Pountain and Robins 26)[27] of a cool persona, to show that he is utterly disaffected by other people's judgments of him. He thus refuses to be wounded by Leonard's insult and rejects the negative connotations of 'creepy' by purporting to endorse it. His strategy of claiming a depreciative attribute for himself and thus

26 See chapter 2 of this study as well as Pountain and Robins 34-42, Thompson, "Aesthetic" 41-42, Danesi 42-46, and Clarke par. 13-17.

27 According to Pountain and Robins, the other two cool personality traits are narcissism and hedonism (26-29). See also chapter 2 of this study.

neutralizing its power to offend him heavily draws on traditional forms of Jewish self-mockery and has been classified as a vital component in the "maintenance of a protective Cool persona" (Pountain and Robins 27).[28] In keeping with the comic intent of the sitcom, his last remark is, of course, ultimately ironic, as the viewer is meant to interpret it as affirming, rather than refuting Leonard's verdict. Nonetheless, Howard's alleged indifference to social judgments can easily be read as a form of emotional mask intended to avert any abuse from the outset. It is thus highly reminiscent of how cool "serves, like a suit of armor, the prevention of mental and physical damage. […] To be cool means […] not to be injured if you do not want to be" (Poschardt 11).[29] On the outside, Howard purports that he is "fine" and thus keeps a cool mask, seemingly unaffected by the self-imposed insult. This dimension of cool has been central in its conceptualization as a countercultural attitude adopted by the underdogs, rather than the darlings of society: Commonly defined as an art of indifference, it actually requires the investment of quite a considerable amount of thought, time, and/or money in order to accomplish this high degree of seeming complacency.

In that sense, the characters of Leonard, Sheldon, Howard, and Raj display typical traits of what Dar-Nimrod's study has identified as contrarian coolness, which is to be distinguished from cachet coolness. While the latter strives for peer approval and greatly correlates with social desirability, the former is based on unconventionality, difference, and, as the key virtue, rebelliousness. It is, in essence, a detached, passive, and camouflaged cool that exceeds the desire to be popular among and accepted by one's peers. Hence, "[c]ontrarian coolness is a kind of coolness that potentially transcends conventional norms of valuation" (Dar-Nimrod 181), which corresponds to Harris's notion of coolness deriding dominant standards of beauty. Sheldon and Leonard's overtly anti-social behavior and their refusal to conform to social expectations align them with the "detached-from-the-mainstream style of coolness" (Dar-Nimrod 183) that has figured prominently in the cultural history of cool. In that sense, the distinction between cachet and contrarian coolness helps to explain how the protagonists are able to wield particular notions of cool despite

28 The use of self-mockery as a defense strategy is reminiscent, as Pountain and Robins note, of the so-called 'sick humor' popular in the 1960s, which was characterized by "deliberately using racial epithets as a way of defusing their power to hurt" (27) and in return draws on traditional Jewish humor. Coincidentally, Howard Wolowitz is the only Jewish character of the show, a fact which is repeatedly resorted to for politically incorrect laughter. See Mills, *Sitcom* 88-92, for a discussion of racist and similar anti-PC humor in terms of the relief theory.

29 Original wording: "Als individuelle Praxis dient die Haltung des 'Cool' wie eine Rüstung der Abwendung von Unheil auf psychischer und körperlicher Ebene. […] 'Cool' sein heißt, […] nicht verletzt werden können, wenn man es nicht will."

their repeated display of socially undesirable traits and a prominent distance from the larger peer community. Additionally, such a distinction between different forms of coolness and their divergent correspondence to notions of social desirability can serve as a feasible strategy to account for the co-existence (both within the show and American popular culture in general) of standard notions of cool and the protagonists' alternative, geeky version of it. Hence, it may help to understand how both the scientists and the scientifically neutral or adverse characters, represented by Penny, her boyfriends, and other members of the cast, display aspects associated with coolness, though on diametrically opposed ends of the spectrum: While Penny and her boyfriends can be understood as cool by their display of socially desirable qualities such as attractiveness, pro-social values, and trendiness, the scientists are presented as rather wielding contrarian values of coolness, such as unconventionality, individualism, and emotional control[30]—all of which are, directly or tacitly, induced by their devotion to science and various geek subcultures.

Hence, the (contrarian) coolness of *TBBT* is directly linked to, or even indeed induced by the protagonists' identity as scientists and geeks. Similar to how techno-scientific expertise in *CSI* allows the scientist-detectives to emerge as superior and cool, it is *because of*, rather than despite the association with science and all the codes and insignia this entails that the sitcom's portrayal of its protagonists allows for the detection of typically contrarian traits of coolness. In that sense, science can be understood as a direct source of, rather than impediment to cool. This is often directly thematized by the narrative of the show, for instance when Sheldon justifies his stubborn refusal to apologize to a judge he insulted and thus avoid imprisonment as follows: "I'm a scientist. I never apologize for the truth" ("The Excelsior Acquisition"). Equally, the above listed anti-social acts establishing Sheldon as (contrarian) cool, rebellious, and individualistic can be directly accounted for by his rational mind-set and/or his obsession with sci-fi geek culture, such as when he stalks famous comic book authors or remains utterly oblivious to Penny's well-meant attempts at small talk. His lack of emotional empathy and expressiveness, two vital aspects in the traditional conceptualization of cool detachment, can be directly related to his stout adherence to (what he perceives as) scientific principles: After all, Sheldon is "a man of science, not someone's snuggle-bunny" ("The Isolation Permutation") and must accordingly find it absurd, childish, and inappropriate to express and share his feelings. Interestingly indeed, many of the principles of proper scientific devotion that Sheldon reiterates in the show, such as indifference to social norms, emotional detachment, aloofness, and refusal to follow the lead of others, concur with some of the most central aspects of cool. This also extends to all

30 See Dar-Nimrod for correlations and differences between ratings for coolness and social desirability respectively (179-181), as well Pountain and Robins 13-33 and Majors and Mancini Billson 1-2.

the other scientific characters of the show, as the following conversation between Penny and the generally more moderate and socially assimilated Leonard illustrates:

Penny: So, you know, isn't there maybe some way you and Sheldon could compromise on this whole presentation thing?
Leonard: No. Scientists do not compromise. Our minds are trained to synthesize facts and come to inarguable conclusions.

("The Cooper-Hofstadter Polarization")

In this case, the refusal to compromise and thereby sacrifice one's beliefs can be read as directly fulfilling cool's credo of nonconformity, self-assurance, and individuality. Indeed, Leonard's reply suggests that scientists are different from the 'normal' people Penny might have in mind and can thus not be expected to live up to social expectations. The refusal to yield to societal pressure, a vital facet of contrarian coolness (Dar-Nimrod 179-180), is hence portrayed as a central aspect of what it means to be a scientist. Similar to how their scientific profession makes the protagonists stand out visually from the other characters in terms of looks and attire, it also constructs them as singular in their conduct and attitude.

Instances of Sheldon's and Leonard's eccentric, unusual behavior thus reflect how science can effectively be portrayed as cool in the show: It provides the underlying rationale for an extreme form of nonconformity and difference, highlighting the protagonists' individuality and uniqueness in terms of behavior, dress, and occupation. In that respect, the above delineated process of othering, which becomes most manifest in the show's treatment of science, in fact serves the production of a very distinct, alternative notion of cool. The "revenge of the nerds" ("Death of the Cool", 15) announced by Gioia is thus not one that derides cool altogether, but adroitly updates it to the concerns and demands of the twenty-first century, in which individualization and nonconformity have gained unprecedented significance for contemporary youths faced with the increasing demands of mass culture and consumerism (Harris 47-49, Pountain and Robins 165-170). Rather than suffering a premature death, cool can be regarded as having redrawn, once again, the boundaries of its vast semantic field to align itself with the changed premises of contemporary knowledge society. In *TBBT*, the semantic change currently undergone by cool is additionally complicated by the comedy of the show: As a pivotal source of laughter, science provides the motivation for a cool but simultaneously comic difference which, to varying degrees, affects all of its protagonists. As the dualistic structure of the sitcom repeatedly draws on the dichotomy between scientists and non-scientists for jokes and comic situations, the protagonists' inability and/or unwillingness to abide by the norms presented by scientifically unmarked characters is a key strategy for the humor (and success) of the show. Additionally, however, this pronounced focus on difference for an ostensibly comic effect allows particular constructions of contrarian coolness, based on notions of rebelliousness, nonconformity, and affective indifference, to emerge. In that sense, the show's representation of science, one of its main thematic concerns, as comic and Other simultaneously contributes to the portrayal of its protagonists as proponents of an alternative, geeky cool. Taking these reflections into consideration and observing the particularities of the genre as a comic, rather than serious treatment of its subject matter, the following section will now conclude this chapter with a close reading of one complete episode, which will help to join and apply the notions explored so far.

"WANNA SEE SOMETHING COOL?": A CLOSE READING OF THE GEEK IN ACTION

In line with Feuer's argumentation that due to their extensive run time "US sitcoms usually reveal their ideological frames much more clearly in the first season than in ensuing ones" ("Situation Comedy" 70), the episode chosen for close analysis is taken from the first season of *TBBT*. In narratological terms, the first few installments of a sitcom or, more generally, a television drama can be understood as primarily fulfilling an expository function by introducing the main characters, their relationship to each other, and the overall conflict or dualism on which the show is based. Following this premise, it seems safe to assume that episode three, "The Fuzzy Boots Corollary," constitutes a fairly prototypical installment concerned with establishing the characters' personalities and the overall thematic orientation of the show rather than aiming for diversion through the appearance of guest stars or the use of outdoor scenes, as is increasingly the case in later seasons. Additionally, the episode displays scenes of actual scientific practice, even if only in its application to household tasks, which become rarer in later seasons as the emphasis notably shifts from science to the protagonists' (unlucky) relationship with their girlfriends. At this point of the narrative, all four scientists are still single, although Leonard especially is portrayed as being desperate to change this circumstance as fast as possible. The episode hence revolves around his comic failures at dating, first with his co-worker Leslie (Sara Gilbert), and later with his new neighbor Penny. These and other comic situations, such as the scientists playing a video game or discussing the possibility of Sheldon being a robot, help to showcase their eccentric behavior and lack of social skills, which will be treated as a given in later seasons.

For the purpose of exposition, the episode not only compares and contrasts the characters through dialogue and behavior, but also draws on a number of stereotypes commonly employed for the popular cultural portrayal of scientists. Most conspicuously, the scientists are modeled on what Roslynn Haynes has identified as the 'silly scientist,' who is "totally ridiculous, gullible, and out of touch with the 'real world'" ("Scientist in Literature" 387). This stereotype, as Haynes perceptively demonstrates, originates in seventeenth-century (predominantly British) satire of natural philosophers[31] and has been updated in a range of twenty- and twenty-first-century feature films in the figure of the absent-minded professor.[32] The following utterly absurd, yet deadly serious conversation between Sheldon, Howard, and Raj seems to allude to this common stereotype:

Howard: Sheldon, if you were a robot, and I knew and you didn't, would you want me to tell you?
Sheldon: That depends. When I learn that I'm a robot, will I be able to handle it?
Howard: Maybe, although the history of science fiction is not on your side.
Sheldon: Uh, let me ask you this. When I learn that I'm a robot, would I be bound by Asimov's three laws of robotics?
Raj: You might be bound by them right now.
Howard: That's true. Have you ever harmed a human being, or through inaction allowed a human being to come to harm?
Sheldon: Of course not.
Howard: Have you ever harmed yourself, or allowed yourself to be harmed except in cases where a human being would have been endangered?
Sheldon: Well, no.
Howard: I smell robot.[33]

Applying a common strategy used throughout the show, this scene additionally emphasizes the scientists' foolishness through their participation in various geek fan cultures, as the above discussion of robotics has probably not only been inspired by Howard's work as an aerospace engineer, but also by the guys' ardent veneration for science fiction and comic books. Additionally, their geeky ridiculousness is demonstrated through physical comedy, for instance when they are all wired and covered in sophisticated headsets for a video game or when Leonard perspires heavily before his date with Penny. Their portrayal as being "preoccupied with some minute triviality" (Haynes, "Scientist in Literature" 388), typical of the stereotype of the foolish and confused scientist, is thus effectively teamed with geek identity, which is characterized by an obsessive devotion to odd and seemingly irrelevant pursuits.

31 Famous examples are Jonathan Swift's portrayal of scientific virtuosi in *Gulliver's Travels* (1726) and Thomas Shadwell's play *The Virtuoso* (1676).

32 See, for instance, *Back to the Future* (1985), *I.Q.* (1994), and *The Nutty Professor* (1996).

33 All direct quotations in this section are taken from the episode "The Fuzzy Boots Corollary" unless indicated otherwise.

Through the addition of geek insignia, the centuries-old stereotype of the 'silly scientist' mobilized in the show is thus updated to the concerns of the information age. Indeed, repeated references to modern information and communication technologies, such as social networks or virtual identities, cannot only be read as confirming traditional stock character portrayals of the scientist by adding some geek chic, but also as modernizing these stereotypes and thus potentially attracting more positive connotations. The following conversation between Sheldon and Leonard, which takes place immediately after the successful completion of a video game quest, is paradigmatic of their portrayal as digital citizens fluent in the codes and conventions of web 2.0.

Sheldon: Wooh, I'm all sweaty, anybody want to log on to second life and go swimming? I just built a virtual pool.
Leonard: No, I can't look at you or your avatar right now.

Thanks to the possibilities of the information age, Sheldon and Leonard are literally living in a different world, or rather a place removed from the 'real world,' which in this case is the virtual world of the Internet. Accordingly, the geeky occupation of creating online personas teams up smoothly with historically inherited stereotypes of the scientist, who was equally ridiculed for being "out of touch with the real world of social intercourse" (Haynes, *From Faust to Strangelove* 3).

Alternatively, however, the repeated references to geek culture can also be read as prompting a thorough revision of and challenge to traditional stereotypes. As young, imaginative, and unconventional men in touch with popular culture (comic books, video games, film, TV series) and modern technology (virtual life, eBay, online blogging, social networks), Sheldon, Leonard, Howard, and Raj present a viable alternative to the white-coated, spectacled, bald, and middle-aged male who still constitutes one of the most common popular images of the scientist (Mead and Métraux 384-388). The protagonists' young age (they are in their mid to late twenties) serves as an unusual spin on the conventional portrayal of scientists. While their professions would rather align them with adulthood and middle age, it is the recurring references to geek culture which accentuate their youth and, consequently, their (partial) deviation from the stock image of the elderly, old-fashioned scientist. In that respect, *TBBT* challenges the pervasive claim that "[t]he youth cult that dominates other genres has not penetrated the images of science" as concerns its depiction in fiction and film (Weingart, Muhl, and Pansegrau 282). Not coincidentally, of course, youth and youthfulness are strongly associated with the multifaceted lexical field of cool (Mentges 25, Pountain and Robins 21-22, Danesi 37-42).[34]

---

34 Indeed, scholarship on (modern day) cool often conceptualizes it as a by-product of the rise of youth culture(s) from the 1950s onwards. Several youth subcultures, be it the hip-

In addition to the notion of the 'silly scientist,' which the episode both draws on and challenges, one may also observe other stock characteristics in the portrayal of the show's protagonists, which can be understood as orientation guides for the audience in this early expository episode. Hence, although Leonard is highly devoted to his dating endeavors, he nonetheless displays traits of the Romantic stereotype of the inhumane, cold, and emotionally dysfunctional researcher. This is particularly alluded to in his 'romantic' encounter with his co-worker Leslie, who, as an apparently equally rational-minded character, could be understood as a catalyst (or, in literary terms, a foil) to his abstract thinking and pragmatic approach to all aspects of life. The following scene takes place after Leonard discovers that Penny might have found a new boyfriend and is advised by his friends to "go after someone [his] speed," which makes him seek out Leslie, an experimental physicist, in her lab:

Leonard [enters]: Hello Leslie. [...] I would like to propose an experiment.
Leslie: Hang on. I'm trying to see how long it takes a 500-kilowatt oxygen iodine laser to heat up my Cup o' Noodles.
Leonard: I've done it. About two seconds, 2.6 for minestrone. Anyway, I was thinking more of a bio-social exploration with a neuro-chemical overlay.
Leslie: Wait, are you asking me out?
Leonard: I was going to characterize it as the modification of our colleague-slash-friendship paradigm with the addition of a date-like component, but we don't need to quibble over terminology.
Leslie: What sort of experiment?
Leonard: There's a generally accepted pattern in this area. I would pick you up. Take you to a restaurant. Then we would see a movie, probably a romantic comedy featuring the talents of Hugh Grant or Sandra Bullock.
Leslie: Interesting. And would you agree that the primary way we would evaluate either the success or failure of the date would be based on the biochemical reaction during the good night kiss?

---

pies of the 1970s or the punks of the 1980s, made cool an integral part of their philosophy, although they frequently re-baptized it in the wake of it. On the other hand, however, it must be noted that "Cool is equally about teenagers behaving with precocious maturity" (Pountain and Robins 21), hence striving to refute any association with (too youthful) childishness. Accordingly, the protagonists' obsession with video games could also be read as infantilizing them, resulting in a de-masculinization that underlines their comic function. Indeed, scientists' portrayal as immature and childish has served as a popular strategy for emphasizing their "retardation in all areas other than the intellectual" (Haynes, *From Faust to Strangelove* 220). See this chapter's section on "Laughing At Science or Laughing With Science?" and Butsch (2003), who explores this strategy with regard to the sitcom's representation of working-class men.

Leonard: Heart rate, pheromones, et cetera. Yes.

Leslie: Why don't we just stipulate that the date goes well and move to the key variable?

Leonard: You mean kiss you now?

Leslie: Yes.

Leonard: Can you define the parameters of the kiss?

Leslie: Closed-mouth but romantic. Mint? [They kiss]

Leslie: What do you think?

Leonard: You proposed the experiment, I think you should present your findings first.

Leslie: Fair enough. On the plus side, it was a good kiss, reasonable technique, no extraneous spittle. On the other hand, no arousal.

Leonard: None?

Leslie: None.

Leonard: Ah. Well, thank you for your time.

Leslie: Thank you.

Within the thematic framework of this study, the encounter between Leonard and Leslie is interesting for several reasons. On the one hand, the scene (as well as the episode on the whole) could be read as "exemplifying the destructive effects of an obsession with science to the exclusion of human relationships" (Haynes, "Scientist in Literature" 391), as Leonard's and Leslie's extremely scientific mind-set impedes them from following the "generally established pattern in this area" and allowing their mutual feelings to slowly develop during a conventional date. Rather, for the purposes of temporal and emotional efficiency, they pragmatically choose to apply scientific methods, thus illustrating how the stereotypical scientist is so devoted to research that he or she "sacrifices his or her emotions and human relationships in an obsessive pursuit of scientific materialism" (Haynes, "From Alchemy to Artificial Intelligence" 249). Accordingly, the whole encounter is categorized as an 'experiment' and conducted in the required formal jargon. Leslie is portrayed as the more extreme of the two, since Leonard only seems to disguise his invitation to dinner as a research proposal because she is more likely to accept the latter. Furthermore, it is Leslie who proposes an instant good-night kiss in order to avoid what she apparently regards as a waste of time and effort. However, as the outcome reflects, such a purely scientific approach to romance is not destined for success. In that respect, the whole episode can be read as an illustration of (and warning against) scientists' alleged "arrogance of the intellect" (Haynes, "Scientist in Literature" 391) and its (self-)destructive consequences. The scene thus also stands as a prime instance of how science repeatedly comes to assume the role of the Other through overt, comic deviations from the norm. In accordance with Romantic literature perpetuating this stereotype, Leonard as well as most of the other male characters will be presented as more sensible, modest, and moderate when their respective girlfriends are introduced to the show, thus demonstrating the capacity to reform "through the power of love" (Haynes, "Scientist in Literature" 391).

With regard to genre conventions which are, generally speaking, closely followed by the show, the scene in the lab offers a noteworthy contrast to most other situations in which science provides for a comic effect. In this case, it is not Penny, the most prominent scientifically neutral character, who encounters one of the scientifically marked characters, but two scientists on an equal level. Accordingly, laughter cannot be interpreted as happening at the expense of one scientist, who would usually offer the comic deviation from the norm presented by Penny. As the canned laughter and the reaction from the studio audience suggests, however, the scene is nonetheless intended to be funny. Hence, one could argue that the 'norm' in this case is implicitly provided by the viewer, who may, despite the absence of Penny as a "surrogate for 'us' in the audience" (Wiatrowski n.p.), still recognize the situation as comically different from what one would expect to happen in a romantic encounter. Similar to the show at large, it is science which fuels the obligatory opposition to social expectations that is prerequisite for incongruity and, consequently, humor to emerge. Again, this positioning of science as Other serves as a vital prerequisite for the viewer to laugh *at* the inaptness of scientific principles for the evaluation of human relationships. In accordance with the superiority theory, laughter could hence be read as an expression of derision at their emotional retardation, which is not only signaled, but indeed justified by the canned laughter of the studio audience.

In view of the above explored subversive potential of humor, the comedy of this scene may, however, also be evaluated the opposite way. For instance, humor also derives from Leonard expertly dissecting the socio-cultural components of a traditional date: "I would pick you up. Take you to a restaurant. Then we would see a movie, probably a romantic comedy featuring the talents of Hugh Grant or Sandra Bullock." Laughter at the end of this line is not, I would argue, directed at the scientific analysis, but rather at the absurdity of such a typical social encounter. Hence, the scene provides a prime illustration of how comedy questions social norms and conventions by highlighting their arbitrariness and, in this case, predictability, which are only revealed by analytical reflection. In that sense, laughter can be read as appreciating and supporting, rather than undermining the scientists' perspective. The pragmatic approach to dating is offered as an unusual, yet potentially worthwhile alternative to conventional procedures, which are, as Leonard's analysis reveals, no less absurd or comic.

With regard to the thematic scope of the present study, Leonard's and Leslie's emotionally restrained attempts at courtship can, of course, also be understood as (their version of) playing it cool. Indeed, the scene could stand as an exemplary rendering of cool affective control, as both parties demonstrate "the ability to maintain detachment, even during tense encounters" (Majors and Mancini Billson 2).

Considering Leonard's repeated failures and the enormous pressure he places on himself, the awkward encounter with Leslie can certainly be classified as such. The above scene in which Sheldon, Howard, and Raj are compared to soulless robots allows for a similar interpretation: Their aloofness and emotional deficiency, which "has been leveled more frequently [...] [and] more vigorously" (Haynes, *From Faust to Strangelove* 211) than any other charge against scientists, can alternatively also be understood as a display of cool, which has traditionally been conceptualized as founded upon similar associations. In that sense, the show's representation of science as evidently Other, marked through the scientists' deviation from generally received norms regulating social intercourse, contributes to their portrayal as proponents of an alternative, detached cool thriving on connotations of emotional restraint, complacency, and disaffectedness. This effect is even enhanced by the fact that both situations, Leonard and Leslie's kiss and the robotics discussion, heavily rely on the use of specialized terminology and insider codes. As neither of them involves a non-scientist character as stand-in for the audience and generally accepted norms, many references can only be decoded after sufficient explanation, such as Leonard's obscure way of asking Leslie out for dinner:

Leonard: Anyway, I was thinking more of a bio-social exploration with a neuro-chemical overlay.
Leslie: Wait, are you asking me out?

Laughter from the studio audience only occurs after Leslie's reply, which clarifies (or indeed translates) Leonard's statement for the viewer. This suggests that the audience is not even expected to understand the highly specialized jargon and is, initially, deliberately excluded from the exchange. The situation can thus be read as deliberately contrived to convey an impression of exclusiveness and obscurity, appearing mysterious and obscure to the slightly bedazzled, yet admiring onlooker. As the above sections have demonstrated, cool depends on exactly such a form of secret (verbal or visual) code only shared by an elite set of insiders. Similar to how various youth subcultures have developed their own visual and verbal symbols to both signal affiliation with each other and exclude outsiders (Harris 46-48), the scientists' utterances are impenetrable from the outside. In the present scene, this is, of course, only maintained for a brief second, as Leslie's comment immediately clarifies Leonard's intent. Nonetheless, it offers a glimpse into the sustained unequal distribution of information that necessarily arises from the sitcom's use of scientific and non-scientific characters, who hold highly discrete sets of knowledge. As in the discussion of robotics or similar geeky pastimes, the scientists are continually portrayed as sharing an exclusive set of references that must always already be translated to the audience in order to be perceived as funny and entertaining. Initially, coolness obscures rather than reveals the show's comedy, as Wiatrowski also points out: "While much of the comedic production takes place at the expense of Geek culture, it simultaneously requires an intimate knowledge of geek culture to be truly funny" (n.p.). In that respect, the addition of scientifically unmarked characters is a vital strategy for decoding the inherent humor of such situations.

On the other hand, however, unmarked (side) characters are repeatedly used to offer contrasting versions of cool. Above all, this is one of the primary functions of the character of Penny, who acts as a foil not only with regard to the protagonists' superior scientific (and geeky) knowledge, but also in terms of their failures at wielding mainstream notions of coolness. In the present episode, however, this contrast becomes even more pronounced through the appearance of Penny's lover Dough. When Leonard unexpectedly meets Penny and Dough in a tight embrace outside of her apartment, he is introduced to his rival, who nonchalantly quips: "What's up, bro?" Leonard's reply is uttered in a shaking voice, demonstrating unfamiliarity with this sort of jargon: "Not much," he says, before tentatively adding: "...bro." While Dough acts self-assured, laid-back, and in utter control of the situation, Leonard is portrayed as insecure, shy, and over-whelmed. In Dough's case, cool emerges as a "flamboyant charade of toughness and authority" (Harris 40). He effectively wields what has earlier been described as mainstream or cachet cool, thriving on notions of popularity, peer approval, trendiness, and (physical) attractiveness (Dar-Nimrod 180), which Leonard and his friends have no chance of participating in. However, the humor of the scene does not allow for an uncomplicated dismissal of Leonard as the inferior party. While Dough's question is uttered in an overtly confident and aggressive manner, laughter occurring after Leonard's attempt to emulate his opponent's jargon is not only derisive of the scientist. Conversely, it is Dough who is subtly mocked by Leonard's repetition of his phrase, which cannot only be read as admiring and emulating, but also satirizing Dough's street jargon. This effect becomes even more prominent when Leonard, after some minutes of awkward small talk with Penny, makes to return to his apartment. He waves goodbye to Penny and, after a second of contemplation, adds: "Oh, and bye, bro!" Laughter after Leonard's utterance suggests that even more so than with his first 'bro,' it is Dough (and, more generally, the sort of street talk he employs) that is ridiculed. Through the sitcom's comedy, Dough's way of talking, a pervasive and normative specimen of teenage argot, is revealed as immature, pretentious, and completely inappropriate in the present situation. The fact that Leonard takes Dough's quip "What's up?", after all merely a figure of speech for casual greeting, as a literal question and answers it accordingly adds to this effect by revealing the phrase's inherent illogicality.[35] The scene is hence a prime illustration of how hu-

35 On the other hand, of course, it just represents another instance of how Leonard and his friends repeatedly fail at correctly decoding and employing standard social protocol such

mor can question and critique social (or, in the present case, linguistic) conventions by exposing their absurdity and irrationality. For the present situation, this also entails that Dough's coolness is shown to be contrived and flawed, while Leonard, despite his timidity, emerges as superior. Again, his individualistic, unconventional attitude that does not bow to peer pressure or passing trends more overtly aligns him with contrarian coolness, though it may be less socially desirable (Dar-Nimrod 180-183).

The notion of contrarian coolness, however, emerges most conspicuously in the character of Sheldon, which is line with the overall structure of the sitcom. Though in the present episode, which revolves around Leonard's trials and tribulations concerning dating, Sheldon only plays a secondary role, the scenes in which he does appear establish him as the most outspoken, unconventional, and emotionally deficient of the four scientists. Right from the opening sequence, which shows the protagonists playing a multi-player online game, he emerges as ruthless, egotistical, and anti-social, as he deserts his virtual team members in order to win the main prize, which he promptly sells on eBay. This sort of behavior aligns Sheldon with traditional notions of cool as an "attitude of narcissistic self-absorption" (Pountain and Robins 62) that is dismissive of or even destructive to human relationships. Equally, the episode strives to establish Sheldon as utterly indifferent to social décor, as he is brutally honest when Leonard asks for his opinion on the upcoming date with Penny.

Leonard: Oh, you know what, maybe this isn't such a good idea…
Sheldon: Oh, no, no, no, well now, there's always the possibility that alcohol and poor judgment on her part might lead to a nice romantic evening.
Leonard: You're right, alcohol, poor judgment, it could go well.
Sheldon: Of course, there's the other possibility that this date kicks off a rather unpleasant six months of the two of you passing awkwardly in the hall until one of you breaks down and moves to another zip code.
Leonard: You could have stopped at "it could go well."
Sheldon: If I could have, I would have.

Again, Sheldon's inappropriate reaction can be explained by his stout adherence to scientific methodology, which in this case means that he has to take all possible outcomes into consideration. Concealing half of Leonard's options would, in his opinion, amount to sloppiness or scientific fraud. The result of such an exceedingly rational outlook is, of course, a comprehensive violation of social expectations, as Leonard would have hoped for encouragement, rather than the plain truth. Accord-

as rhetorical questions. Again, the scene illustrates how comic representations frequently allow for conflicting interpretations.

ingly, it is his scientific profession that motivates Sheldon to adopt typically contrarian values of cool, such as nonconformity and indifference to social norms. At the same time, however, his utter disregard for his flat mate's emotional turmoil provides for the humor of the scene, just as science generally comes to serve as a pivotal source of comedy in the sitcom. Through the character of Sheldon, science, coolness, and comedy are thus intricately joined to serve the purposes of the genre.

THE LEGACY OF THE GEEK

"Wanna see something cool?" Leonard asks Penny during their date in a fancy restaurant before he demonstrates how one can make an olive drop into a water glass without touching it. The solution to how "something cool" like this magic trick can possibly happen is, of course, science, as Leonard seizes the opportunity to demonstrate centripetal force to Penny. While Penny is reasonably impressed, the experiment eventually also propels the breakdown of the date: The olive drops and Leonard, in an attempt to retrieve it from under the table, hits his head and suffers a concussion. In a nutshell, this little scene sums up which roles science comes to assume in the analyzed episode and the show in general. On the one hand, science is the underlying factor that allows all four of the protagonists to showcase different aspects of (contrarian) cool, based on their unconventionality, emotional restraint, individualistic attitude, and indifference to social norms. In most instances, these traits are, explicitly or tacitly, ascribed to the protagonists' rational mind-set and stout adherence to scientific principles (even in private matters), which the show presents as an inevitable consequence of their purely intellectual occupation as physicists. On the other hand, this violation of social expectations due to science also serves as an (in)direct source of comedy throughout the show, which in this scene is realized through Leonard's unlucky injury during the date. Their adherence to science emphasizes the protagonists' utter indifference to generally received conventions and regulations that govern social intercourse. This underlying dualism between science and what the plotlines consistently establish as 'the real world' forms the narrative basis of the sitcom and gives rise to a variety of comic situations that highlight the otherness of science and of those who perform it. In *TBBT*, science thus acts as a catalyst for the protagonists' display of (contrarian) coolness as well as a reliable source of comedy that draws on the incongruity of two highly divergent lifestyles. In both cases, the protagonists' overt construction as representatives of and participants in a range of geek subcultures fundamentally contributes to the effect: Their geek identity additionally highlights their divergence from social expectations and marks them as different from the mainstream, which is a vital factor in the display of alternative forms of cool as well as the basis for the show's comedy. This not only becomes prominent in terms of behavior and mannerisms, accentuating their eccentricities and quirks, but also with regard to looks. The scientists' idiosyncratic clothing style, involving colorful hooded sweaters, black-rimmed glasses, and backpacks, is both a comic deviation from mainstream trends and a viable instance of 'geek chic.' It is in this respect that *TBBT* comes to "highlight the collapsing of cultural boundaries in an information age and focus on the relevance of the geek" (n.p.), as Wiatrowski argues. Through the geeky characters of the scientists, the sitcom smoothly joins science, coolness, and comedy while remaining faithful to its generic conventions.

TBBT hence reflects a feasible way in which science can be represented as cool in popular cultural contexts, namely by foregrounding its comic potential through underlining the deviation from mainstream social protocols, which the show codes as more prototypically 'normal' than the ones followed by the scientific characters. Similar to its function in *CSI*, science is effectively stripped of its cognitive authority, in this case by making it comic, bizarre, and laughable. The comic derision that the protagonists repeatedly have to suffer is thus key to defusing the perceived cognitive superiority of science, which enables the viewer to no longer perceive the subject as intimidating or threatening. While the previous chapter has demonstrated how *CSI* foregrounds the aesthetic value of science through a focus on technoscientific machinery, *TBBT* can be read as capitalizing on its inherent comic potential through an effective juxtaposition with diametrically opposed lifestyles, resulting in a dualistic narrative structure that aligns the show with the great tradition of the American domestic sitcom. As one of the first prime time comedy programs revolving around a group of scientists, *TBBT* hence reflects how the focus on this particular profession and the eccentricities evidently caused by it is not only made tolerable, but indeed entertaining to the general sitcom audience. Its proficient use of scientific references for the production of jokes, punch lines, and comic situations suggests that the legitimation of science within the narrative confines of the sitcom must, first and foremost, be attributed to its comic deviation from whatever norms the plot affirms, rather than its cognitive or ethical value.

It is also for this reason that *TBBT* ought to be understood as being essentially about the performers, i.e. the scientists, rather than the actual practice of science. As the above analysis of one of its expository episodes has demonstrated, the viewer will hardly ever witness the performance of science, especially not in an academic context or for professional purposes, which mirrors the irrelevance of science for (direct) plot development. In this the show notably digresses from the other texts analyzed for this study, in which science does serve as a vital factor for narrative progression and eventual resolution. While other popular cultural formations may equally refrain from directly displaying (the practice of) science, they nonetheless emphasize its centrality for propelling the plot toward its successful dénouement. For the purposes of television comedy, the actual portrayal of science appears to be less than effective, which is why the focus is clearly placed on the quirks and comic misfortunes of the protagonists, only tacitly motivated by their scientific occupations. What the sitcom, however, clearly shares with other popular cultural productions thematizing science is that coolness can be understood as a direct consequence of, rather than antonym to, the use of scientifically informed storylines and the resulting rational, detached worldview of its main characters. Hence, it seems safe to conclude that it is not only Leonard's amazing science trick that makes the audience tune in on *TBBT* when they 'wanna see something cool.'

Only The Cool Survive

The Science of Disaster in Roland Emmerich's *The Day After Tomorrow*

> Only the day after tomorrow belongs to me. Some are born posthumously.
> FRIEDRICH NIETZSCHE/*THE ANTICHRIST*[1]

In a recent episode of the popular animated sitcom *South Park* (1997-present), the inhabitants are faced with an unexpected natural disaster that leads to a series of satirical misunderstandings. "Two Days Before the Day After Tomorrow" (2005) opens with Stan and Cartman stealing a motorboat, thereby destroying a dam and causing the flooding of a nearby village. The incident is vastly exaggerated by media reports, suggesting that the population is facing a national, if not indeed global crisis. Mass hysteria ensues, while scientists try to determine the cause of the disaster, which they soon attribute to global warming. The blame for the sudden out-

1 This quote from the preface to Nietzsche's *Antichrist* allegedly inspired the movie's enigmatic title (*The Internet Movie Database*, <http://www.imdb.com/title/tt0319262/trivia>, accessed 24 Jan. 2013). Nietzsche's oeuvre is, in fact, briefly discussed in the movie when the survivors gathering in the public library in NYC collect books to burn in order to keep them warm during the blizzard. A debate ensues between two of them as to whether or not his books should be burned because "[h]e was a chauvinist pig in love with his sister" (*TDAT*), as one of the survivors argues.

break of flooding and cold weather is consecutively put on the government, FEMA, terrorist attacks and, eventually, alien crab people. While there is no actual cause for alarm, South Park's inhabitants start to act out their own emergency plan, gather in the local community center, and dress in multiple layers of coats despite the warm and sunny weather. Eventually, Stan and Cartman admit that they have broken the dam, but the deluded villagers interpret this as a merely symbolic confession, suggesting that everyone ought to feel responsible for the alleged ecological crisis.

The episode is noteworthy insomuch as it succinctly pinpoints, despite necessary comic exaggerations, the most vital ingredients of a commercially successful disaster movie and attests to the continuing popularity of the genre, which firmly revolves around a recurring central theme: The salvation of the American nation and indeed the whole planet through scientific ingenuity and technological progress. The staples of the disaster genre parodied in the above episode, which includes scenes of panic-mongering and evacuation, exaggerated media coverage, a latent conspiracy motif, warnings that are not taken seriously but turn out to be (more than) accurate, bizarre infographics suggesting the application of cutting-edge science, and a final heroic rescue which restores the protagonist's nuclear family, all of which have appeared repeatedly since the genre's inception in the 1970s.

The present chapter is intended to explore the enduring success of the genre and examine one of its central thematic concerns, i.e. its representation(s) of science as a vital factor in the evolution and/or solution of the disaster. As the following pages will show, the genre's stance toward science has been an inherently ambiguous one, especially so in its early days: While disaster narratives tend to endorse the humane and ethical use of science, they simultaneously condemn its application to immoral causes. Accordingly, early disaster movies were populated by unscrupulous scientists whose unethical and/or failed experiments directly caused or, at the very least, majorly contributed to the outbreak of the disaster, but seldom averted it. This ambivalent image of science and its practitioners, as I argue, has been perceptively updated and ameliorated in recent instances of the genre, roughly starting in the early years of the new millennium.

As a case in point, the present chapter will examine Roland Emmerich's climate drama *The Day After Tomorrow* (2004, henceforth in this chapter *TDAT*), which confronts its audience with a disaster scenario inspired by current debates on global warming. In contrast to earlier representations, the contemporary disaster movie scientist has been elevated from comic or tragic sidekick to single-handed hero of the story, who not only accurately predicts the disaster by virtue of his scientific ingenuity, but also applies his physical potency to save the lives of others. In all of his endeavors, this updated figure of the scientist, exemplarily represented by *TDAT*'s young, handsome and athletic paleoclimatologist Prof. Hall, retains an indestructible cool pose, facing the impending end of the world with equanimity, composure, and an undisturbed exterior. This cool pose, as I will show in the following sections, is not adverse to his scientific profession, but rather acutely dependent on it: It is only by virtue of his direct access to and unshakable faith in scientifically consolidated information about the disaster (and apposite technological gadgetry to cope with it) that the scientist remains cool and detached in the face of the utmost horror and destruction.

The present chapter will show that Prof. Hall has every reason to counter chaos and destruction with immanent composure: In the contemporary disaster movie, those characters most closely and immediately associated with science stand the best chances of surviving the catastrophe almost entirely unscathed. Science is hence overtly constructed as an immediate precondition for the salvation of the human race, as it endows its allies with the indispensable mind-set of emotional detachment: In the midst of havoc and destruction, only the cool survive. The scientist-hero of the contemporary disaster movie thus emerges as another central character of the information age who adopts a cool stance because of, rather than despite his association with Western science and technology. Part of my argument will thus also be to propose a radical break between earlier disaster movie cycles and the present one, which is majorly hinged on a changed perspective of science and the role(s) of the scientist. The following pages will examine the negotiation of these concerns via *TDAT*'s narrative structure and cinematography, including the restricted use of facial close-ups and a dedication to the visualization of the natural and technological sublime via computer-generated imagery (CGI). First of all, however, a brief overview of the movie's central themes and its position in the generic tradition of the disaster story will provide background information for the subsequent analysis.

Disaster for the Masses: *The Day After Tomorrow* and Generic Tradition

The disaster movie has established itself as a highly popular and commercially profitable genre, which has continuously had to suffer allegations of perpetuating reactionary cultural values through predictable plot lines, stereotypical character portrayal, and simplistic dualisms. Due to this often hasty dismissal, disaster movies have seldom been subjected to serious academic critique and "have remained relatively neglected within film studies" (1), as Stephen Keane observes in the early 2000s. Although this situation has changed in the last decade as a range of studies from various academic vantage points, including ecocriticism and gender studies, emerged, there still seems to be a widespread urge to justify one's professional engagement with the popular genre. Despina Kakoudaki's insightful analysis of race relations in recent disaster movies, for instance, starts with the apologetic note that "[t]he visual landscape created by the science fiction/disaster films of the last decade was pervasive, obsessive, and, in my view, politically eloquent despite its lack of depth" (110), while Franziska Seewald detects a "strained divergence" (9)[2] between cinematographic practice and academic critique when it comes to filmic representations of disaster. The present analysis of a recent disaster movie of the last decade, *The Day After Tomorrow* (2004), suggests that this cautious stance proves largely unnecessary: Despite their clichéd narrative orientation and an evident commitment to nationalistic positions, disaster movies present their audience with viable insights into the circulation of dominant cultural discourses, including, but certainly not limited to notions of science and technology, which establishes them as highly resourceful objects of academic critique. The present section will briefly review the historical development and general properties of the genre with reference to *TDAT* as one of its most successful instances to date.

Frequently cited as the highest grossing Hollywood film ever produced in Canada,[3] Roland Emmerich's climate drama *TDAT* depicts the catastrophic events evolving from abrupt global warming. The story revolves around Prof. Jack Hall (Dennis Quaid), a distinguished paleoclimatologist working for the NOAA (National Oceanic and Atmospheric Administration), who during an expedition to Antarctica discovers that the polar ice caps are melting rapidly. At a United Nations conference in New Delhi, he tries to warn government officials, including a hardheaded vice president (Kenneth Welsh) with a striking visual resemblance to Dick Cheney, that this worrisome development will cause a new ice age: The cold stream from the north pole will disrupt the North Atlantic current that brings warm weather to the Northern hemisphere. While none of the diplomats believes in this alarming hypothesis, it soon becomes dire reality, faster than even Hall expected it to happen. All around the globe, extreme weather phenomena such as huge tidal waves, tsunamis, tornados, hail, and blizzards are recorded. Together with Prof. Rapson (Ian Holm), a senior researcher from Scotland, Hall builds a forecast model which predicts the development of several massive super storms that will instantly freeze anything they get into contact with. Meanwhile, Jack's son Sam (Jake Gyllenhaal) has been flying from Washington to New York for an academic competition with his classmates. As one of the super storms is approaching New York and the weather worsens, Sam and his friends seek shelter in a public library, where they start burning books in order to keep warm. They are joined by other survivors, including the librarian and a homeless man with his dog. Sam manages to call his father in Washington, who promises to come and save him. While waiting for the rescue, Sam and his classmate Laura (Emmy Rossum) grow closer to each other and share a kiss.

2 Original wording: "gespannte[...] Divergenz."

3 See, for instance, O'Connell (2004).

Meanwhile, the United States undergo massive evacuation as American refugees flee south to Mexico. Together with two of his assistants (one of whom perishes on the way), Jack makes it to New York almost completely on foot and finds Sam's group alive in the snow-buried library. In the meantime, the storm has cleared so that the survivors can be rescued by helicopters and brought to safety across the Mexican border, where Jack reunites with his wife.

Just like its parody in *South Park*, the movie's plot summary already reflects several standard narrative devices typically employed by the genre. From a narratological point of view, it exemplarily shows how "the representation of disaster offers another ideal arena for the combination of narrative and spectacular dynamics" (King 15), which is illustrated by the effective entwinement of melodrama with action-movie elements and elaborate special effects. In *TDAT*, this is successfully realized through the developing love story between Sam and Laura in the midst of havoc and destruction. Despite these sentimental elements, the movie is overtly dedicated to realism, capitalizing on a potentially possible, rather than completely fictional phenomenon. Indeed, the story was directly inspired by several recent publications on climate change and global warming, above all Art Bell's and Whitley Strieber's *The Coming Global Superstorm* (1999).[4] It feeds on the scientifically proven fact of global warming, but vastly accelerates its potential effects so that the climate shift happens almost overnight, rather than through the course of a century.[5] Despite these obvious sensationalist overtones, *TDAT* can be identified as the most outspoken critical disaster movie to date, as it openly addresses pollution through the burning of fossil fuels and seems dedicated to delivering the message that "our climate is fragile" (*TDAT*).[6]

4 The book is explicitly acknowledged as a source in the movie's closing credits and repeatedly referenced by Emmerich in interviews (Bowles 2004). Incidentally, however, *TDAT*'s plotline echoes a much earlier (fictional) source, namely Rudolf Daumann's *Protuberanzen*, a 1940 German science-fiction novel in which a scientist predicts the coming of a new ice age, but is ignored by the general public.

5 In that respect, the movie unwittingly undermines dominant notions of reproductive futurism (Edelman 2004) that are usually voiced in connection with environmental debates. It is, in contrast to the actual threat, not our children or grandchildren for whom the planet needs to be protected, but the current generation that is immediately endangered by ongoing pollution.

6 Emmerich repeatedly stressed that his motivation in making *TDAT* was more than merely financial, as he wanted "to give people a message" (qtd. in Gilchrist 2004). The German newspaper *Der Spiegel* reports that he donated 150.000 dollars to the organization Future Forest to make the production carbon neutral (Becker 2004). This alleged commitment to environmental issues also surfaces in the movie: While *TDAT* aims for spectacular scenery and goose-bumping action, the narrative is far from the ironic tone which made Em-

Regardless of its arguably laudable intentions and the consultation of leading climatologists (Hodgson 14), the movie has received mixed reviews and suffered repeated allegations of inaccurate science.[7] Its representations of science are, however, remarkable for their continuous glorification of technoscientific progress as the potential savior of mankind, which is most distinctly personified by the prodigious character of Prof. Jack Hall. Given the necessary exaggeration of events for cinematographic effect, the movie incited a series of political and ecological debates, particularly revolving around the American refusal to sign the Kyoto protocol.[8] Regardless of some of the movie's clumsy scientific plotlines, both critics and audience praised its special effects[9] and "complete digital environments" (Emmerich n.p.), which were unsurpassed at the time and, according to Emmerich, involved the work of twelve independent postproduction companies (Gilchrist 2004). Its devotion to graphic virtuosity and elaborate CGI (computer-generated imagery) shots stands as one of the most appreciated and redeeming qualities of the film and even induced *The Guardian* to label *TDAT* "one of the best disaster movies ever released" (Monbiot par. 6). Hardly surprisingly, the movie's iconic visuality inspired a myriad of rip-offs and parodies, of which the above introduced *South Park* episode only represents the most prominent one. *TDAT* thus remains one of the most commercially successful instances of a genre that has captivated Hollywood's imagination for the last five decades.

---

merich's previous advances into the genre, above all *Independence Day* (1996) and *Godzilla* (1998), an easy laugh despite the havoc.

7 *TDAT* was, in fact, named one of the 'Top 10 Scientifically Inaccurate Movies' in 2008. See <http://movies.yahoo.com/photos/top-10-scientifically-inaccurate-movies-slideshow-903/>, accessed 22 Jan. 2013. A persistent rumor states that NASA scientists received official directions to not talk to the media about the movie, a story which, however, most likely originated in a satiric *New York Times* article on April 1 (Bowles par. 10, Becker n.p.). Indeed, several governmental organizations did react to the film, such as the NSIDC (National Snow and Ice Data Center), which posted its own FAQ for moviegoers' questions on their website. The answers are partly based on an official NASA note to the public, which contradicts the rumor of a gag order allegedly issued by the White House. See <http://nsidc.org/news/press/day_after/>, accessed 23 Jan. 2013.

8 The movie was endorsed, among others, by climate activist and politician Al Gore, who found the depicted scenario highly plausible and organized a series of town hall meetings in response to it. Equally, several environmental activist groups handed out leaflets with information on climate change and its prevention to moviegoers (Bowles par. 9). *Environment*, the official journal of the School of Forestry and Environmental Science at Yale University, dedicated a special issue to *TDAT*, exploring the political impact of the movie on its viewers (Leiserowitz 2004).

9 See, for instance, Bowles (2004) and Hoff (2004).

There is widespread critical consensus that the disaster movie emerged as a distinct genre in the early to mid-1970s.[10] This so-called 'first cycle' or wave saw the production of a range of aesthetically similar, highly formulaic Hollywood feature films facilitated at the time by new technological and cinematic advances. Famous examples include *Airport* (1970), *Earthquake* (1974), and *The Poseidon Adventure* (1972), starring top Hollywood actors such as Paul Newman, Charlton Heston, Ava Gardner, Faye Dunaway, and Fred Astaire. The movies of the "savage seventies" (19), as Keane dubs them in his critical overview, almost invariably turned out to be major box office hits, despite their frequent dismissal by critics. In the late 1970s, a period of transition began, featuring films that would be equally heavy on violence and disaster, but could be more clearly identified as action movies.

It was only in the mid-1990s that disaster movies, for various industrial, political and ideological reasons, witnessed a glorious revival, sparked off by major blockbusters like *Independence Day* (1996). One main difference to the earlier cycle is the fact that millennial movies were almost always conceptualized as comprehensive 'end-of-the-world' scenarios, depicting nothing less than the potential eradication of the whole planet. While there are obvious and commercially highly successful exceptions to the rule such as *Twister* (1996), *Titanic* (1997), and *Volcano* (1997), the major representatives of this cycle featured existential threats caused by aliens, meteors, or geological irregularities with global consequences: *Twelve Monkeys (*1995), *Godzilla (1998), Last Night* (1998), and *Deep Blue Sea* (1999) are well-known representatives of the overall apocalyptic vision that dominated the 1990s cycle. In contrast to the character-driven dramas dominating the 1970s cycle, the movies produced two decades later capitalized on big budgets, action-driven plots, and breathtaking CGI effects,[11] which apparently fared well with the intended

10 See Roddick (1980) for a detailed survey. There are also earlier feature films, such as Roman epics of the 1910s to 1930s, which deal with disaster as a central narrative device. For an overview, see Keane 6-11.

11 This circumstance has to be attributed, among other factors, to the fact that the 1970s simply lacked the technology to present the disaster as such, so that the movies accordingly diverted the focus to interpersonal relations and social consequences of the catastrophe. This has been explored by Nessel, who argues that "[t]he fact that the focus of the 1970s cycle was placed on the social consequences of the catastrophe can be explained by the contemporary political climate, but may also be read as the effect of a void: in the 70s, one could simply not yet command the image" (3-4, original wording: "Daß der Fokus in den 70er Jahren auf den sozialen Konsequenzen der Katastrophe lag, läßt sich mit der damaligen politischen Stimmung erklären, ist jedoch ebenfalls lesbar als der Effekt einer Leerstelle: in den 70er Jahren verfügte man noch nicht über das Bild."). A similar point is made by Belmont, who, despite of acknowledging the substantial ideological dimension of the genre, concedes that "developments in film technologies are the

audience: The majority of disaster movies in this decade rank among the 100 highest grossing American films to date.[12] Despite the overwhelming commercial success, the millennial cycle quickly "reach[ed] a saturation point in the summer of 1998" (Keane 105) with *Armageddon* and *Deep Impact*.

The genre was, however, only granted a short break. Potentially incited by the emergence of a series of new real-life threats, the 2000s also witnessed a range of apocalyptic movies, including *The Core* (2003), *Sunshine* (2007), *2012* (2009), *Knowing* (2009), and the movie that will be more closely explored in this chapter, *TDAT* (2004). Whether these recent films constitute their own cycle or ought to be treated as a continuation of the millennial phase in the 1990s still remains to be seen. While the majority of studies do not explicitly comment on this categorization and tend to tacitly treat the millennial cycle as continuing well up to the present (e.g. N. Schröder 2010, Seewald 2011),[13] I do discern a more or less radical break between 1990s and post-2000 disaster movies. Their disparate treatments of and ideological perspectives on science, which the following sections will explore more closely, constitute the most crucial factors for this distinction, but there are several other properties which, as I contest, firmly set contemporary Hollywood disaster narratives apart from earlier instances.

Generally speaking, the films of one cycle can be expected to "show a remarkable consistency in terms of themes, narrative structures and, above all, ideological content" (244), as Nick Roddick points out in one of the earliest explorations of the genre. His observation suggests that disaster movie cycles are caused by and depict current societal, financial or ecological crises (see also D. Rosen 1975, Fricke 2001). Accordingly, the first cycle has often been interpreted as an expression of the social and political unrest of the late 1960s, a tense economic situation after Watergate, national disillusionment (or, conversely, reactionary patriotic backlash) following the war in Vietnam, and several other contemporary sociopolitical events.[14] Appropriately, early movies invariably employ strong, uniformed leadership figures, such as the young fire fighter played by Steve McQueen in *The Towering Inferno* (1974). Movies of the millennial cycle, on the other hand, frequently thematize new information technologies and their potentially destructive potential. They hence tend to offer a blend of disaster movie and science fiction aesthetics, as

driving force behind the subjects and styles of disaster films as well as their number" (349).

12 See Belmont (2007), who cites *Box Office Mojo* as a source: <http://www.boxofficemojo.com/alltime/domestic.htm>, accessed 18 Jan. 2013.

13 N. Schröder (2010) even explicitly suggests that the last "two decades [...] witnessed what one could call a revival of the disaster-genre" (290).

14 See Roddick (1980) and Art Ross (1975) for these and other possible explanations that have been put forth to account for the popularity of 1970s disaster movies.

critics have pointed out (see Loretan and Martig 1999, Martig 2000). Finally, contemporary examples can be read in view of the approaching year 2012, a global financial crisis, 9/11 and the war on terror, as well as a heightened consciousness for ecological issues. The last point in particular is significant, given the fact that recent films increasingly blame humans for generating disaster, as exemplified by *TDAT* in terms of pollution and the unchecked burning of fossil fuels. For the sake of comprehensiveness, it should also be noted that the immediate connection between the upsurge or recession of disaster movies and current political, economic and social trends has been heatedly disputed by a range of critics (see Kaplan 1975, Keane 2001).[15]

What unites movies of all cycles despite significant ideological and aesthetic differences, however, is the fact that the disaster is central to the story. While catastrophe, destruction, and spectacular visions have been staples in cinema since its inception and continue to exude a "cross-generic appeal" (Keane 4), to which the multitude of action movies, thrillers, and Westerns testify, the genre of the disaster movie is generally classified as being exclusively 'about' the disaster, i.e. dealing with or depicting the reasons, aftermath and/or actual occurrence of the catastrophe by focusing on a group of potential survivors. Generally speaking, there are two main categories the disaster can be classified in: 'natural' disaster, including typical 'weather-porn'[16] movies dealing with earthquakes, volcanic eruptions, and storms,

15 It has been argued that consecutive cycles are simply sparked off by one commercially successful movie, such as Emmerich's *Independence Day* (1996), which incited the 1990s cycle. The industry then quickly jumps on the bandwagon, hoping to capitalize on the public's apparent interest in the topic (Keane 2001, Altmann 1999). Additionally, technological progress is a vital factor in Hollywood production cycles, especially for a genre that majorly depends on expensive equipment and lavish mise-en-scène. These industrial factors, however, are not inextricable from wider ideological causes, as the movie discussed in this chapter also suggests: Whether covertly or explicitly, disaster movies do reflect current anxieties and phobias from HIV-inspired virus pandemics to terrorist attacks.

16 For the use of this term, see the media coverage for *TDAT*, e.g. Hoff (2004) for the *Pittsburgh City Paper*. In her article on "Ecocriticism and the Natural Disaster Heroine" (2007), Belmont explores this phenomenon in great depth and cites John Seabrook from *The New Yorker*, who defines 'weather porn' as "the voyeuristic weather-watching experience that has become a condition of modern life" (44). In investigating the growing audience of *The Weather Channel*, Seabrook links the fascination with and interest in the weather to notions of biblical guilt, suggesting that "extreme weather is taken as a sign of cosmic displeasure for our failure as stewards of the earth" (45). However, as Belmont's analysis shows, weather porn also negotiates issues of alienation, (in)visibility, and control, just like actual porn (353-355).

or 'man-made' disaster, such as scientific experiments gone awry or pollution and its consequences (Kakoudaki 122, Roddick 247-8, Ch. Mitchell xiv). *TDAT* clearly represents a combination of both factors, i.e. a natural catastrophe essentially fuelled or at least accelerated by (both unlawful and unethical) human interference.

While the depicted disaster can hence transcend the binary of being either 'natural' or 'mad-made,' there are determinate criteria it has to fulfill in order for the production to qualify as a distinct example of the genre: According to Roddick (1980), the disaster must be diegetically central, factually possible, indiscriminate (i.e. it affects the population of a given place as a whole), largely unexpected, all-encompassing, and ahistorical (246). Several of these characteristics, such as ahistoricity and factuality, are certainly problematic, as the movie discussed in this chapter demonstrates: *TDAT* is tied to a very specific point in time as global warming did not register as an actual threat to human existence only fifty years earlier. Similarly, a cataclysmic climate shift occurring over the time span of mere eight days can hardly be classified as factual. In the same vein, millennial studies such as Keane (2001) predicted that global scenarios would be substituted by more localized disaster dramas, such as Kathryn Bigelow's *K19: The Widowmaker* (2002) about a nuclear submarine suffering a radiation leak, which is neither all-encompassing nor indiscriminate. More recent examples, including *TDAT*, however contradict this prognosis. Finally, one vital factor that appears to be missing in the above list of criteria but is implicitly acknowledged in Roddick's article is that the genre is committed to dramatic realism rather than the fantastic sci-fi tradition, thus confronting the viewer with what might be possible now or in the near future.[17] *TDAT* clearly fulfills this basic criterion, as the depicted disaster, despite the evident exaggeration of events, is essentially based on a plausible future scenario.[18]

Since the genre functions by attributing agency to natural and/or technological forces, the actual antagonist of the story remains unspecified and diffuse, which

17 According to Roddick, this criterion disqualifies movies featuring attacks from outer space. Other critics, such as Ch. Mitchell (2001) and Keane (2001), do include more overt sci-fi scenarios in their discussion of the genre. The question of what constitutes a "credible threat" (Ch. Mitchell xi) is thus clearly subject to debate.

18 Additionally, as N. Schröder (2010) shows, the reality effect of the movie is heightened by its generically typical meditation of the disaster via news reports and intra-diegetic media images, such as when several massive tornados hit Los Angeles during live broadcasts. While these scenes can, on the one hand, be seen as underlining the artificiality of the production (and its completely computer-generated disaster shots), they can also, on the other hand, be read as "confirmations of the 'truth'-quality of such representations" (N. Schröder 101) and the movie per se. Indeed, *TDAT* has several deaths happening 'live' in front of the news camera, emphasizing the immediacy and severity of the situation.

makes the catastrophe all the more threatening and allows for a variety of symbolic inferences.[19] In the disaster-driven movie, "the enemy is now 'everywhere'" (71), as Wheeler Winston Dixon argues, a trope which clearly delineates the genre from the "readily identifiable antagonist who is the staple of the contemporary action film" (96). The unspecified enemy, however, is also the most dreaded one, as it is elusive and hard to combat.[20] Consequently, every action taken by anyone of the characters is usually only a *reaction* to the disaster (Fricke 118): The characters are hardly ever portrayed as reacting preemptively or with foresight, but always seem to be waiting for the disaster to hit. In *TDAT*, this emerges as a particularly noticeable trope, as humanity, despite climatologists' repeated warnings, does not act in time to avert the climate shift, but only becomes active in dealing with its aftermath. This also applies to the story's hero, who will typically be depicted as an ordinary guy rising to the challenge posed by the crisis. In narratological terms, the disaster thus acts as a form of catalyst (Roddick 258), initiating the heroic transformation of the (usually male) protagonist from common man to heroic leader and propelling his final catharsis.

This basic plot formula endows the classic disaster movie with an essentially mythological pattern, as biblical pictures of apocalypse and destruction are repeatedly evoked. Michael Bay's *Armageddon* (1998), for instance, refers to the Old Testament's place of final battle between good and evil in its title.[21] In the Christian sense, disasters hence often announce the possibility of redemption and the dawning

19 I here strongly diverge from Kakoudaki's point of view, who argues that the "allegorical dimension is seemingly lost when the disaster film [...] does not propose an agent or reason for the destruction, now delivered unintentionally by nature or accident" (120). While the symbolic level is certainly more diffuse and less didactic in natural disaster movies than (post)apocalyptic films exploring issues of political responsibility (e.g. *Planet of the Apes*, 1968, reflecting contemporary Cold War paranoia), they still allow for a range of symbolic allegations, as the present example reflects: Although *TDAT* focuses on a more or less 'natural' disaster, it contains thinly veiled allusions to members of the Bush administration (above all the uncanny resemblance of actor Kenneth Welsh to Dick Cheney) and hence clearly positions itself with regard to political investment.

20 At the same time, it will serve as a crucial narrative device to propel the pseudo-liberal overcoming of ethnic, religious or class differences and celebrate human community across the globe in the fight against one collective enemy, as *TDAT*'s tear-jerking final minutes exemplarily reflect.

21 Hollywood's visualization of mass destruction through gigantic fire balls, lightning, and storm is often directly inspired by St. John's "Book of Revelation," which describes the battle at Armageddon in vivid colors. Equally, other predictions of the 'end of days,' such as those by the prophetic seer Nostradamus or, more recently, the Mayan calendar have repeatedly provided fodder for apocalyptic scenarios on the big screen.

of a new beginning, purging civilization of evil, decadence, and moral decay. In the American context, such a dichotomous structure of punishment and redemption becomes particularly acute, suggesting that the 'Old World' makes room for a new, ostensibly better place (see King 2000, N. Schröder 2010). Particularly movies depicting grand-scale disasters that are neither spatially nor temporally contained but threaten the global extinction of mankind can typically be associated with visions of utopia or, conversely, dystopia. Disaster movies engage with this common trope by showing that "[d]estruction also means liberation" (Kakoudaki 113), as one first needs to go about "destroying what looked like home [...] in order to reaffirm what home really means" (Kakoudaki 113). While the stories may, at first glance, present dystopian visions of punishment and destruction, their final promise for a new, better future has an optimistic tone.[22]

In keeping with this tradition, *TDAT* ends with a rather inappropriately cheerful vision of the American population's future in a country covered in ice and snow: The helicopter that rescues Sam and his friends trapped in the New York library is seen flying over a plane of glittering snow, offering a view of eerie beauty as the death-bringing storm has finally subdued and the gray sky cleared. The scene echoes the opening shot of the movie which displayed the idyllic world before the disaster, only that it is now, in keeping with the story's circular structure, the area around New York City rather than the middle of Antarctica that is offering this image of icy bliss. The utopian and, given the scale of destruction, inexplicably optimistic tone of the movie becomes even more pronounced in the final shot depicting a group of international astronauts in a space station. Beholding his home planet from up above in a reiteration of the Romantic motif of the remote spectator, one of them exclaims in awe: "Have you ever seen an air so clear?" (*TDAT*) This is supported by a final view of the blue marble from outer space, now ostensibly more radiant and fresher than before, with hardly a cloud formation to disturb the sublime view. The earth, these images seem to say, has been purged and the future looks bright for the survivors.

Indeed, the movie's ending suggests a new age of brotherhood and solidarity: Given the changed climatic conditions, the former developing countries are now the only place where human life is possible and where the First World inhabitants seek refuge. The geopolitical situation has been ironically reversed by the disaster, but humanity, as the movie implies, is altruistic enough to act in dignity and unity, even

22 However, not all disaster movies must necessarily be apocalyptic in the sense that an imminent catastrophe threatens to eradicate all life on the planet and will pave the way for a new beginning, as one of the most successful representatives of the genre, James Cameron's *Titanic* (1997), reflects. Conversely, not all apocalyptic movies feature scenes of mass destruction, since the apocalypse may also be signified as an internal event, as, for instance, in Lars von Trier's *Melancholia* (2011).

if resources have become rarer than ever. The disaster has acted as a great equalizer: Every American is homeless now, and the original bum who survived in the New York library is even better equipped to deal with the changed situation than most of his fellow survivors, teaching them how to find food in the garbage and stay warm by insulating their bodies with newspaper. Even the formerly evil vice president, who has risen to presidency in the meantime, has changed into an ardent environmentalist and openly regrets his mistakes.[23]

PRESERVING AMERICA AND THE REST: NATION, MASCULINITY, AND SCIENCE IN THE AMERICAN DISASTER MOVIE

> Science fiction films are not about science. They are about disaster, which is one of the oldest subjects of art.
> SUSAN SONTAG/"THE IMAGINATION OF DISASTER"

As several critics have remarked (e.g. Roddick 1980, Keane 2001, Belmont 2007), the disaster movie is an essentially conservative genre in the most literal sense: The status quo, threatened by the imminent arrival of disaster, must be saved and preserved. Accordingly, Hollywood disaster narratives offer a particularly viable form of escapism by perpetuating dominant conceptions of gender and race relations, social class, military power, and, last but not least, science. This is typically supported

23 The highly optimistic, almost utopian ending reflects how *TDAT* refrains from evoking the present generation's responsibility for global warming too ardently in order not to make the audience looking for easy pastime entertainment uncomfortable. In this vein, Peter Bürger convincingly argues that "Emmerich's 'ecological movie' *The Day After Tomorrow* (2004) does not show which strategies and models humanity may employ *preemptively* in order to avert the impending destruction. Only *post*-apocalypses are shown" (367, italics in the original, original wording: "Gezeigt wird im Kino bis hin zu Emmerichs 'Ökologiefilm' *The Day After Tomorrow* (2004) keineswegs, mit welchen Strategien und Modellen die Menschheit *vorbeugend* einem drohenden Untergang entkommen könnte. Gezeigt werden vielmehr *Post*-Apokalypsen"). Thus, the disaster is both inevitable and ultimately beneficial for the planet, offering the welcome possibility of a new and fresh start by ridding decadent civilization of its most poignant problems. Preventing the catastrophe from the outset is not an option, as it would deprive the movie of its central spectacular potential.

by a linear plotline, circular narrative structures including the heavy use of foreshadowing, a fast and predictable dénouement, and clichéd character drawing.[24] As a result, the comforting and morally reassuring tone of the standard disaster movie contributes vitally to its continuing attraction: "Part of the pleasure of the spectacle of disaster," as Kakoudaki argues, "derive[s] from the perceived safety and comfort of the world" (111). Not only does the audience know that the disaster is safely contained in the fictional world of the movie, but they can also be reassured, thanks to the formulaic structure of the genre, that the characters they are emotionally invested in, i.e. the protagonist, his family and allies, will live to see a happy ending. Conversely, the disaster will have conveniently eliminated all the bad guys by the end, or, in the case of *TDAT*'s stubborn vice president, converted them to the right side of the battle. Closure, whether or not it comes about naturally, is an imperative of the genre. The following pages will show how conservative conceptualizations of nation, gender, and science, in combination with the characteristic plot devices of the Hollywood disaster movie, support the celebration of a scientific mindset of reason and rationality and promote the adoption of a 'cool pose' as an effective response to danger and existential threat.

Generally speaking, the moral reassurance offered by the standard disaster narrative tends to be supported by a strong nationalistic undercurrent. Disaster movies typically strive to affirm to their audience that they are concerned with distinctly American values. To clarify this proposition from the outset, the very first shot in *TDAT* following the opening credits depicts the American flag in extreme close-up, filling up the whole frame.[25] It flaps agitatedly in the strong wind, already foreshadowing that there is heavy weather ahead for the American nation. During the movie, the Stars and Stripes even help to save lives: Prof. Hall sees the flag above his head freeze instantly, which indicates that the super storm bringing extremely cold temperature is approaching fast and his team needs to take shelter. Another shot of the flag shows it completely covered in ice, symbolizing the state of the nation during the arctic snowstorm. Eventually, it quivers gaily and confidently in the soft breeze at the end of the movie to announce the dawn of a new American civilization,

---

24 However, filmmakers usually ensure that the easy and predictable plotline is not realized at the expense of narrative suspense. Part of the pleasure of watching a catastrophe unfold on the big screen certainly stems from guessing who will survive and who will perish in it. See also Keane 6-7 as well as the final section of this chapter, which explores these issues with regard to the disaster movie's politics of information distribution.

25 This initial sequence also already reflects how in the disaster genre generally and *TDAT* in particular, close-ups tend to be reserved for objects and displays of the disaster rather than human actors. See Seewald (2011) and the subsequent section in this chapter, which explores this issue with regard to affective audience involvement.

which may well, as the heroic background tune suggests, outshine the previous one.[26]

The incessant display of the flag is indeed symbolic of the movie's unmistakable patriotic orientation. Although most recent disaster movies revolve around disaster on a global scale (allegedly paying tribute to current processes of globalization and diversity), both the crisis and its solution are typically depicted from an exclusively American vantage point. In *TDAT*, the audience is initially shown minor scenes occurring outside of the United States, such as massive hail in China or heavy snow in New Delhi, which are intended to firmly establish the climate shift as a global problem. After these initial sequences, heavy weather phenomena all around the globe are only mediated verbally, rather than visually: Hence, weather specialists report that Australia is afflicted by gigantic typhoons, while Russia suffers from the freezing eye of a hurricane. By the end of the movie, even these brief reports are suspended, as the rest of the world seems to have become utterly irrelevant: Two of the three major super storms generated by the abrupt climate shift conveniently form right above Scotland and Siberia respectively, suggesting that most of the population of Central Europe and East Asia is probably already dead, which is apparently supposed to justify the complete disregard for continents other than the American.[27] Another one of Emmerich's signature pieces, *Independence Day* (1996), even elevates the 4th of July to a global holiday that will from now on commemorate the world's survival amidst an alien attack. The whole globe has, thus, become American. In his address to the nation—a stock element in the disaster genre—a grave, yet confident American president not only addresses his people, but also speaks to and for everyone on the planet.[28]

---

26 The frequent flaunting of the American flag or, more generally, the Stars and Stripes has been identified as one of Emmerich's trademarks and earned the native German the reputation of being "more American than many a US patriot" (Becker n.p., original wording: "amerikanischer als mancher US-Patriot"). Other films, however, also make ample use of nationalistic imagery, such as Michael Bay's blockbuster *Armageddon* (1998), in which the flag is even hissed on the asteroid that threatens to eradicate all life on the planet. See also Bürger 386-390 and Fricke 125.

27 The only exceptions are Prof. Rapson and his team in northern Scotland, whose fate is also traced in the movie (although they do suffer an early death as the eye of the storm forms right above the UK). Despite the inclusion of countries other than the United States, the movie thereby remains firmly grounded in the Anglophone world.

28 Coincidentally, disaster movies have traditionally cast (male) African-American actors as American presidents well before other genres (or reality, for that matter). Movies like *Deep Impact* (1998), *Polar Storm* (2009), or *2012* (2009) show that despite the indisputable conservative orientation of the genre, "these films use the space of fantasy to allow potential action, potential visibility to the politically unrepresented" (Kakoudaki 135).

It is also for this reason that mainstream disaster movies are almost inevitably set in an American urban environment, mostly well-known metropolises such as Los Angeles, New York, or Las Vegas. By capitalizing on immediately recognizable sights, the movie immediately affirms its Americocentric perspective by means of narrative setting and simultaneously suggests to the average American viewer that the fictional disaster presented on screen could well occur in their more or less immediate neighborhood.[29] This "feeling of 'involvement' on part of the audience" (N. Schröder 292) is key to the attraction of the disaster genre, counterbalanced by the comfort we take in the (still) intact world around us. The destruction of famous sights, be it New York's Statue of Liberty or San Francisco's Golden Gate Bridge, provides a perfect canvas for appeasing the viewer's appetite for spine-tingling scenes of large-scale destruction, as tumbling skyscrapers offer a more impressive scenario than crashing farmhouses or suburban picket fences. Movies set in the American heartland, such as the box office hit *Twister* (1996), are an obvious exception. Generally speaking, however, the Midwest, stereotyped as both flyover country and the hotbed of America's true values, is routinely spared the large-scale destruction experienced by the fringes, be it the decadent East or the libertine West Coast. Global metropolises as the alleged centers of social coldness seem to offer a much more viable canvas for cool destruction scenarios; *TDAT*'s deluge of ice and snow engulfing the supposedly coldhearted population of New York City perfectly exemplifies this tendency.

Due to their loaded scenery, urban spaces are appreciated by filmmakers for their blatant symbolic value as national icons. As the "material and symbolic centres of 'civilization' and modernity" (Keane 101), cities offer a perfect playground for the cinematographic exploration and ensuing castigation of the abysses of Western, and particularly American culture. Both historical and contemporary hedonistic centers, be it biblical Babylon or modern-day Hollywood, allow for simplistic moral judgment when their sites of decadence and digression are leveled to the ground.[30] Accordingly, the spectacular destruction of the Hollywood sign in *TDAT*

29 Keane calls this the "postcard theory of disaster movies" (83), stressing the significance of immediately recognizable buildings for the effective execution of the disaster mode. He also points out, however, that this trope has only been initiated in the 1990s cycle (he specifically mentions *Independence Day* as its chief incubator), since earlier disaster movies usually only featured small-scale catastrophes, such as plane crashes or sunken ships, which did not necessitate scenes of urban destruction.

30 Incidentally, those most closely associated with decadent civilization are usually the first to die (King 152), such as the employee at the L.A. weather center in *TDAT*, who uses his night shift for a tête-à-tête with his girlfriend and is fatally interrupted by the approaching tornado, suggesting that he must be immediately punished for neglecting his duties. Conversely, a cleaner who is diligently pursuing his job in the same building miraculously

by a massive tornado can be read as an ecological comment on the industry's contribution to global warming. This adds a curious layer of meta-commentary to the movie, suggesting that Hollywood as the place where these images of large-scale destruction are made is, in fact, destroyed by its own creation in a Frankensteinian manner, similarly to how humanity is now forced to pay the prize for the decadent and unnatural civilization it created. In contrast to the subtle critique offered in contemporary movies, the 1990s cycle plays with the demolition or, conversely, survival of celebrated landmarks in a tongue-in-cheek fashion, as the implausibly imperishable Statue of Liberty reflects in many blockbusters of the era.[31] In *TDAT*, Miss Liberty similarly withstands the massive Atlantic tidal waves and remains largely intact, apart from a thick coat of ice and snow, symbolizing the state of the nation at large.

The United States, however, not only provide the city upon a hill for the rest of the world to behold, empathize with and emulate in times of crisis, but are also depicted as the only nation capable of offering a practicable solution, be it in the form of nuclear weaponry, military strategies, or space travel. US superiority in terms of military and technological power is thus casually reasserted.[32] Curiously, assistance from other, equally advanced countries is seldom required in this endeavor, neither

survives the tornado. In the streets, several curious onlookers and sensationalist Hollywood reporters have to perish. Generally speaking, reporters and journalists tend to lead a precarious existence in the disaster genre, as they are frequently employed to represent a condemnable passivity that is juxtaposed with the brave and noble hero who does not stand by and watch (and possibly capitalize on the misery of others), but act. The disparate group gathering at the NYC library displays similar dynamics of life and death: While Sam and his classmates, the librarian and a homeless guy with dog survive, notably more well-off residents in fur coats and a police officer freeze to death in the snowstorm.

31 See, for instance, *Godzilla* and *Independence Day*. While in the latter instance, Miss Liberty is blown off her pedestal and comes to rest horizontally in the Hudson River, the statue miraculously remains in one piece.

32 Many disaster movies explicitly acknowledge the cooperation of state and military institutions in their end titles, such as the NASA (*Armageddon, Deep Impact*), the Department of Defense (*Armageddon, The Core*), the U.S. Army (*Deep Impact*), and the U.S. Marine Corps (*The Core*). In reference to the continuing cooperation between Hollywood and the military, Bürger claims that many disaster movies (he specifically names those produced by Hollywood mastermind Jerry Bruckheimer) "increase their budget by benefitting from the U.S. Army's 'advertising costs'" (388, original wording: "Dass einige seiner erfolgreichen Titel ihr Budget aus dem Etat 'Werbungskosten' der U.S. Army aufbessern"). While *TDAT* does not officially acknowledge any cooperation with state services, the production did borrow several Blackhawk helicopters from the U.S. Army for the shooting of the final rescue scene (Bürger 396).

in terms of resources nor warfare. The world is saved by and for America alone, a recurrent plot line which offers an evident recourse to and modern interpretation of the frontier myth. This "ideological fantasy of omnipotence in terms of a *pax americana*" ("Weltuntergang" 54, italics in the original),[33] as Charles Martig calls it, renders the classic disaster movie a powerful vehicle for the nationalistic enterprise. In this light, it does not seem too far-fetched to read the genre's formulaic ending as the twenty-first-century rendering of social Darwinism (Roddick 260), since the American race is routinely portrayed as fitter for survival than any other people, while being simultaneously purged of unfit and degenerate elements in their own ranks. The already introduced final sequence of *TDAT*, depicting two astronauts rejoicing over an air that has never been so clear, can hence also be interpreted as a sardonic comment on the number of casualties claimed by the disaster: The USA have literally been purified.

Similarly to how America will be routinely depicted as ensuring the survival of the human race almost single-handedly, the solution of the crisis does usually not stem from collaborative efforts or team work, but requires the "ritualized legitimation of strong male leadership" (23), as Michael Ryan and Douglas Kellner observe. Traditional virtues of male strength, moral integrity and professional competence are thus teamed with an almost superhuman courage to save the beloved at the risk of one's own life. *TDAT* is an exemplary representative of this formulaic characterization, as it features two central male heroes: Prof. Hall and, with regard to the second narrative thread situated at the library, his son Sam, who despite his young age quickly emerges as the natural leader of the highly disparate group. Through the figure of the strong male specialist, whose moral authority is majorly enhanced by technoscientific expertise, the disaster genre conveys "a sense of unity out of deeply disparate factions" (Dixon 68), which is also exemplified through *TDAT*'s final reconciliation scene between the obstinate government and the NOAA science squad. In another one of Emmerich's commercial successes, *Independence Day* (1996), the disaster even propels the overcoming of ethnic differences, above all through the joint rescue mission of a white Anglo-Saxon president, a Jewish satellite technician, and an African-American jet pilot.[34] Each of them represents another central male virtue, from political power to technological prowess to physical strength. Their chances of combating the hostile aliens are noticeably increased by their teaming

33 Original wording: "ideologische[...] Allmachtsfantasie einer *pax americana*."

34 The blissful neglect of ethnic differences for the greater American good, however, may easily result in a problematic cultural appropriation of minorities as well as the strategic refusal to thematize race and ethnicity as actual problems in the United States. See especially Kakoudaki (2002), who explores race relations in science fiction and disaster movies.

together and thus combining their individual qualities. Contributions by female heroines, on the other hand, are rarely required.

In accordance with its visions of male potency, it is thus hardly surprising that the disaster movie perpetuates conservative conceptualizations of gender and femininity. In her exploration of the figure of the disaster movie heroine in recent blockbusters, Belmont (2007) observes that "heroines who are initially characterized as 'modern women'—capable, intelligent, and employed—are quickly returned to the domestic sphere and to helpless dependence on masculine physical prowess and technological know-how" (350). *TDAT*, however, does not even feature an initially strong female character. From the outset, the heroic role is clearly attributed to Prof. Hall and, to a lesser degree, his son Sam. Although Hall's wife Lucy (Sela Ward) is a hard-working professional, employed as an oncologist at a local hospital, the movie noticeably foregrounds her motherly qualities as she nurses a young cancer patient who was left behind by the rescue team. The only female scientist in the movie, NASA hurricane specialist Janet Tokada (Tamlyn Tomita), seems to be primarily included as a love interest for Jack's younger colleague Jason (Dash Mihok). The same applies to Sam's classmate Laura, whose function in the story, despite her evident intellectual abilities, is to fuel the romantic subplot.

Ultimately, the future envisioned in the movie's buoyant ending is a distinctly heteronormative and patriarchal one, primarily hinged on the survival of the nuclear family: Not only is the first generation of the family reunited by returning Sam to his parents, the audience is also given to understand that Sam and Laura will form another central heterosexual couple that shall preserve the survival of the human race in the world after the storm. This is implied in the final rescue scene, when Sam and Laura sit next to each other in the helicopter and lock hands, which Sam's father, seated opposite of them, condones with a faint smile. The patriarchal order is thus clearly handed down from father to son, from one generation to the next. Laura is officially instigated as Sam's potential future wife, which she also seems to acknowledge in the library during the storm: While Laura is a straight-A student and has often been ridiculed for her strong academic ambition, she now intimates to Sam that the disaster has caused her to revise her notions of what is really important in life:

> "How am I supposed to adjust, Sam? Everything I've ever cared about, everything I've worked for has all been preparation for a future that no longer exists. I know you always thought I took the competition too seriously. You were right. It was all for nothing." (*TDAT*)

What is more important now than the career she envisioned for herself, as Laura instinctively comprehends while the world outside the library changes for good, is the mere survival of the human race, to which she can contribute by virtue of her sex, rather than her intellect or ambition. Not by chance, Laura and Sam's first kiss occurs immediately after this confession, signaling to Sam that she is no longer manically obsessed with her career plans and lost in academic abstraction, but ready to succumb to her natural role of love interest, wife and, eventually, mother to help repopulate the barren planet. *TDAT* is thus exemplary of the way in which disaster movies show that "the catastrophe ends [...] precisely in the moment when the 'imbalance' of traditional gender roles has been reversed and traditional (family) values have been reinstated" (N. Schröder 99). The conservative gender politics of the genre emerge most conspicuously in its idealized and highly formulaic endings.

Additionally, it is noteworthy that although Sam, Laura and Brian (Arjay Smith) participate in the same academic competition as a team, the boys are depicted as more versed in mathematics, engineering, and natural sciences, while Laura's talent seems to lie in the humanities, as she helps a family in the storm by translating into French or prevents Brian from getting a history question wrong during the competition.[35] These skills are, however, depicted as far less valuable during the disaster than technological prowess, such as when Brian repairs an old radio to receive information about the approaching blizzard. By this time, Laura has already succumbed to the role of 'damsel in distress,' as she suffers from severe blood poisoning and high fever induced by her injured leg. Consequently, Sam evolves into a modern knight in shining armor, risking his life for his beloved by retrieving medicine from a stranded ship in the storm outside the library. This is the second time he saves her life within the mere 120 minutes of the movie, having previously pulled her out of a parked taxi caught in the flooding. Laura's physical and symbolic enfeeblement thus enables Sam's transformation from the cowardly teenager who panicked and grabbed her hand during the bumpy flight to New York to the brave leader of a group of survivors, whose authority is supported by his technoscientific prowess. Science in *TDAT* thus repeatedly serves to support and legitimize conservative notions of gender by naturalizing the connection between technological expertise and dominant conceptualizations of masculinity.

The overall conservative tone of the genre also pervades the images of science reiterated by the standard disaster movie. Evidently, the above delineated major at-

35 Interestingly, Laura only knows that Brian's proposed answer is wrong; she still needs Sam to wage in and tell her the correct response, which she herself could not come up with. In the same vein, the situation that allows her to apply her language skills seems rather arbitrary, as it is highly unlikely that middle-aged French tourists in New York City of the twenty-first century do not even speak rudimentary English. Hence, her knowledge of French emerges as an archaic skill that proves rather insignificant in an age of highly developed technological gadgetry that could have equally done the job. Technological prowess and the natural sciences are thus constructed as hierarchically higher paradigms of knowledge in such a time of crisis; the humanities, it seems, can hardly contribute anything at all to the survival of the human race.

traction of the genre, i.e. reassuring the audience that the scenes of mass destruction are safely contained in the fictional world of the movie, depends on an element of science, at least implicitly so. Being safe, as especially recent disaster movies seem to tell us, acutely relies on an advanced level of science, which the United States, as a technologically progressive economy, are able to offer. A world in which Western industrialized science can no longer provide definite answers or where its warnings are, conversely, ignored by the political elite of the country is invariably headed for disaster, in quite a literal sense. On the most general level, disaster movies can hence be understood as "boost[ing] confidence in technological progress, even when it is directly responsible for the disaster in the first place" (N. Schröder 292). This assessment can clearly be applied to *TDAT*, which explicitly blames the unrestricted burning of fossil fuels, driven by industrial progress, as the direct source of the disaster. On the other hand, although scientific knowledge may not, in the present case, avert the catastrophe, it is indubitable that it would have been able to: "You didn't want to hear about the science when it could have made a difference" (*TDAT*), as a NOAA official accuses the vice president. The failure to prevent the catastrophe cannot, as this quote suggests, be attributed to science, but to exterior forces such as political or economic prerogatives that ignore the science that could have saved the day. Indeed, science will still prove to be the central savior of the story, as it helps to predict the crisis in advance and hence diminish its consequences, organize nation-wide evacuations in time, and enable the population to make comprehensive provisions for dealing with the aftermath.

A key critical text for the analysis of cinematic disaster and its representations of science is Susan Sontag's "The Imagination of Disaster" (1965). Sontag does not, strictly speaking, focus on the coherent set of feature films that were later popularized under the umbrella term 'disaster movie,' but discusses the particular motifs used in traditional science fiction films that involve an element of external catastrophe, which is loosely identified as "the arrival of the thing" (207). Most often, this involves an alien invasion, the landing of a foreign spaceship, or a similar extra-terrestrial event. As Sontag perceptively notes, such staged disasters are "concerned with the aesthetics of destruction, with the peculiar beauties to be found in wreaking havoc, making a mess" (213). While the films discussed by Sontag, such as *When Worlds Collide* (1951) or *The Time Machine* (1960), are weak on diegesis and character development, they offer breathtaking special effects and sublime scenery. The representation of science in these early movies is essentially one of ambivalence: While the story lines clearly endorse the humane and ethical use of science, they simultaneously condemn its application to immoral causes. Science hence functions "as both satanist and savior" (Sontag 218) and will prove to be immensely dangerous when abused by willful individuals.

In accordance with Sontag's vantage point in the late 1960s, her observation remains particularly accurate for earlier movies of the genre: Science was either represented as the direct cause of the disaster, such as through unethical experiments or scientific miscalculations (Ch. Mitchell xiv), or the indirect root of it, such as when techno-scientific innovations, like the airplane in *Airport 1975* (1974) or the electric wiring of the world's largest building in *The Towering Inferno* (1974), malfunction and cause havoc. Manhood is thus punished for its hubris and presumption of divine omnipotence through high technology, which, being man-made and 'unnatural,' will produce fatal liabilities like mutations or disturbances of the planet's stability. The films hence express, as Martig observes, "a fundamental distrust that natural forces can be kept in check through technological means" ("Weltuntergang" 51).[36] Frequent topics, among other concerns, are the limits of science and technological progress and the societal responsibility of scientists for their creation, a recurrent trope through which the genre is clearly indebted to Mary Shelley's *Frankenstein*. In response to the illicit usurpation of divine rights by fallible and reckless humans, nature will be compelled to leave its assigned territory and break into the clearly bounded, civilized, and artificial world of the protagonists, who have seemingly lost the connection to their mythical roots in the natural world.[37]

Overall, however, science functions as a "great unifier" (220) in an overtly utopian sense, as Sontag convincingly argues. For that reason, there seems to be no logical contradiction when the solution to the technological disaster is brought about by the very technology that initiated the catastrophe. While unchecked scientific progress is usually condemned as unnatural, sinister, or immoral,[38] science will eventually emerge as an idealized, omnipotent and almost magical force, providing an "antidote to the human imperfection that has destroyed the Earth" (Yergensen 156) or, simply, to the adversity of nature. In the disaster genre, science is thus frequently endowed with religious insignia (Yergensen 163-164) and presented in a biblical rhetoric, as *TDAT* exemplarily reflects: In accordance with earlier examples of the genre, the rescuer appears "from on high like an angel" (Roddick 254), since Prof. Hall and his assistant have to climb in through one of the library's high windows in order to get into the snow-covered building. Descending from a snowy

36 Original wording: "das grundlegende Mißtrauen gegenüber der technischen Kontrollierbarkeit der Naturgewalten."

37 See King 143-160, N. Schröder 92-94, and the final section of this chapter for the thematic opposition between nature and civilization in movies featuring spectacular disasters.

38 What qualifies as the 'ethical' and 'correct' use of technical innovation is subject to debate: While many disaster movies feature nuclear weaponry as the ultimate solution to all kinds of threats, they seldom thematize its consequences such as nuclear fallout or radiation risks. See also Bürger (2007), who shows how disaster movies endow nuclear weapons with "the most possible *positive* image" (392, italics in the original, original wording: "denkbar positivem Image") and Fricke (2001), who explores these issues with regard to *Armageddon*.

slope, they enter a completely white room whose furniture and chandeliers are covered in ice, exuding a heavenly atmosphere. The sunlight emanating from the window through which they are entering endows the two rescuers with wings of light. Through the personification by its central practitioner, science thus appears like a *deus ex machine*, unexpectedly and in the very last second, when all hope has already been lost. The final rescue via helicopter reiterates this sudden and somewhat contrived resolution, with the minor modification that it is now the technological forefront of the U.S. government and military that descends upon the survivors like a guardian angel.

According to Sontag, however, it is not only technoscientific progress per se which will bring salvation and redemption, but also the according scientific mindset of reason and rationality that will prevent society from succumbing to chaos and anarchy in times of crisis. This "regime of emotionlessness" (Sontag 221) or the imperative of 'striking a cool pose,' as one might call it today, is promoted as an effective response to all kinds of danger. In *TDAT*, rational thought and abstract knowledge are indeed endowed with a halo of cultural reverence, above all through portraying the public library where Sam and his classmates seek shelter as a safe haven in the midst of chaos and disaster. With a wild super storm roaring outside, the survivors huddle cozily around the library's fireplace. Quite literally, knowledge fuels their survival, as they start burning books to keep warm. The motif of book burning, an emotionally highly invested and suggestive gesture, reflects the severity of the situation and gives rise to the question of what kind of knowledge needs to be preserved for future generations.[39] According to the movie's underlying rationale, the books found at the library crystallize everything that is worthwhile saving in Western society. This is evoked by one of the survivors at the library (coincidentally the spitting image of Hollywood intellectual icon Woody Allen) who takes a Gutenberg Bible with him when they are finally rescued by the helicopters:

> "This Bible is the first book ever printed. It represents the dawn of the age of reason. As far as I'm concerned, the written word is mankind's greatest achievement. You can laugh. But if Western civilization is finished, I'm gonna save at least one little piece of it." (*TDAT*)

The peak of civilization, this quote implies, is codified knowledge representing the chief industrial virtues of reason, rationality, and literacy. This exhilarating moment celebrating abstract thought as mankind's most precious achievement cannot even be tainted by the fact that the Bible as a religious document could equally be read as

39 At a closer inspection, the reason for burning books seems to be somewhat arbitrary, as the survivors could equally use the wooden chairs, doors, desks, or paneling of the library, as Hyde (2004) observes.

representing an alternative explanatory system that is both culturally and epistemologically diametrically opposed to science.[40]

Indeed, one might equally interpret the scenes at the library as an expression of "science's hierarchical position of worth during humanity's tribulation" (Yergensen 161), since the potential contribution of other paradigms and academic disciplines to the survival of mankind is negated from the outset by readily surrendering them to the fire. The humanistic scholarship thus offered is not deemed relevant, so that these books can, at best, contribute to the group's survival by providing fuel, rather than exploitable knowledge. The depiction of Sam and his classmates scavenging through the library's shelves can hence also be read as a final decision to act, rather than sit around and think (or read): The gathering and burning of the books is presented as a physically challenging endeavor, requiring them to employ trolleys, ladders, and an intricate organizational system, since the fire must be constantly rekindled with new paper. Just as his father in the primary narrative strand of the movie needs to become physically, rather than merely intellectually active to save his son by virtue of technological progress, Sam as the natural leader induces the group of survivors to action, suggesting that the time for introspection and intellectual occupation has passed.

On a metaphorical level, this may be read as an entreaty for technoscience to eventually move from theory to action and venture out of its neatly bounded cognitive sphere in order to unlock new territories in need of scientific guidance. Rather than books, it is through the technological devices of the information age that science can truly flourish and effect viable change. The texts in this study, whether it be *CSI*'s aestheticization of the technological tool box or the cool charm of personal computing as exemplified in the biographical movies *Jobs* and *The Social Network* explored in the following and final chapter of this book, to varying degrees all support this observation. The contemporary disaster movie, however, can be read as representing the most overt promotion of an active technoscientific responsibility in lieu of poring over books, which can much more meaningfully be put to use by surrendering them to the fire.

40 Given the strong patriotic undercurrent of the movie, it is indeed surprising that religion is hardly ever evoked as offering help and comfort during the crisis. It is only in one minor instance that Prof. Hall, acknowledging that the nation-wide evacuation plans have been initiated too late for the northern states, advises the remaining population to "pray" (*TDAT*). Religion thus only takes over when all political and scientific resources have been thoroughly exploited.

STRADDLING THE ULTIMATE FRONTIER: THE SCIENTIST AS HERO IN THE POST-2000 DISASTER MOVIE

While technoscientific knowledge, as the above pages have shown, will ultimately be redeemed as a highly worthwhile human endeavor and the most efficient response to danger, the image of the scientist perpetuated by the genre is a much less affirmative one. As Sontag already notes in the late 1960s, the overall ambiguous stance toward science is primarily cast as an ambivalence toward the obscure, hermit-like figure of its practitioners: "In science fiction films the antithesis of black magic and white is ran as a split between technology, which is beneficent, and the errant individual will of a lone intellectual" (223). It is not technology or Western civilization per se that must be condemned, but the irresponsible employment of it by unskilled, unworthy and morally deficient disciples. I would, however, argue that this assessment primarily applies to the early cycles of the genre, especially the 1970s and, to a lesser degree, the 1990s variety. Contemporary examples, roughly starting around 2000, display a much more encouraging, unambiguously positive image of the scientist as ingenious hero and dauntless modern frontiersman. Far from "[b]eing a clearly labeled species of intellectual, [...] always liable to crack up or go off the deep end" (Sontag 217), the scientist of the post-2000 disaster movie has taken center stage as heroic protagonist, handsome love interest and powerful auxiliary in the fight against the disaster. As the virile, young and adventurous hero of the story who almost single-handedly saves the planet from the imminent apocalypse, he counters maximum physical danger with an indestructible cool pose, remaining self-assured, detached and in utmost control when awaiting the presumable end of the world. The following pages will trace the scientist's development from sinister and/or comic sidekick to heroic leader of the story and demonstrate how he employs cool as an indispensable suit of armor against the impending disaster, aided by his unshakable faith in the powers of modern science and technology.

Scientific characters in early disaster movies, especially those of the 1970s cycle, were seldom depicted as likable, trustworthy or even indeed heroic. Their image was, at best, a highly contradictory one, endowing them with both major intellectual prowess to respond to the disaster and sinister intentions directed against the very survival of the human race. *Earthquake* (1974), for instance, reveals an initially similar narrative structure to *TDAT* in the sense that a group of scientists discover heightened seismic activity, suggesting that a major earthquake will hit Los Angeles in the next couple of days. In order not to jeopardize their funding if their predictions turn out to be inaccurate and to avoid a nationwide panic, the scientists decide against alerting the public and only notify the National Guard of the potential crisis. In contrast to Prof. Hall, thus, who immediately tries to warn the U.S. government but fails to have his voice heard, *Earthquake*'s group of scientists are depicted as selfish, greedy, and unethical, as they are indirectly to blame for the disastrous effects of the quake that could have been prevented or attenuated. Further movies of the 1970s cycle and examples from other genres support this reading: The scientist character will appear either sinister and literally or figuratively 'mad,' or, conversely, as a harmless sidekick that provides for the necessary comic contrast with the adventurous protagonist, such as Dr. Brown in the vastly successful *Back to the Future* (1985-1990) trilogy.

The actual hero of the early disaster movie is a figure diametrically opposed to this image of the nerdy and unlucky scientist: the uniformed professional (Roddick 256). While this character clearly also possesses major technological prowess, the storylines usually painstakingly strive to assert that his is, above all, a hands-on approach: 1970s disaster heroes are no quixotic academics lost in theories or endless rows of calculations, but technologically educated specialists, such as pilots, architects or engineers, who are familiar with the everyday world of practice and ready to get their hands dirty. In essence, this tradition is continued in the millennial cycle, which features satellite technicians (*Independence Day*), oil drillers (*Armageddon*), and emergency managers (*Volcano*) as male leads. The central virtue which all of these heroic protagonists necessarily have to display is their practical ability and enthusiasm for manual labor, as King notes with regard to the role of Mike Roark, head of the city's emergency management in *Volcano* (1997): "[T]he major male hero can be permitted other roles, such as leadership or authority, but he must also display a strong degree of direct hands-on activity" (149). As a result, the audience will seldom be induced to appreciate his proficiency in coordinating troops and supervising the rescue team, after all vital activities in a city struck by heavy disaster, but rather see him saving afflicted individuals, drilling trenches to redirect the flow of lava, or preventing lootings almost single-handedly. His intellectual and administrative qualities are notably downplayed or entirely elided for the sake of action-heavy plotlines. The perfect hero does not think, but act, seems to the underlying message.

In *Volcano*, the appreciation of physical strength at the expense of intellectual ability is emphasized through Roark's juxtaposition with the geologist Dr. Amy Barnes, whose gender and profession already suggest that her role is primarily that of a potential love interest for Roark rather than that of the empowered heroine. Indeed, the scientist fails to provide necessary proof for her (accurate) hypothesis that a dangerous volcano is forming near the city and can hence hardly contribute anything at all to the solution of the crisis or the population's rescue. When she tries to take soil samples with her assistant Rachel, the latter is killed in a seismic quake, suggesting that the science squad may be intellectually, but definitely not physically equipped to perform well in an emergency situation. The character of Dr. Barnes hence illustrates "the inadequacy or failure of technology and its predictive powers" (Doane 231), as do the majority of scientist representations in the 1970s and the

1990s cycle. All heroic qualities are relegated to the man of action Roark, who may consult Dr. Barnes for scientific assessment, but is ultimately the one who engineers an apposite plan and carries it out. A similar dichotomy emerges in another central movie of the millennial cycle, *Armageddon*, in which the heroic deep core drillers are juxtaposed with NASA scientists. The latter are presented as scientifically educated, disciplined, and highly knowledgeable, while at the same time pasty and spineless.[41] They provide the necessary scientific data for the extraterrestrial mission, but it is eventually the untrained drillers who will execute it, also because their approach is rough, eccentric, and much less academic. Although ultimately "some of the differences [are] reconciled" (King 85), it is clearly the hands-on mentality of the drillers and their years of experience that save the day, not the scientific superiority of the NASA squad, whose mere function is to provide the adequate hardware.

Both 1970s and 1990s disaster movies hence demonstrate that the role of heroic rescuer is reserved for the technologically solid, but practical leader, whose physical strength and vigor must necessarily outshine his undoubtedly existing intellectual prowess. This type of disaster movie hero, as Roddick writes in the early 1980s, "is not to be confused with the scientist who, in Hollywood imagery, is either a dangerous fanatic, a potential traitor liable to share his knowledge with 'the other side,' or, where he represents no direct threat to the community, a dotty boffin devoid of all normal masculine attributes" (257). As the examples from *Armageddon* and *Volcano* show, one will usually encounter inept scientists, who must, almost rightfully, meet their fate in an overtly silly and absurd fashion, caused by their physical inferiority. 1970s movies are even more outspoken in their wholesale dismissal of scientist characters, ranking them among the first victims of the disaster for which they are directly responsible. Survival or, at the very least, heroic death is reserved for the practical heroes. In the most positive representations of the era, the

41 King (2000) attributes this ambiguous, less than ideal representation of NASA to the fact that Hollywood action blockbusters frequently display a general distrust of the government and state-funded organizations, which are juxtaposed with the hard-working individual everyman. State institutions will hence be represented as essentially suspicious entities, possibly associated with conspiratorial forces. In *TDAT*, the government is portrayed as foolish and directly responsible for the disaster as it refused to sign any climate protection measures in the past and continues to ignore Hall's scientific data. During the disaster, the audience is shown a deserted Oval Office depicting the government's impotency in the face of disaster and suggesting that the power needed to cope with this national crisis is no longer the one formerly located in Washington. Indeed, the US president (Perry King) is killed when his motorcade gets caught in the storm. In contrast to Prof. Hall, none of the government officials, including the vice president, are directly involved in the action but coordinate the evacuation from a safe distance in their Mexican exile.

man (or, less frequently, woman) of knowledge may contribute to the solution of the crisis by providing relevant background information by virtue of dedicated (or even indeed obsessive) scientific research, but he or she is certainly not the one who participates in or even heads the heroic rescue mission in an overtly physical fashion.

These stereotypes stand in stark contrast to contemporary movies, in which the image of the scientist, as I contest, is a much more positive, powerful, and unambiguously heroic one. Just like the notions of science and technology have become inextricably fused in modern information societies, so the technologically skilled adventure hero of the 1970s and 1990s cycles has joined forces with the traditional scientist, whose image has been significantly ameliorated as a result: Not only does Prof. Hall detect the impending disaster well before its time, he also invests a large amount of his own time and resources to warn the world, even though his attempts are continually frustrated. While *TDAT* confronts the viewer with the generically rather rare scenario that the disaster cannot, despite the great powers of Western science, be averted or mitigated, it is still technoscientific research that contributes vitally to humanity's survival by accurately predicting the crisis and suggesting ways of dealing with its aftermath. Accordingly, Prof. Hall's research model, which he designed in order to visualize prehistoric climate shifts, becomes a "prophetic tool that will be the means that enables the Hall research team to call on Americans to evacuate the United States" (Yergensen 160). Indeed, Hall is the only character who correctly attributes the increasingly extreme weather to the right source, while his colleagues remain baffled by the situation and propose a variety of outlandish theories, such as changed solar output via sunspots.

The fact that his call on the public is initially not heeded by those in power adds a sense of noble failure to his character, turning him into a form of modern Cassandra. This fate makes him all the more likeable and relatable to the audience: The viewer knows that Prof. Hall is right, and is impelled to experience the same feeling of outrage at the politicians' ignorance and self-importance. Although he is initially labeled "sensationalist" (*TDAT*) and asked by the vice president if he has "lost [his] mind" (*TDAT*), the scientist rests self-assured and confident that his theory will eventually be listened to, which it is. The 1970s disaster movie image of the scientist as 'dotty boffin' and evil maniac is thus firmly turned around. While Prof. Hall does get labeled "obsessive" (*TDAT*) by his colleagues and is accused of never lightening up, he is clearly not portrayed as the traditional 'mad scientist' lost in obscure, untenable theories. Rather, it is the government officials who act foolishly and, given the overwhelming evidence, against reason. The character of the climatologist hence more overtly aligns with a stock figure of the sci-fi genre, namely "the dissident scientist battling against the odds to persuade politicians to take his data seriously" (Monbiot par. 6). Despite repeated rebuffs by the vice president, Hall continues to act heroically and selflessly, gathering allies to work underground in order to be ready as soon as the government is prepared to listen to his hypothesis. In contrast to the depiction of scientists in earlier disaster movies, *TDAT* thus presents a notably more altruistic and humane image of scientific ability.

What primarily distinguishes the contemporary disaster movie scientist from previous representations of the genre, however, is the peculiar fusion of the traditional adventure hero with an academic background: Prof. Hall's role in the movie is not only to predict the crisis and make scientifically grounded suggestions on how to survive it, but he must also embark on a seemingly suicidal foot march from Washington, D.C., to New York to save his stranded son. The appeal made by his senior colleague, Prof. Rapson of the Hedland Climate Research Centre in Scotland, is thus clearly directed to him (and the power of science) rather than political or military authorities: "Save as many as you can!" (*TDAT*), the experienced researcher commands Jack, which the latter seems to take literally when marching to his son's rescue. In contrast to the secondary figure of *Volcano*'s Dr. Barnes, who remains inept and powerless throughout the story, Prof. Hall takes center stage by not only providing academic analysis, but also bodily action.[42] The modern disaster scientist hence combines intellectual and physical strength, which is also expressed through his physique: As a young, athletic, and virile hero, whose sexual attractiveness is repeatedly underlined through the narrative addition of various female foils and potential love interests (and, incidentally, the casting of burly action movie star Dennis Quaid), he appears more than apt for all sorts of physical adventures. *TDAT* emphasizes his bodily ability through a series of (narratively rather arbitrary) scenes, depicting Prof. Hall topless or in bed, thus showcasing his muscular chest and broad shoulders.[43] Glimpses of his domestic life, which the viewer is granted at the beginning of the movie, suggest that he is capable of a functional marriage and has well-developed social skills.[44] This is clearly a hero, the underlying message seems to say, who distinguishes himself by both his intellectual and sexual potency.

42 Not coincidentally, of course, the gender of the scientist has also been adjusted, as the masculine and muscular Prof. Hall is portrayed as much more prone to physical action.

43 The same narrative construction applies to his teenage son Sam, who needs to strip down in the library to get rid of his wet clothes, which allows him to get physically closer to Laura. Incidentally, the other survivors at the library are also completely soaked by the storm, but none of them changes into dry clothes. The scene hence clearly functions as a way to emphasize Sam's physical attractiveness in addition to his intellectual prowess, again suggesting that the scientist characters are adequate action hero material.

44 The audience does, however, learn that Jack frequently neglected his family for his research, which both his wife and his son repeatedly blame him for. The march to New York is thus also cast as a form of ultimate retribution for Jack, who may have let his son down in the past, but will make up for it now by coming for him in this hour of need.

What he certainly is not, however, is a nerdy, pathetic scientist failing in his professional and/or private endeavors.[45]

The emergence of this new, adventurous type of scientist, as I argue, can be attributed to the shifting values of the information age, which also seem to have affected the image and scope of the classic American movie hero. The standard Hollywood blockbuster, including the disaster movie, essentially presents a continuation of the American frontier myth that requires the adequate type of hero. In his exploration of *Spectacular Narratives* (2000), Geoff King emphasizes "the continued saliency of something akin to the myth or ideology of the American 'frontier'" (2) in contemporary Hollywood cinema. In *TDAT* and other recent disaster movies, which must be classified as 'spectacular narratives' *par excellence*, the frontier myth is, however, not juxtaposed with "an articulation of technological or bureaucratic modernity" (King 2) as in many other recent blockbusters, but rather closely entwined with and augmented through it. This entails that constructions of technoscience and, by analogy, modern industrialized civilization are no longer presented as being diametrically opposed to the frontier myth as used to be the case in its traditional conceptualization. Rather, science itself becomes an important frontier with the promise of rich and glorious territories awaiting the brave frontiersman beyond. This trope stands in seeming unison with Vannevar Bush's well-known 1945 report to the government, which understood science as 'the ultimate frontier' that the American nation needed to explore and conquer in order to maintain its leading position in the world.[46] In the disaster situation presented by *TDAT* and related movies, technoscientific knowledge is accordingly put under duress to push the limits and forge into hitherto unknown territories in order to safeguard the survival of the human race. Indeed, many disaster movies feature a solution to the imminent crisis that applies novel and yet untested research which may potentially prove fatal, such as nuclear weaponry or unsafe vaccinations. The disaster thus literally accelerates the expansion into unknown intellectual territories and the closing of one of science's many frontiers, for better or for worse.[47]

---

45 There are, however, also more stereotypical depictions of scientists in *TDAT*, such as Jack's senior colleague Prof. Rapson, who represents the white-bearded, elderly, and physically inept researcher. While the audience learns that he is married and has grandchildren, his private life is, in contrast to Jack's, never depicted in the story. Tellingly, he remains only a minor character who, in contrast to Jack and despite his scientific education, is not able to save himself.

46 See chapter 3 for more background information on Bush's report.

47 In accordance with many other instances of the genre, *TDAT* is indeed saturated by overt attempts to restage the frontier experience, as, for instance, through a series of aerial shots, which are supposed to simulate "the freedom, mobility and space within which the heroes move as they race around the countryside" (34), as King argues.

The appropriate hero for a frontier narrative of the information age would by necessity be a fearless cowboy with a tenure track or, more easily feasible, a scientist with a taste for physical, rather than purely intellectual adventure. Importantly, this does not, as one might assume, include humanities scholars, who have emerged time and again as adventurers in popular culture, including Indiana Jones and, more recently, Dr. Robert Langdon from Dan Brown's mystery novel trilogy. While these characters certainly combine intellectual ingenuity with physical ability, they lack any form of connection to natural sciences and technological gadgetry that proves to be vital in disaster narratives. In contrast to archeologists, linguists, and other scholarly adventurers whose book smartness contributes less to the story's happy ending than do their mere physical capabilities and taste for adventure, *TDAT*'s Prof. Hall is an academic who rises to the challenge with the appropriate technoscientific means. It is not only his bulky frame that makes him the perfect hero for the information age, as the movie's plotline is quick to assert. Rather than transcending his scholarly background, as is the case with Indiana Jones, who has to leave behind the safety of the university walls to test his survival capacities in the real world, Prof. Hall metaphorically takes his laboratory with him when embarking on his heroic foot march to New York. Vital technoscientific equipment, such as GPS tracking, snow samples, and a scientifically-grounded knowledge of projected weather phenomena enabled by his elaborate forecast model, allow Hall and his assistant to evade or, at the very least, deal with the disaster. Technoscience as practiced and personified by Prof. Hall is thus immediately relevant to the resolution of the movie's central conflict—his son's being trapped in the NYC library—and cannot, as in movies featuring humanities scholars like Dr. Langdon and Indiana Jones, be replaced and superseded by mere bodily strength. Rather, Prof. Hall's impressive physique and irresistible sex appeal are portrayed as factors that are equal to his academic training and technological equipment for the solution of the crisis.

This is an underlying message which the movie appears determined to deliver from the very first scene: Rather than pursuing intellectual occupations in his office or a lecture hall (settings that are hardly explored at all in the film), Jack is seen collecting samples in Antarctica and is almost immediately put in a hazardous situation that requires full body action, among others a courageous leap across a quickly widening crack in the ice to save his precious box of ice samples. By bridging the gap between two opposing ends, he is literally cast as a hero "who straddle[s] the border or 'frontier' between the two [poles], offering the best of both worlds, or at least avoiding the worst extremes of either" (6), as King argues with regard to the ideal protagonist of the frontier narrative.[48] The short sequence contains everything

48 Furthermore, Jack clearly portrays traits of the "reluctant hero" (King 11) that is equally characteristic of frontier situations: He only rises to heroic action because his previous attempts of notifying the government (and thus relegating all further action to them) have

the audience needs to know for the further progression of the story: Jack Hall is a noble, athletic man of principles who will risk his life for the things he loves and works for.[49] His physical strength appears almost superhuman, as both his colleagues, one of them markedly brawnier and younger than Hall himself, try to prevent him from jumping because he is "not gonna make it" (*TDAT*) across the crevasse. Hall ignores their advice and prefers to listen to his gut instinct, proving his opponents wrong. Both visibly and psychologically, Hall is thus modeled on the figure of the lone ranger, posing as one man against the rest of the world (in this case, above all, the foolish government), who acts on his convictions and is not intimidated by exterior forces. His character thus overtly aligns with conventional notions of cool comportment, characterized by his "nonchalant, unruffled refusal to play by the man's rules" (Pountain and Robins 28). As a modern frontiersman, Jack unites notions of countercultural, cool rebellion against misguided authorities with the more conservative masculine virtues of courage and physical strength, making him the perfect choice for the job at hand.

The above described scene depicting a disintegrating shelf in the Antarctic ice contains several layers of foreshadowing as concerns the personal development of the story's hero, since Jack will soon engage in equally demanding adventures when trying to save his son who is waiting out the blizzard in New York City. Both the crevasse in the ice and later hazardous situations can essentially be traced back to the same source, i.e. global warming and the cataclysmic climate shift it causes.[50]

---

proven futile. Equally, his march to New York is not motivated by his willingness to prove himself as a heroic adventurer, but necessitated by his son's involuntary stay at the library.

49 Additionally, the opening scene endows the audience with superior narrative knowledge, since they are given a chance to witness, in contrast to the characters, the actual effects of global warning and are thus presented with indisputable visual 'proof' for Jack's hypothesis, a privilege which neither his colleagues nor the government officials enjoy. The latter must depend on abstract graphs and figures to base their consequential decisions on, while the average viewer can easily deduct from this initial sequence what to anticipate in the next hour and a half.

50 Apart from the cracking ice depicting the actual consequences of global warming, scientific knowledge in *TDAT* is primarily made available via cartographic evidence (Seewald 156), such as diagrams, simulations, or satellite images, mostly of the United States. These pictographic representations allow a high degree of simplification, making it clear to the viewer from the beginning how the threat will evolve. One famous scene shows Prof. Hall spontaneously drawing a horizontal line on a map of North America, indicating that everyone above this line will stand little to no chance of surviving the new ice age. This seemingly random line, which does not seem to be backed up by any calculations or coordinates, helps to dramatize the pseudo-scientific background of the movie and pro-

The scene hence proves vital in mediating the psychological motivation of Hall's character, as the viewer will be immediately led to understand why the scientist remains dedicated to and passionate about his hypothesis despite the initially overwhelming number of skeptics: He has already been personally affected by it, and barely survived. Indeed, the only reason Hall gives for embarking on a seemingly suicidal mission in the midst of an approaching super storm are his ethics: "I made my son a promise. I am going to keep it" (*TDAT*). The audience is thus already informed from the outset that his true calling is not to be safely buried in a pile of books at his university desk, but that he is destined to warn the world and take action as a sort of modern martyr figure of science. The mythical origins of the genre emerge conspicuously in this opening scene, which can be read as visualizing Hall's initiation into the world of adventure. His willingness to heroic self-sacrifice, thus already determined in the first few seconds of the movie, will be echoed later when he works manically to create an accurate forecast model to inform evacuation decisions: "Jack, you've been working for 24 hours straight. You're the only one who hasn't taken a break" (*TDAT*), as his worried colleagues try to appeal to him. Both literally and figuratively, Jack will soon exchange his lab coat with arctic gear and become not only intellectually, but physically proactive in the search for his son. He thus seems to translate into literal terms what Ulf Poschardt defines as the underlying functionality of a cool and composed attitude, namely "the attempt to counter the ice ages of our existence with affirmative strategies" (10),[51] which makes Jack the best possible character to deal with the crisis at hand.

In that sense, the contemporary disaster movie can be regarded as resurrecting positive fictional scientist figures from the nineteenth and early twentieth century, as, for instance, exemplified in Jules Verne's oeuvre. Verne's scientists are presented as heroes, saviors, and fearless adventurers, who are "vivacious, humorous, young at heart and optimistic" ("Scientist in Literature" 393) and "embarked on a physical journey, a metaphor for a mental journey which, in turn, will reveal a marvelous future" ("Scientist in Literature" 393), as Roslynn Haynes observes in her exploration of the scientist character in Western literature. Just like Prof. Hall, nineteenth-century scientist figures become empowered through the seemingly infinite possibilities of modern science and remain utterly devoted to the idea of progress, which will ultimately always prove beneficial for the human race. Similar to the

---

vides an effective form of visualization for conveying necessary plot information to the audience. Not coincidentally, the scene has been the subject of several parodies (including the initially described episode from *South Park*). It also stands as a prime instance of the highly simplified and frequently inaccurate science employed in the disaster genre, despite the standard consultation of specialists.

51 Original wording: "der Versuch, den Kältepassagen der Existenz affirmative Strategien entgegenzusetzen."

utopian scenario conjured up in *TDAT*, Verne's novels construct the scientist as the new leader for a new (ice) age, which requires scientific ingenuity to adapt to the fundamentally changed circumstances of human existence. The scientist encounters these new living conditions with the conviction that "bravery, endurance, optimism, and reverence for scientific knowledge would overcome all difficulties" (Haynes, "From Alchemy to Artificial Intelligence" 250). This belief in the seemingly unlimited powers of modern science and technology allows him to face all adversities with "*sang-froid*" (Haynes, *From Faust to Strangelove* 136), the nineteenth-century equivalent to the cool pose: Supported by his utter confidence in scientific knowledge as a symbolic safety net for the most hazardous situations, he may encounter danger unabashedly, composedly, and in 'cool blood.' Given this cool mindset, it is hardly surprising that the scientist characters emerge as "technological knights" (Haynes, *From Faust to Strangelove* 129) prepared to risk their lives for the greater cause. Prof. Hall's construction as the idealized martyr of climate change distinctly mirrors these concerns by virtue of his commitment to the combat against one of the most immediate threats of the twenty-first century.

Selflessness, self-sacrifice, and heroism are, however, not only linked to the male protagonist, but also emerge conspicuously in the portrayal of other scientific characters. While Prof. Rapson is well aware that he will not be able to survive the storm in his research center at the northern tip of Scotland, he faces his fate with dignity and urges Jack to save others, rather than himself. Through Rapson's death, science is effectively passed on from one generation to the next, rejuvenating it in the process. Even more dramatic than this dignified acceptance of fate, however, is the scene in which one of Jack's research assistants who offered to accompany him on his foot march to New York sacrifices himself when he breaks through the glass roof of a snow-covered building by cutting the rope that connects him to the rest of the crew. This selfless stance, which all science-related characters curiously seem to be endowed with and which, conversely, is completely absent in any of the government officials, also emerges conspicuously in Sam and his scholastic decathlon team. The teenagers help others trapped at the library to survive by passing on information, organizing food and medicine, and repairing the radio. Sam's friend Laura saves a family by translating for them into French, while Sam rescues her when she is stuck in a parked taxi in the approaching tidal wave. Rather than the other, apparently disaffected characters, whose professions range from librarian to police officer and homeless person, it is the self-proclaimed nerds who take action and emerge as courageous, selfless, and strong. Sam's brave deeds even manage to impress Laura, so that eventually, science not only saves humanity from extinction, but also helps him to get the girl. In this sense, Jack's son undergoes a similar development to himself: While Sam is initially cast as a typical high school nerd, who even goes to New York to participate in an academic competition, the disaster propels his transformation into the heroic (and scientifically most literate) leader of the survivors at the library. His superior knowledge of the situation, gained in a short phone conversation with his father, endows him with the unusual courage to speak in front of the large crowd of people gathered at the library. The utmost faith in the scientific data he is about to relate to them acts as a boost to his self-confidence and makes him deliver an ardent speech worthy of a true leader, even though the majority of survivors do not heed his advice. It is not only in words, but also deeds that Sam acts heroically, as he later even embarks on a dangerous and physically challenging mission to retrieve antibiotics from a stranded ship near the library for his love interest Laura. For both father and son, the disaster hence serves as a catalyst propelling their transformation into strong leadership figures (and attractive romantic partners) who readily undertake the noble rescue of the weak and needy.

In contrast to the unambiguously positive portrayal of science figures, characters opposed to scientific knowledge are portrayed in a comparatively negative light, which may be understood as a direct reversal of priorities compared to the 1970s and 1990s cycle: Not only will non-scientists be punished for their lack of faith in scientific data by making their survival difficult or indeed impossible, they are also openly belittled for their ignorance when it comes to empirical knowledge. When asked why he aggravated the vice president, after all the second most powerful man in the state, Jack justifies his disrespectful behavior by the allegedly low intellectual capacity of the diplomat: "Because my 17-year old kid knows more science" (*TDAT*), he angrily quips. Science is thus presented as a superior form of knowledge when compared to alternative explanatory systems, such as politics, economics, or the military. This is also symbolized by the split in the group of survivors trapped at the library: While one (major) part of them, guided by a policeman as representative of state authority, decides to leave the improvised shelter and march south for rescue, Sam as the scientific leader of the gathering manages to convince a small minority to stay at the library and wait until the storm has passed, as his father advised him to do. As the movie later depicts, all the members of the group who left the library froze to death in the blizzard, while all those who followed Sam's advice are rescued by his father. Science is thus clearly constructed as the better counselor in times of crisis and an unfailing source of certified information.

The grave consequences afflicting those who fatefully choose to ignore a scientific warning are suggestive of the moral burden carried by the movie's science characters: Both father and son are consistently depicted as struggling with scientifically illiterate people who prove to be a hazard to themselves and others. While Jack has trouble getting his voice heard by the government, Sam is failed by his mathematics teacher, who, unable to believe the boy solved a complex mathematical problem in his head, accuses him of cheating. Both are repeatedly misunderstood by their peers because society, as the storyline implies, lacks basic science education and fails to comprehend their reasoning, which will invariably turn out to be accurate. Nonetheless, the movie is very cautious not to present their superior knowledge as scientific arrogance: In contrast to the geeks of *The Big Bang Theory* (chapter 6), the science characters in *TDAT* more overtly align with the socially desirable traits of coolness, such as popularity, courage, attractiveness, prosocial values, and confidence (Dar-Nimrod 179). These laudable and overtly positive qualities are effortlessly combined with, or indeed reliant on the scientific profession, as Prof. Hall's bold leap across the widening crack in the ice implies: It is only due to his strong dedication to scientific principles and his consequently high research ethics that he is willing to risk his life for the well-being of others, in this case by saving the precious ice cores needed for predicting the disaster. The selfless rescue mission for his son reiterates this heroic stance on the personal level.

In accordance with *TDAT*'s overtly positive representations of the scientist, his key role in the resolution of the central narrative conflict, i.e. the disaster brought about by the cataclysmic climate shift, is repeatedly reaffirmed by the narratological function attributed to the scientific knowledge he harbors. Just like disaster itself, science emerges as a vital narrative element in the contemporary disaster movie, propelling both the main conflict of the story as well as its eventual dénouement. Indeed, the power of science is not just reasserted in the final happy ending generated by technoscientific innovation, but already pervades the narrative from the very beginning as the imminent disaster can only be meaningfully presented and identified as such (even if not prevented) through scientific discourse. This observation concurs with an assessment by Keane, who detects a recent tendency in the disaster genre of "deferring [the] developing disaster until the second half of the film" (120), thus indulging in a relatively slow exposition with the aim of "making the viewer wait for the final spectacular release" (King 165). Indeed, the 'developing disaster' in post-2000 movies depends upon competent and fast scientists to detect its potential oncoming and enlighten the audience early on in the narrative as to why geomagnetic storms/a minor rise in temperature/the disruption of the gulf stream may prove to be dangerous. Accordingly, recent films commonly start with professional or amateur scientists intensely staring into a microscope or a telescope, detecting a potential threat. All of the consequences thus predicted will, as the viewer versed in disaster rhetoric is well aware, soon become dire reality.

In that respect, science is often used to add a momentum of suspense to the plot, as the present example reflects: The central story in *TDAT* revolves around a scientific warning that is fatefully ignored by politicians[52] and inevitably leads to disas-

52 This narrative structure of a scientific warning being ignored emerges as a recurrent theme in Western literature, as Haynes (1994) observes. A famous example is Sir Arthur Conan Doyle's novella "The Poison Belt" (1913), in which Prof. Challenger accurately predicts the end of humanity due to the planet's impending collision with a poisonous ether belt. However, it turns out that the effect of ether on the world's population is only

ter. Indeed, Prof. Hall accurately predicts that "the weather anomalies are gonna continue" and the world is "on the verge of a major climate shift" (*TDAT*), a dreaded scenario which arrives even sooner than expected. Scientific explanations hence function as a viable form of foreshadowing since the audience already knows from the very beginning that the prognosis will prove to be accurate.[53] Other disaster movies only make the information about a coming crisis available to a fraction of the characters and, necessarily, the audience (who, of course, knows what to expect from the mere fact of being familiar with the movie poster). Consequently, the viewer watches the characters in blissful ignorance of the impending end of the world, which adds an element of highly effective and at times amusing dramatic irony. When a family unsuspectingly goes about their everyday routine or a young couple makes plans for the weekend ahead, the audience is well aware of the fact that their petty problems will soon be overshadowed by a much more substantial one. *TDAT* also contains several scenes that work along the lines of this intentionally unequal information distribution between characters and audience: For instance, when the Hedland Centre research team notices that one of the buoys they are surveying via satellite shows an unusual drop in ocean temperature, they attribute this to the rough weather at sea which must have caused a malfunction. The audience, however, well aware of the impending disaster, will read the research assistants' downplaying of the temperature indication as a wrong and ultimately fatal reaction, and will thrill at their utter obliviousness of the actual danger ahead. The typical narrative structure of the post-2000 disaster movie, driven by the unequal distribution of scientific information, hence translates into a "systematic divergence of knowledge between characters and the audience" (Seewald 163),[54] placing the viewer in an enjoyable position of omniscience and foresight. On the extra-diegetic level, science hence adds to the audience's viewing pleasure by enabling them to anticipate events and to enjoy a privileged subject position, experiencing suspense for the main characters' struggle for survival as well as indignation, laughter, or malicious glee for those ignoring the impending catastrophe. A vital part of the dis-

temporary and merely causes a coma, from which everyone awakes. Prof. Challenger not only struggles to have his theory taken seriously before, but also after the catastrophe, as the Londoners have no memory of the time passed during their coma.

53 A similar instance in which science serves to foreshadow future events is when Sam and his classmates visit the Natural History Museum and come across a mammoth which "was found perfectly preserved in the Siberian tundra with food still in its mouth and stomach, indicating that it froze instantly while grazing" (*TDAT*). The same fate, as the viewer may easily infer, will await the characters if they do not heed Prof. Hall's scientific advice. Indeed, several minor characters will be found frozen instantly just like the prehistoric mammoth.

54 Original wording: "systematische Wissensdivergenz zwischen Figuren und Zuschauer."

aster movie's continuing attraction in the new millennium must thus be attributed, perhaps rather unexpectedly, to its capitalization on scientific plotlines for generating a privileged viewing situation.[55]

A similar expository structure in terms of imbalanced information distribution can be detected among the characters: Apart from the omniscient audience, access to scientifically consolidated information about the disaster is only granted to an elite set of characters, typically including the scientist and his family or friends. Science is thus employed as a tool of narrative empowerment, as usually only those with access to it will a) know about the threat in good time to make provisions and b) find a scientifically inspired way of averting or, at the very least, surviving it. In *TDAT*, it is the government via its obstinate refusal to listen to and, later on, to publicize Prof. Hall's research results which is directly to blame for the resulting hierarchy among the characters as regards the amount of information they receive. This clear-cut hierarchy directly translates into the characters' likelihoods of surviving: As a scientist, Jack Hall emerges at the top of the pyramid, while those characters most immediately associated with him or with science in general, such as his colleagues, the NOAA squad, his wife, and several external NASA officials, watch the disaster from afar in relative safety. Similarly, his son Sam is notably better informed than any of his classmates due to his father's profession and will become an important source of information for the other survivors at the library, as he represents their only gateway to scientifically consolidated knowledge and, by analogy, their only hope for survival.[56] Accordingly, the "informative advantage of the view-

55 The highly appreciative stance toward science is also echoed, it appears, in the overall narrative orientation of the disaster movie post-2000: While the 1970s witnessed the emergence of character-focused dramas exploring the personal and social consequences of disaster and the 1990s saw the establishment of special-effects-driven blockbusters with a not too serious, tongue-in-cheek undercurrent, the most recent wave seems to aim for much more solemn realism. Not only are the scenarios threatening human existence now much more plausible than in the 1990s (global warming and geomagnetic storms vs. alien invasions and super volcanoes), they also tend to explore human responsibility for these disasters much more critically. The changed narrative orientation and the overall more serious tone can, I would argue, certainly be linked to the novel role that science comes to play in post-2000 disaster movies, which will almost invariably offer a final reassertion of technoscientific powers for saving or, at the very least, restoring the world.

56 Interestingly, Sam's love interest Laura initially offers an even better advice than he himself, suggesting that they "should just stay" (*TDAT*) at J.D.'s place, their new friend in New York (Austin Nichols). As J.D. lives on the sixth floor in an Upper Eastside apartment, they would not only gain cozier lodgings and more provisions, but would also be better protected from the tidal wave and massive flooding than in the much lower building of the public library. Since it is not Laura but Sam who must emerge as the story's in-

er is directly and intra-diegetically mirrored" (Seewald 157)[57] by *TDAT*'s character constellations. Indeed, scientific knowledge not only saves the day when it comes to the big catastrophe and its aftermath, but also administers help in the small details: Fortunately for the survivors at the library, they are trapped with a bunch of self-proclaimed science nerds chairing in "the Electronics Club, the Math Club and the Chess Club" (*TDAT*). Consequently, Sam comes up with the idea of using an old payphone that is still in power to call his father to rescue, while his classmate Brian repairs an old radio to receive the latest updates on evacuation plans and weather forecast. Additionally, a medicine book that has fortunately not yet been burned helps to determine what causes Laura's fever. Their privileged access to codified, scientifically consolidated knowledge, whether by virtue of their academic education or the books found at the library, directly increases their chances of survival, which gives them every reason to remain cool in the face of imminent danger.

WHEN THE PLANET COOLS DOWN: SUBLIME DISASTER AND TECHNOSCIENTIFIC SALVATION

This Year, A Sweater Won't Do.
TAGLINE FOR *THE DAY AFTER TOMORROW*

In accordance with the overtly positive image of science perpetuated by the genre, the classic disaster movie acutely depends on an adverse image of nature, continually stressing its cruelty and otherness to human civilization and technological progress. Science and nature are hence tacitly defined as clear adversaries. Ecocritical studies of disaster movies, such as Greg Garrad (2004), Cynthia Belmont (2007), and Nicole Schröder (2010), have shown that the majority of films "uphold 'nature' as the threatening 'Other' that has to be kept under control" (N. Schröder 290). Since disaster movies, as established above, capitalize on the enemy being unspecified and obscure, the realm of nature offers a perfect canvas to which to ascribe antagonistic forces. In order to fulfill this vital narrative function, nature is frequently personified and endowed with emotions and desires, as Belmont observes: "[D]isaster films often envision nature as female, as a mother—in this case, not a

tuitive hero, however, her reasonable advice is ignored by the group and will never again be thematized in the movie. In the same scene, Sam's order to take the stairs rather than the elevator because of the failing electricity is portrayed as much more consequential and valuable.

57 Original wording: "finden die Wissensvorsprünge des Zuschauers hier klar erkennbar eine innerdiegetische Spiegelung."

loving one who deserves our love in return, but rather an alienated adversary in a battle of good and evil" (358). The seemingly determined, adverse attitude of nature as a moody woman consequently justifies the use of heavy weaponry, above all nuclear bombs, against or at the very least despite nature's fragility and need of protection.[58] Therefore, the weather or natural phenomena such as earthquakes and volcanic eruptions function, however indefinite and diffuse they may be, as the immediate antagonist of the movie, providing a viable form of opposition for the protagonist and his allies to battle against and conquer.

This adverse image is frequently combined with a sublime and awe-inspiring view of nature, which both contrasts with, yet ultimately highlights the destructive and vicious forces of the earth. In *TDAT*, long-angle aerial shots, almost exclusively computer-generated, dominate the camera work, an effect which is enhanced by the chant-like, menacing signature tune sounded off every time the eerie beauty and destructive potential of nature are depicted.[59] These dynamics become particularly prominent as the diffuse threat of 'global warming' is manifested through myriad overwhelming natural phenomena, including blizzards, tornados, tidal waves, hail, and a dangerous drop in temperature. The fact that the catastrophe consequently does not appear as "a unified scenario" (Seewald 156),[60] but continuously changes its shape and form makes the disaster all the more threatening, insurmountable, and unpredictable, both for the characters and the audience. The global crisis is excessively re-produced and re-imagined, with every new image even more breath-taking and petrifying than the previous one.[61] Consequently, an easily discernible choreography emerges, as the continuous amplification of the visions of disaster and destruction can certainly be understood as working toward an inevitable climax,

58 Not coincidentally, thus, the application of phallic weaponry against nature, such as nuclear missiles or ice drillers, tends to be visualized along the lines of a symbolic sexual penetration.

59 The same tune is heard during scenes of heavy foreshadowing, mostly abetted by scientific data, such as when the majority of the people who sought shelter at the library decide to leave and march south, a clearly doomed endeavor. The voice of reason, as these scenes and the eerie music suggest, is fatefully ignored.

60 Original wording: "ein einheitliches Szenario." Most disaster movies do capitalize on one specific threat, such as lava (*Volcano*) or wind (*Twister*). The combination of several natural catastrophes in *TDAT* can, however, not only be explained in terms of ideology, but must also be linked to the fact that contemporary CGI technology allows for a much easier production of these effects than merely a decade ago.

61 As Seewald also shows, this is augmented by the fact that images of the blizzard or the tidal wave are usually presented from more diverse perspectives than any other subject (161-162). Additionally, the disaster scenes will be accompanied by a faster pace and more abrupt editing.

which could be identified as the moment in which the wall of cold air, which instantly freezes everything it touches, reaches New York and thus Sam's hiding place.

TDAT's spectacular images hence resonate particularly well with Sontag's 'aesthetics of destruction,' especially when they do not occur on their own, but are effectively supplemented and contrasted with a form of technological sublime, such as huge helicopters, airplanes, dams, or skyscrapers. While such scenes clearly serve to demonstrate how technology and man-made civilization may easily equal or even indeed excel nature's grandeur (N. Schröder 299), they can also, on the other hand, offer a crucial complement to and enhancement of the natural sublime: The awe-inspiring forces of nature appear even more breath-taking and terrifying when they seemingly effortlessly bury modern skyscrapers and ocean liners underneath them. In *TDAT*, the clash of the natural with the technological sublime emerges as a recurring motif, which is already introduced in the first few seconds of the movie and will be echoed in the final closing sequence: The long, completely computer-generated[62] aerial shot of the opening credits, depicting a seemingly endless plane of Antarctic ice with an impressive sunrise and cloud formations in the background, is followed by an abrupt cut to a heavy ice driller in close-up, which is about to be mounted by Prof. Hall's expedition crew.[63] Both shots are impressive on their own, but their effect is clearly augmented by the immediate juxtaposition. For a similar reason, the movie later even stages the highly unlikely stranding of a Russian cargo ship in the water-flooded streets of Manhattan. This bizarre set-up exemplarily reflects how in the visually most spectacular scenes of *TDAT*, the natural sublime will repeatedly collide with the technological, resulting in a doubling of the original effect. The natural and the technological elements augment and mutually inform each other, as the one could not be effectively staged without the other.[64]

Whether natural or technological sublime, however, both poles originate from the same source, namely computer-based postproduction, and are thus essentially founded on technoscientific innovation. The actual 'disaster' presented in the clas-

62 According to Emmerich, the opening credits scene alone took six months to create (Emmerich 2004).

63 In this shot, the view of the breathtakingly beautiful ice also has the function of depicting the state prior to the catastrophe, i.e. idyllic and unthreatening nature. The aftermath emerges as all the more horrendous when compared to this pastoral bliss.

64 In correspondence with the information age, a sublime and almost magical view of technology also emerges in the form of small ingenious gadgetry, instant data processing, or highly accurate forecast models that are seemingly made from scratch with barely any data or calculations needed. While the scenes depicting these operations are far from visually overwhelming, they nonetheless present a highly impressive display of technology's seemingly unlimited abilities and leave the viewer in awe of its magic powers.

sic disaster movie is, according to this argumentation, a visualization (if not indeed personification, considering the way it is endowed with apparent emotions and will) of technoscientific ability. This is also pointed out by Kakoudaki, who argues that "'seeing' computer space and computer power in some concrete form should [...] be considered one of the main unconscious desires of the disaster narrative" (143). Through the engagement of special effects for the production of sublime aesthetics, the technologized image is naturalized, as it is no longer 'outside' of or adverse to nature, but indeed appears to reenact nature, whether as computer-generated tidal wave, tornado, or hail. This perspective aligns with David Nye's notion of the *American Technological Sublime* (1994), to which Kakoudaki refers. According to Nye, both technology and the sublime serve as unifying elements in the American popular imagination, welding society together. Since "[t]he sublime underlies th[e] enthusiasm for technology" (Nye xiii) and vice versa, the audience's appreciation of the spectacle of disaster in the Sontagian sense can also be understood as an appreciation of the technology making it happen. In the case of disaster movies, this assessment seems to be particularly accurate, as a large amount of the story is indeed produced on the computer screen in meticulous post-production and editing. It is hence not only at the level of narrative that technoscience will literally save the world and restore human community, dignity, and integrity, but also at the level of production, as the viewer is constantly confronted with and immersed in digitally produced images generating a sublime effect and hence promising, in line with Nye's argument, a new moment of (American) community building through radical emotional experience.

The disaster movie's capitalization on the sublime as "the strongest emotion which the mind is capable of feeling" (36), as Edmund Burke phrases it in the Romantic era,[65] must not, however, be equated with outbreaks of passion or an agitated exterior. On the affective level, the disaster genre clearly values rationality over panic, which reiterates the dichotomy between untamed, wild nature and detached, reasonable science.[66] Seewald connects this to the fact that disaster movies eschew facial close-ups in favor of multiple views of the natural catastrophe. In lieu of the classic framing experience, she hence detects "a shift in the aesthetic function of the

65 See Burke's *Philosophical Enquiry into the Origin of our Ideas of the Sublime and Beautiful* (1756), the first major European treatise on aesthetic theory and, to this date, the central work in the Anglophone world.

66 This presents no conflict with the actual properties of the sublime, as Edmund Burke explicates in his treatise: "Thus are two ideas as opposite as can be imagined reconciled in the extremes of both; and both in spite of their opposite nature brought to concur in producing the sublime" (74). In this sense, the sublime is an essentially ambiguous quality, fundamentally based on the negotiation of polarities, particularly that of pleasure and pain. See also Adam Phillips's 1990 introduction to Burke's *Enquiry*.

facial close-up, the emergence of an experience of empathy that is not strictly grounded in a particular character" (Seewald 178).[67] Close-ups do occur in the disaster genre, but instead of tightly framing characters, they are employed to provide minute views of the natural phenomenon at stake, such as details of an ocean wave or cars frozen by the blizzard in *TDAT*. In contrast, facial close-ups depicting emotional states and hence inducing the viewer to form a sentimental relationship with the character are perceptibly rare. In that sense, affect is clearly located in nature and, by analogy, the CGI imagery of the disaster genre, rather than the actors' facial expressions. It is, in other words, the technologically produced, yet natural disaster that moves the audience, rather than those affected by it. The absence of facially motivated close-ups hence fails to provoke what Seewald terms a "visual deficit" (164),[68] since images of nature substitute the human face as the most immediate screen for displays of emotionality. Even though affect may hence not necessarily be associated with individual characters, it remains a key element for the disaster movie's continuing popularity. It is in the breathtaking images of destructive natural forces that the emotional level emerges most clearly; accordingly, the movie's close-ups capitalize on depictions of the catastrophe to closely and immediately capture its sublime scenery for audience involvement. The indispensable affective impact of the genre is hence inextricably linked to its scenarios of mass destruction visualized through CGI imagery.

In the most basic sense, thus, technologically generated images of the disaster effectively replace the human actor as the central agent of the story. It is the natural forces of destruction, such as tidal waves and blizzards, which drive the story. This is also observed by Sabine Nessel, who similarly argues that the actual star of the disaster movie since the 1990s is the technology behind it:

"It is not without pride that [the cinema of the 1990s] presents a new generation of images that are not produced cinematographically, but calculated and designed on a computer: the visual effects. The stars of *Twister*, *Volcano*, *Deep Impact*—the disaster movies of the 90s—are natural phenomena without a stable semantic sex." (1)[69]

67 Original wording: "ästhetischen Funktionsverschiebung der Großaufnahme des Gesicht, die Emergenz eines nicht streng figürlich fundierten Erlebens von Empathie."

68 Original wording: "defizitäre Bildlichkeit."

69 Original wording: "Nicht ohne Stolz präsentiert es aber eine neue Generation von Bildern, die nicht mehr filmisch hergestellt werden, sondern per Computer errechnete und designte Effects sind, die Visual Effects. Die Stars aus *Twister*, *Volcano*, *Deep Impact*—den Katastrophenfilmen der 90er Jahre—sind Naturphänomene ohne stabiles semantisches Geschlecht."

Although Nessel's final verdict would have to be revised with regard to the clear sexualization and feminization of natural forces as shown by Belmont (2007), one may safely add *TDAT* to the list of movies noticeably foregrounding the CGI image as the driving force of the movie. King supports this premise, as he convincingly argues that the disaster genre "exists up to a point precisely for the purpose of spectacular display" (161). This corresponds to the audience's desire when watching the disaster on screen: The viewing pleasure is directed at (and, at the reception level, made to derive from) the underlying technology offering sublime and cool images of the catastrophe, rather than the human performance accompanying it. Indeed, the actors always only act *in reaction to* the disaster, rather than independently and self-determinately: "Everything that happens only happens in reaction to the expected arrival of the comet" (188),[70] as Harald Fricke succinctly sums up the plotline of *Armageddon*. All of the actions taken by the characters are directly or implicitly initiated through the possibility of crisis, with the disaster as the quite literal hub of the narrative world.

Directing affective involvement and narrative agency through artificially produced images of natural phenomena, however, entails that the human actors, on the other hand, seem to remain rather unimpressed by their surroundings: Given the paucity of facial close-ups, excessive emotional involvement is discouraged on the level of character. This does not entail, however, that emotionality remains utterly absent in the movie: While the protagonist and his allies are not affect-free and can, in most cases, be assumed to move in the usual emotional spectrum, this is, due to the lack of close-ups, not as closely and immediately conveyed to the audience as in other genres, such as melodrama or soap operas. Indeed, not even in the most overtly sentimental scenes of the movie, such as the climactic kiss scene between Sam and Laura or Jack's eventual reunion with his son, will the viewer be presented with actual close-ups; rather, they will have to make do with medium shots of the face. As a result, the characters are, despite dramatic events and storylines, perceived as cool, laid-back, and emotionally reserved. Having to fight ice and cold from the very beginning, they seem to serve as illustrations of Poschardt's argument "that a cool way of life is required by coldness both in symbolic and literal terms. 'Coolness' enables humans to live with coldness instead of freezing in it" (11).[71] In *TDAT*, the absence or presence of a cool stance in the face of the dawning ice age quite literally decides the question of life or death: Only the cool survive, as its storyline repeatedly reaffirms.

---

70 Original wording: "Alles was geschieht, geschieht nur als Reaktion auf den zu erwartenden Kometeneinschlag."

71 Original wording: "daß eine 'coole' Lebenspraxis herausgefordert wird durch die Kälte im symbolischen wie im realen Sinn. 'Coolness' ermöglicht den Menschen mit der Kälte zu leben, statt in ihr zu erfrieren."

The coolest of all potential survivors is, of course, the protagonist, who as the world's leading paleoclimatologist has every reason to stay relaxed: His immediate access to both scientifically consolidated information about the disaster and the technoscientific means to avert it (or cope with its aftermath) provide him with the best possible conditions to survive. Hence, he displays *par excellence* the list of attributes with which Lewis MacAdams's *The Birth of Cool* (2001) characterizes the early paragons and progenitors of cool, including "intellectual clarity, [...] physical bravery, and [...] studied indifference to the vagaries of fate" (97). Keeping a calm and undisturbed exterior in the face of imminent danger is central to contemporary popular notions of cool, and indeed, Jack remains calm and composed throughout the crisis despite the severity of the situation. When the expedition's truck crashes into a snow-covered train on their way to New York, making it necessary to continue the march on foot, Jack seems unimpressed by the accident, demands his arctic gear, and nonchalantly quips: "I've walked that far before in the snow" (*TDAT*), a fairly confident assertion in the midst of one of the heaviest blizzards the world has ever seen. Supported by his faith in the abilities of modern science and technology, he manages to look completely detached, laid-back, and in utmost control when facing what might, after all, turn out to be the end of the world. Indeed, as Bert Yergensen argues with regard to *The Core* (2003) and *TDAT*, "scientists become, because of their technology, immune to the Earth's fury" (162), which accounts for their otherwise inexplicable composure during the global disaster. Prof. Hall's coolness is thus acutely dependent on the advantage of knowledge allowed by his scientific profession: He can only remain relaxed in this time of crisis because his immediate access to scientific methodology and data allows him to learn instantly which actions to take at which point of time in order to achieve the best results, be it the rescue of survivors or the calculation of an accurate forecast model. Consequently, the fact that he manages to maintain "poise under pressure and the ability to maintain detachment, even during tense encounters" (2), as Richard Majors and Janet Mancini Billson define the underlying functionality of cool, is directly linked to his status as the world's leading paleoclimatologist, who is necessarily best equipped to deal with the situation at hand because of his intimate knowledge of it, backed up by years of research and indisputable scientific data. The viewer is hence again faced with a character who emerges as cool because of, rather than despite his association with standard Western science.

The seemingly indestructible cool stance of the modern disaster movie hero is even further enhanced by juxtaposing him with scenes of panic and the frenzied masses. Additionally, he will frequently be given a sidekick who represents the more willful, emotional part of the duo and contrasts with the protagonist's thoughtfulness and calm. In *TDAT*, this role is played by Jack's brawny but clumsy assistant Jason, whose life he saves during their march to New York. Although Jason thus ultimately survives the blizzard because of Jack's help, his figure still exemplarily reflects how irrational, impulsive characters tend to be much more at risk of perishing in the disaster than the scientist himself. The effective narrative juxtaposition of two emotionally antithetic characters such as Prof. Hall and his assistant suggests that "humanistic emotion and expression are helpless in relation to the pursuit of answers to the destiny of mankind. Such answers are apparently found only in the zealous pursuit and employment of technology, of course bridled [sic] by the trustworthy perspective of scientific rationalism" (Yergensen 164). Eventually, the distinction between cool rationality and dangerous impulsiveness results in a clear hierarchy between the characters, with the scientist at the very top, which is reflected by Hall's seemingly natural usurpation of the role of the leader, formerly reserved for political, rather than intellectual power.

Through his rational and scientifically informed attitude toward the disaster, the scientist emerges as the only logical hero for the post-2000 disaster movie that is fundamentally based on technological advances: As a professionally detached and emotionally subdued protagonist, he personifies the computerized rationality and efficiency required for producing the tumbling skyscrapers and overpowering water masses surrounding him. He thus offers an ideal space of identification for the viewer, who, according to Nye's notion of the 'technological sublime,' will appreciate the latter for its promise of community building through technologized encounters. The elevated and overtly heroic figure of the scientist hence comes to embody the association of science with the absence of extreme emotionality (including panic and fear), which is constructed as the direct counterpart to and worst enemy of the impending disaster, as several notably more agitated characters reflect; according to the narrative logic, science is the only element that can keep the disaster in check and possibly counteract it. The audience is thus left with a culturally pervasive dichotomy, equaling nature with heightened affect and science with detached rationality and the (laudable) absence of emotions. More generally speaking, this corresponds to the underlying binary saturating the majority of popular cultural productions: the unambiguous opposition between nature and science/technology. In the general cinematographic orientation of the disaster genre, this universal dualism is visualized through a natural disaster offering affective involvement and, on the opposite end of the spectrum, emotionally restrained scientist characters counteracting the devastating effects of the disaster with cool rationality.

Hence, the movie can be understood as feeding on the oscillation between two extremes: One the one hand, affective involvement is obstructed by limiting the amount of facial close-ups, while, on the other hand, this cinematographic strategy simultaneously functions as a way of rousing the viewer's desire for exactly this form of involvement. As Seewald observes, facial shots in *TDAT* tend to stop short right before they would turn into close-ups and capture the emotional state of the character (165). The audience is hence visually stimulated to anticipate affect, the satisfaction of which will be categorically denied. It is only in the final sublime close-ups of the disaster that they will be granted an eventual appeasement of the yearning for affective involvement. What hence emerges is an ambiguous form of emotional spectacle as "the viewer undergoes an oscillating interplay between the infinite deferment of a projected fulfillment and the currently experienced overwhelming of a surpassed expectation" (Seewald 165).[72] Consequently, a paradoxical situation emerges, switching between the stimulation of the audience's desire for the display of affect via facial close-ups, the immediate frustration of this desire by disrupting the zoom on a character's face shortly before a close-up, and a final satisfaction through granting sublime visions of the catastrophe in tight framing.

This strategy of juxtaposition characterizing the displays of affect in *TDAT* is mirrored and simultaneously augmented by the aesthetic composition of its scenes. The movie's mise-en-scène will commonly be composed of opposing or clashing images, such as the neon yellow tent used by Prof. Hall's expedition team in the middle of the snowy-white Antarctic shelf (Seewald 159). Another recurrent composition involves the use of single light sources which feebly illuminate an otherwise pitch-dark surrounding, such as the light emanating from Prof. Hall's laptop in his office or the frequent use of flashlights in the blackout caused by the storm. The above described sequences displaying the genre's vital clash between natural and technological sublime, such as the contrast between serene Antarctica and noisy ice drillers, work along similar lines. As the previous chapters of this study have already shown, aesthetic juxtaposition is a vital prerequisite for the production of cool visuality in the broadest sense. As Jeff Rice asserts in the context of electronic representations, cool emerges "as a juxtapositional practice" ("Writing" 224) as well as a creational style whose function it is "to connect ideas and images, [...] motivated by juxtaposition" ("Writing" 226). In *TDAT* and other contemporary disaster scenarios, combination and contrast can certainly be classified as recurrent compositional strategies for emphasizing the above explored clash between natural and technological sublime. The basic ecological theme of the movie, i.e. humanity's responsibility for global warming, is repeatedly played out by colliding images of idyllic, serene nature with symbols of decadent Western civilization, such as overcharged military helicopters getting caught in a massive cloud formation that makes them look like disorientated flies. The same function is essentially fulfilled by the repeated (computer-generated) shots from outer space, depicting the outlandish beauty of the blue marble, and, depending on the point in the narrative, its disturbance by weather anomalies that obstruct the sublime view for the astronauts. The resulting effect is, in accordance with the rather inconsistent conciliatory tone at the

72. Original wording: "das den Zuschauer über die Dauer des Films dem oszillierenden Zwischenspiel von infinitem Aufschub einer projizierten Erfüllung und der gegenwärtigen Überwältigung einer überstiegenen Erwartung aussetzt."

end, a lasting impression of ambiguity, although this is clearly not aimed for by the movie's overt text.

This evident ambiguity informing the movie's aesthetic composition is echoed on its narrative level, as the disaster genre is, in fact, based on a number of unresolved contradictions. As the preceding sections have shown, the central paradox upon which most disaster movies are based (but which they seldom thematize) is the fact that technological progress, in the widest sense, must be held responsible for the outbreak of the crisis, but simultaneously helps to avert or mitigate it. In the case of *TDAT*, this is apparent in the destruction of the Earth's natural climate through greenhouse gases and pollution, which are an immediate result of industrial progress. Both cause and solution of the disaster are thus closely entwined, a circumstance which is not usually commented on explicitly as it would detract from the optimism and dramatic power of the ending. In accordance with the resolution typically employed by the genre, it is the figure of the hero who offers a reconciliation of extremes that makes these inherent paradoxes acceptable (and indeed hardly noticeable) for the audience (King 80). The scientist of the post-2000 cycle is an especially apt example of a disaster movie hero bridging extremes, as he effortlessly combines intellectual prowess with physical strength, common sense and practical know-how. Prof. Hall's courageous leap across the gulf opening in the Antarctic shelf symbolizes this ability: "[T]he movie's first sequence, the long-drawn depiction of a linear crack in the Antarctic ice, [can] be read as a metaphor for the divergence of two standpoints" (158),[73] as Seewald contests. Hall's crossing back and forth between these two is emblematic of the way in which his character consistently and adeptly moves between (extreme) positions, thereby blurring the boundaries between them. His strong heroic ethos as the well-informed, single-handed rescuer of the story, which his bold leap perfectly epitomizes, thus prevents the viewer from looking too closely at any obvious narrative contradictions.

Nonetheless, these contradictions exist and remain largely unresolved, as Hall in fact defies his own orders: While the audience learns early on that staying inside and riding it out is the only way to survive the storm, he starts the rescue mission for his son right at the storm's peak, just after having briefed the government that the extremely cold air generated by the storm's eye that forms right above New York City will instantly kill everybody who comes into contact with it. Equally, the self-proclaimed ecological mission informing *TDAT*'S storyline remains ambiguous, as the actual human failure to *prevent* climate change is presented as both an inevitable and ultimately positive development, celebrated as the only practicable way for humanity to survive in a corrupted world that needs to be purged. Almost

73 Original wording: "[…] die erste Sequenz des Films, die lange Darstellung eines linearen Bruchs im Eis der Antarktis, [kann] als Metaphorisierung der Divergenz zweier Standpunkte gelesen werden."

from the very beginning, climate preservation and protection are ruled out as possibly helpful measures in dealing with the change. After all, the disaster will turn out to be less fatal for the main characters than expected. In accordance with the cheerful closing scene of the movie, the rather more urgent question of survival *after* the storm is downplayed in one single, laconic comment by Prof. Hall, stating that "mankind survived the last ice age. We are certainly capable of surviving this one" (*TDAT*). How exactly this survival in a world deprived of more than half its resources, fresh water supplies, and crucial technology will be organized would be an interesting question, but is not pursued any further. The future that the incoherently optimistic tone of the movie seems to project appears vague and unrealistic, and the audience is further distracted by the happy personal ending of its central characters: The reunion of Prof. Hall, his wife, and their lost son at the end of the movie delivers "emotional satisfaction that distracts from the existence of other, unresolved tensions" (King 25), in this case the high number of casualties claimed by the disaster and the question of future human existence on a planet covered almost entirely in ice and snow.

Consequently, one may detect an "ambiguous sensibility" (43) underneath *TDAT*'s sleek Hollywood surface that noticeably aligns with the contradictory stance that "lies beneath Cool" (43), according to Dick Pountain's and David Robins's analysis of blues culture. This cool ambiguity also emerges conspicuously in moments of the disaster, which, given the icy drop in temperature it involves, stages a quite literal performance of coolness. *TDAT* enhances this 'cool look' of large-scale destruction by endowing the completely computer-generated shots of the blizzard and other natural phenomena with a bluish filter, thereby further augmenting the chilly impression. Appropriately, such CGI shots indeed figuratively 'freeze' the narrative, as they usually disrupt the linear story for displays of sublime scenery that contribute little to nothing to the narrative progression. Catastrophic images hence create narrative space in the otherwise fast advancing plotline of the Hollywood disaster movie to parade the spectacular results of technoscientific production.[74] Similar to how depictions of lab work in *CSI* (chapter 5) foreground the visual dimension by their utter absence of plot-related information or verbal communication, the moviegoers are given a chance to revel in the impressive display of computer-graphic virtuosity without having their attention diverted by complex dramatic developments. Indeed, while the depicted catastrophic event is evidently

74 As its long generic tradition reflects, the disaster movie could also exist without any display of the crisis per se, which was the case throughout the 1970s cycle due to a lack of appropriate technology (Nessel 3-4). Actual depictions of the disaster hence neither constitute a vital prerequisite for the logical unfolding of the storyline nor the commercial success of the production. They do, however, provide vital narrative space to exhibit the high level of digital production and aestheticization.

highly destructive, it usually cannot be determined, from the CGI shots alone, where and what damage has been done. Rather than being immediately presented with images of death or survival, the audience's attention is diverted by close-ups of the disaster, zooming in onto the eye of the storm or magnifying details of the approaching waves. This tight framing precludes any exterior view from the outset.

Paralleling the ambiguous epistemological status of the lab scene in *CSI*, the conspicuous lack of narrative contents that would otherwise deflect from the spectacular visuality of the disaster scenario helps to underscore its aesthetic dimension. While the projected audience can thus safely revel in the graphic abundance of the digital environments on display, the scenery has the welcome side effect of building (or rather augmenting) the movie's crucial tension between the known and the unknown: The "constellation of uncertainty" (Martig, "Weltuntergang" 52),[75] i.e. the question of who survives and who perishes in the disaster, serves as a key narrative device in the disaster genre and plays a major role in its continuing attraction, as it contributes vitally to the viewers' interest in the story unfolding on the screen before them. Again, this evokes the notion of unknowability, which has been defined as central for a cool visual effect. As Alan Liu asserts, cool is an essentially "paradoxical ethos" (72) when it comes to the transmission and/or obstruction of information in the digital age: While being cool relies on a basic fluency in relevant societal codes and demeanors, the intimate knowledge it draws from will never be revealed. The lack of information in the standard disaster shot hence provides ideal conditions for coolness to emerge, which is, in this case, dependent on science both intra- and extra-diegetically: Science not only contributes to the cool CGI shots by predicting and directing the disaster from the outset, but also by providing the underlying technological requirements that fuel the digital environments on display.

CONCLUSION, OR: SCIENTIA EX MACHINA

As I write the conclusion to this chapter, 'Superstorm Sandy' is slowly subduing over the North American East Coast, having caused widespread damage in the United States and beyond. As the largest Atlantic hurricane on record, Sandy brought major flooding, heavy rain, tidal waves, and gale-force winds, inducing the U.S. government to issue a series of evacuation orders, especially for the area around New York City, one of the cities most severely afflicted by the storm. The parallels to the movie analyzed in this chapter are more than evident: Just like the super storms depicted in *TDAT*, the formation of the hurricane was monitored by NOAA, the governmental organization Prof. Hall works for, and swiftly attributed to global warming and an imminent climate shift. Sensationalist media coverage

75 Original wording: "Konstellation der Ungewißheit."

promptly circulated several alleged images of the degree of destruction caused by Sandy which, as it turned out, were directly taken from Emmerich's CGI production, such as the famous shot depicting the Statue of Liberty withstanding a massive storm surge (see Madrigal 2012). In an uncanny twist of fate, life seemed to imitate art for the worse.

While the huge media echo all around the globe in response to the storm attests to the continuing attraction of and fascination with (both real-life and fictional) images of disaster and destruction, the actual situation during and after Sandy's onslaught reflected one major difference to its cinematic sibling: Science did not figure as a major factor in either the prediction or the prevention of the crisis. While modern meteorology provided highly elaborate forecast models to outline the path of the storm as accurately as possible, the contribution of cutting-edge technoscientific innovation was hardly ever thematized in the media coverage of the disaster, which must also be attributed to the fact that climatologists did not reach an agreement over the actual relation of the hurricane to effects of global warming. Hence, the extensive glorification and veneration of technoscience emerges as a distinct element of *cinematic representations* of disaster, especially so in their contemporary variety, as the preceding analysis of *TDAT* has shown: While science may not, after all, succeed in averting the disaster, it contributes vitally to the survival of the main characters and will, as the movie's buoyant ending implies, continue to administer help in dealing with its aftermath in a new (ice) age. Despite the movie's ambivalent (and ultimately unresolved) stance toward industrial growth as the implicit cause of disaster, the salvation of the human race is portrayed as being acutely dependent on technoscientific progress.

These concerns are chiefly negotiated, as the preceding pages have shown, through the heightened narrative status of the scientist, who progressed from comic or sinister sidekick to single-handed hero of the story, combining intellectual competence with physical strength. My analysis of Prof. Hall as a major representative of this type thus suggests that the construction of the modern American movie hero as outlined by King may have undergone serious revision:

> "Underlying the confrontations enacted in these [1990s blockbuster] films is an opposition between two of the most powerful components of dominant American ideology: the myth of the frontier and alternative myths of technological modernity, according to which America is defined as a model of modernizing progress, whether industrial or post-industrial. The two mythologies are in some respects mutually exclusive." (22)

Rather than being 'mutually exclusive,' these myths appear to have seamlessly joined forces in the post-2000 disaster movie as exemplified by *TDAT* and become all the more effective through their intertwining, which seems consistent with the changed premises of the modern American information society. There is no longer any contradiction to be perceived between physical and intellectual heroism, a dichotomy which the brawny, yet non-scientific protagonists of earlier productions cycles still reflected. It is now science that comes to stand as the ultimate frontier to be conquered and overcome, a task which the handsome, athletic and adventurous scientist is ready to undertake. It hence appears sensible to conceptualize the current surge of post-2000 disaster movies as a discrete cycle, rather than the continuation of the millennial variety, which is hinged on a notably different image of technoscientific ability.

The representations of science with which *TDAT* is saturated, however, not only reflect its conspicuous digression from earlier instances of the genre, but also reveal changed priorities when compared with the other cultural productions analyzed for this study. In contrast to the images perpetuated by *CSI* and *The Big Bang Theory* respectively, the cognitive value of science is prominently foregrounded in *TDAT* and emerges as a vital factor in making science and its allies appear cool. It is only by virtue of the knowledge of the disaster produced by technoscientific innovation (including Hall's prehistoric climate shift model and valuable data he collected through the years) that the characters, above all Hall himself, are endowed with a higher chance of survival and may hence face the crisis in seemingly cool blood. On the formal level, this is augmented by the movie's restricted use of facial close-ups and the high degree of digitally produced special effects, which replace the human agent as the site of affective audience involvement and equally depend on a high degree of technoscientific innovation. Conversely, the aesthetic dimension of science, which emerges conspicuously in *CSI*'s long-drawn displays of sleek lab equipment, is notably downplayed for the sake of placing emphasis on its epistemic value for preserving Western civilization in the imminent disaster. The second main narrative strand at the library perceptively crystallizes these concerns, suggesting that codified Western knowledge will come to constitute a safe haven for afflicted refugees seeking help in science's almighty powers. Eventually, this help appears in the shape of Prof. Hall, descending upon the troubled masses like an angel sent by divine intervention. In the meantime, the books they are surrounded with have strengthened the disparate group to endure in a suddenly barren world by nurturing both body and mind. They can hence await their imminent rescue with perfect equanimity, demonstrating that a cool attitude is imperative to avoid freezing in the world's symbolic and actual ice ages.

From Nerds to iCons

Consumer Cool and the Rise of the Scientist-Entrepreneur in *The Social Network* and *Jobs*

> To be cool [:] to be in control of a situation because you have swung where the Square has not.
> NORMAN MAILER/"THE WHITE NEGRO"

This chapter is dedicated to recent cinematic representations of two of the most publicly venerated personalities of the information age. One of them was named 'Person of the Year' in 2010 by *Time Magazine*, four years after 'you,' the individual content creator on the Internet, and almost three decades after the computer had been announced 'Machine of the Year.' The other one was graced with the titles 'Entrepreneur of the Decade' and 'CEO of the Decade' and placed second in a poll that asked young Americans to identify the greatest innovator of all time, surpassing Alexander Bell and Marie Curie and second only to Thomas Edison.[1] The people alluded to are, of course, Facebook founder and chief executive Mark Zuckerberg and late Apple co-founder and CEO Steve Jobs. Both of them have recently been honored with their own biopics, which reveal striking parallels in terms of narrative, characterization, and the representation of the protagonist as a key figure in the advancement of the twenty-first century information revolution. David Fincher's *The Social Network* (2010), based on a screenplay by Aaron Sorkin, and Joshua

1 See Burlingham and Gendron (1989), Lashinsky (2009), and Netburn (2012).

Michael Stern's *Jobs* (2013) both reflect and vitally contribute to the pervasive contemporary cultural imagery and rhetoric of technoscience that constructs information and communication technologies (ICTs) as the prime loci of consumer cool.

The following pages will explore the ostensible politics of representation in both films in greater depth. Through its promotion of self-creation, perseverance, and individualism, the genre of the biopic offers a particularly viable vehicle for perpetuating the myth of the American Dream and the self-made man, both of which are employed as key narrative strategies in the movies under scrutiny. Indeed, Zuckerberg's and Jobs's life stories as publicized by popular media lend themselves well to fitting the mould identified by seminal studies of the biopic (e.g. Custen 1994, Landy 1996, Taylor 2002, Bingham 2010). However, as my analysis will show, both films strive to present their protagonists as scientific innovators, rather than greedy entrepreneurs or shrewd businessmen. This conforms to the national veneration of the figure of the inventor, who as a modern frontiersman embodies the quintessentially American values of progress, resourcefulness, and adventure. In contrast to the previously explored roles of the researcher as scientist-detective (*CSI*) or scientist-hero (*The Day After Tomorrow*), the present chapter hence focuses on yet another prominent hybrid in contemporary American popular culture: the scientist-entrepreneur. Through their cinematic representation as defiant visionaries utterly disinterested in material gain, an image that extends to the respective companies they have come to represent, Jobs and Zuckerberg perpetuate a particularly valuable notion of consumer cool. Referring to Thomas Frank's (1998) and Jim McGuigan's (2009, 2011) insightful studies on the uses of cool in business culture, I will demonstrate that through the dissenting sensibility conveyed by their public persona, the tech founders partake in the pervasive ideology of contemporary 'cool capitalism' to legitimate their corporate ventures. The carefully cultivated rebel image of two of the leading icons of the information age hence crucially abets the neutralization of dissent and opposition to the present economic system.

In contrast to the preceding analyses, which have explored the perpetuation of cool technoscience via the visual-aesthetic and the cognitive dimension respectively, this chapter will demonstrate how contemporary popular culture exploits the historical associations of cool with subversion, anti-establishment, and the American counterculture to present technoscience as a prime marker of individuality, difference, and nonconformity. Drawing on seminal studies of cool rebellion, such as Dick Pountain and David Robins's definition of the concept as a "*permanent* state of *private* rebellion" (19, emphasis in the original), I will demonstrate how popular cultural productions in general and the two biopics under discussion in particular succeed in constructing mass-produced high-tech devices as individualistic, unique, and personal. Underlying the use of countercultural imagery and rhetoric for the purposes of global business culture is hence, once again, the already explored "paradoxical ethos" (Liu 72) exuded by cool. It is also for this reason that the former oxymoron of the 'cool nerd' no longer seems to hold. Paralleling the previous chapters of this book, my analysis will be concluded by a close reading of a signature scene from *Jobs*, which is intended to connect the preceding reflections and exemplify the genesis of the scientist-entrepreneur as an innovative and defining character in the movies under scrutiny. First of all, however, a brief overview of its historical development and aesthetic conventions will facilitate a conceptual approach to the genre of the biopic, which both films discussed in this chapter can be classified as.

PERSONALITIES THAT MOVE THE AGE: CONVENTIONS AND INNOVATIONS IN THE BIOPIC GENRE

> Any biography is a novel
> that dare not speak its name.
> ROLAND BARTHES/"RÉPONSES"

Jobs and *The Social Network* (henceforth in this chapter *Network*) have, more or less comfortably, been declared 'biopics' by producers, critics, and viewers. To a large extent, both films do reflect the traditional hallmarks of the genre, with Stern's *Jobs* adopting a decidedly more conventional approach than Fincher's *Network*. Tracing a young man's rise to power and notoriety in the American (and indeed global) IT industry, the movies subscribe considerably to the Great Man narrative. *Jobs* presents a fast-paced potpourri of largely unconnected milestones in the not always linear career path of the visionary genius (played by Ashton Kutcher), such as his tinkering with Steve Wozniak (Josh Gad) in the family's garage in Los Altos, Apple's first appearance at the 1977 West Coast Computer Faire, or Jobs's introduction of the iPod at the Apple Town Hall Meeting in 2001. The story largely explores the professional life of the tech guru, who is depicted as always having been well ahead of his time: The greater part of the movie's 120 minutes are devoted to his battles with ignorant business partners, employees, and shareholders to convince them of his prescient ideas. Whereas the personal background is largely glanced over in this production, *Network*, on the other hand, places considerable emphasis on the private aspects of its lead character's struggle. While the main narrative strand depicts Zuckerberg's (morally questionable) ascent to fame and wealth in 2003, the frame narrative set several years later reveals that two lawsuits have been filed against him—one by his former business partner and best friend Eduardo, who claims that his shares in the budding company were illegally diluted, and the other one by the Winklevoss twins, two fellow students from Harvard who originally approached Zuckerberg with the idea of an online networking website and now accuse him of having stolen their idea. The movie dramatizes the depositions before the actual trials, which end in a settlement (and a non-disclosure agreement) with Eduardo in order not to jeopardize Facebook's reputation. Despite these setbacks, Zuckerberg's journey to become "the youngest billionaire in the world" (*Network*) is depicted as rather linear and ultimately inevitable, while Jobs's road to success is presented as bumpy and indirect. Both life stories, however, as different as they may seem, lend themselves perfectly to the fictional schemas and tropes employed by the biopic genre.

The biopic has, for a long period of time, suffered a lack of critical attention. The most crucial studies in the field, including Custen (1992), Landy (1996), Anderson (1998), Neale (2000), Taylor (2002), and Bingham (2010), begin their analyses by bemoaning that "the biographical film has gone largely ignored in critical writing" (Anderson 344) and "has lacked critical—rather than industrial—esteem" (Neale, *Genre* 60). Despite the rejuvenation and increasing popularity of the genre since the 1990s, a research gap regarding both its receptive and productive end is still manifest. Dennis Bingham attributes this circumstance to the fact that "the term 'biopic' is frequently used as a pejorative," "spurs an instantly negative reaction," and "is so often thought of as tedious, pedestrian, and fraudulent" (*Whose Lives* 11). Although Bingham's assessment stems from the early 2010s, and biopics seem to have acquired, judging by the rising numbers of Academy Award nominations over the last several years,[2] considerable critical acclaim in the meantime, academic commentary still appears somewhat cautious in its treatment of the genre.

Without justifying the overall disregard for one of the commercially most profitable and stylistically malleable genres of the time, such critical neglect must also be attributed to the fact that biopics are generally understood to be a 'weak' genre, or, as Robert McKee has suggested in more positive terms, a 'supragenre,' which embraces a multitude of other, autonomous genres for plot formula, character typology, and imagery (McKee 83). The mainstream Hollywood biopic, especially in present-day configurations, will typically utilize prominent and easily recognizable genres, such as the western, the epic, the melodrama, or the musical. This overlap with other genres has been duly noted by critics and may be understood as yet another reason why academic commentators have traditionally found it either difficult or plainly irrelevant to conduct a deeper exploration into the large corpus of biographical motion pictures. Indeed, their choice of terms to describe the biopic's tendency to cross-generic or multi-generic construction is far from affirmative: Bingham calls it an "'impure' genre" (*Whose Lives* 26), Carolyn Anderson and Jon Lupo detect "contestable boundaries" (91), and Henry Taylor even introduces the sugges-

2 See, for instance, the commercially successful and critically acclaimed *Lincoln* (2012), *The Iron Lady* (2012), and *Diana* (2013).

tive terms "zombie text" (82) and "chameleon text" (21)[3] to describe the phenomenon, implying that the successful biopic will ferociously devour any given, nurturing element of other genres and easily adapt to the most diverse circumstances of production. Only Marcia Landy's classification of the genre as "polyphonic" (154) seems to strike a more positive chord. The need for auxiliary genres to support narrative development is strikingly evident in the two movies to which this chapter is dedicated. Indeed, both of them may also fare well under different, though related labels: *Jobs* can equally be read as a business film or social drama, exposing the morally corrupt practices of the (IT) business world like *Wall Street* (1987) and *The Secret of My Success* (1987) set out to do, while *Network* contains strong traces of melodrama, romance, and docudrama. The general cross-generic orientation of biographical cinematic approaches hence also accounts for the frequent, a priori dismissal of the label of biopic for films that might appear too grand or artistically valuable for a genre with rather low critical esteem. Classics like *Citizen Kane* (1941) or *Cleopatra* (1963) may therefore rather be subsumed into other, potentially more acclaimed categories than be identified as representatives of the Hollywood biopic.

As a distinct genre designation, the term 'biopic' was first used by the American film magazine *Variety* in 1951 (Taylor 20). In the broadest sense, however, the biographical film dates back to the very beginnings of motion pictures. Indeed, one of the first silent movies ever produced, *The Execution of Mary, Queen of Scots* (1895), may be defined as an early representative of the genre, despite its focus on the subject's spectacular death, rather than life (Custen 5). The golden age of the genre, as critics generally concur, was Hollywood's studio era from 1927 to 1960. Altogether, almost three hundred biographical films were produced in the course of less than forty years, focusing primarily on statesmen, scientists, royalty, and, following the years after the Second World War, entertainers, artists, and movie stars.[4] Screen icons like Bette Davis, Greta Garbo, Gary Grant, and Marlene Dietrich all starred in biopics. Some actors, like George Arliss and Pauli Muni, even became exclusively associated with the genre, just like the celebrated director William Dieterle. His productions at the Warner Brothers studio soon marked the productive and artistic peak of the genre in the 1950s and considerably helped to reposition the studio's image in times of censorship and changing audiences (Elsaesser 21). However, as Rick Altman points out in *Film/Genre* (1999), the distinct genre of the 'biographical film' was not established until the late 1930s: It was due to early successes such as Warner Brothers' *Disraeli* (1929) that producers tried to replicate the evidently successful formula that they had until then not fully understood. Only in

3 Original wording: "Zombie-Text," "Chamäleon-Texte."

4 See Custen's table of biopic productions (242). He lists 291 films produced between 1927 and 1960.

retrospect have critics identified the studio era as the cradle and primary producer of the genre.

Today, biopics have shifted from this exclusive association with specific producers and studios to being defined as an 'auteur' genre, with several prolific directors such as Milos Foreman, Martin Scorsese, and screen writers Scott Alexander and Larry Karaszweksi working almost exclusively in the genre. At the same time, the biopic has also caught the attention of independent filmmakers and low-budget productions, of which *Jobs* can be cited as a prime example: Produced by Entertainment One and distributed by Open Road Films, the movie had a budget of a mere $12 million. Nonetheless, its highly conventional structure provides ample evidence for Anderson's and Lupo's claim that "indie biopics and their protagonists were not necessarily radically different from their Hollywood counterparts" (102). Nonetheless, unconventional approaches to the genre exist, as Todd Haynes's much celebrated biographical exploration of Bob Dylan, *I'm Not There* (2007), exemplarily reflects. Its impressive cast (including Cate Blanchett, Richard Gere, Christian Bale, and Heath Ledger) shows that the biopic is still a star vehicle, even providing opportunity for more comedic actors like Ashton Kutcher to remodel their image in the industry. Despite the above explored dismissal by academic commentators, the supposedly grave contents and consequently adult appeal of the biopic (Anderson and Lupo 93) as a form of popularized history lesson secures connotations of 'high culture' entertainment for both performers and audience.

Hence, laments about the alleged death of the biopic, such as George Custen's famous announcement of "the post-studio era of virtual extinction" (214), have proven inaccurate in retrospect. While the genre did experience a creative slackening with the advent of television, the last two decades have again witnessed a veritable biopic boom. Among other factors, cross-fertilization with the related field of documentary and more postmodern approaches questioning notions of authenticity, life writing, and the coherence of character have contributed to the revival of the genre. This suggests that the biopic, like any other genre dating back to the very beginnings of motion pictures, is highly organic and pliable. Whereas television dramas, docudramas, and made-for-TV movies in the 1960s to 1980s would focus on the lives of ordinary citizens, such as cancer survivors, drug addicts, or charity organizers, contemporary Hollywood has redirected the focus to the Great Man Narrative and thus revived the genre's traditional form. With the notable exception of surprise hits like *Erin Brockovich* (2000), the preferred subjects were musicians (*Ray*, *Walk the Line*, *I'm Not There*), movie stars (*My Week with Marilyn*), and politicians (*J. Edgar*, *The Kennedys, The Iron Lady*). At the time of writing, biopics about Walt Disney (*Saving Mr. Banks*) and Freddie Mercury (*Mercury*) are in production.[5] Accordingly, Bingham's verdict that "the [cinematic] biography has been

5 See, for instance, Danton (2013).

a more prestigious genre since 1980 than it was in the studio period" (*Whose Lives* 21) seems accurate, particularly if one considers Hollywood's heightened output since that period and the accolades its productions receive.[6] Fincher's *Network*, for instance, was nominated for eight Academy Awards and won three of them as well as four Golden Globes.[7]

What might hence prove more accurate than Custen's hasty obituary for the biopic is a conceptualization of the genre as passing through different phases over the last decades. As a case in point, Bingham identifies altogether seven "developmental stages" (*Whose Lives* 17), each of which endowed the genre with specific modes and forms of narration that added to the material filmmakers can employ today. Without reiterating Bingham's list, it appears noteworthy that he defines the current cycle of biopics in the new millennium as neoclassical, i.e. depicting a return to the originally celebratory stance of the genre and its close ties to melodrama. At the same time, the neoclassical biopic will eclectically integrate elements of all previous phases, so that a clear-cut definition will ultimately prove impracticable. The two movies in this study support this point of view, particularly the hagiographical *Jobs*. The latter must certainly be defined as the more conventional of the two, although it does strive to depict the protagonist's flaws and shortcomings to a considerable degree. It can thus serve as a prime instance of how the neoclassical biopic revival of the 2000s incorporates the heightened realism of the 'warts-and-all' style to such an extent that it will "almost defiantly mak[e] its subject unlikeable" (Bingham, "Living Stories" 90). Another discernible phase in the development of the genre is marked by the current tendency of self-reflexivity and the fragmentation of the unified subject position. Contemporary biopics feature a complex, postmodern self and often introduce a multiplicity of perspectives, as, for instance, *Network* through its frame narrations by Zuckerberg's friend and business partner Eduardo (Andrew Garfield) and the Winklevoss twins (Armie Hammer and Josh Pence) respectively.[8] In reference to this recent trend, Taylor (2002) broadly divides the gen-

6 Neale finds that in the 1980s, which Custen would have still defined as a historical low point for the big screen biopic, the genre accounts for 5.3% of all Hollywood productions, its highest percentage so far (Neale, *Genre* 65).

7 In 2011, *Network* won three Academy Awards for Best Adapted Screenplay, Best Original Score, and Best Film Editing as well as the Golden Globe for Best Motion Picture (Drama), Best Director, Best Screenplay, and Best Original Score.

8 With regard to contemporary biopics' tendency to portray their subjects as fragmented, complex, and multi-facetted, one could argue that the genre has implicitly always supported such a viewpoint. Custen already notes that "the self depicted in these films [...] is full of contradictions" (175), since the classic biopic protagonist is often both self-made and reliant on heterosexual romance, naturally gifted yet rising through hard work, dedicated to a community yet highly individualistic. Similarly, the appeal of the biopic sub-

re into a classical and a modern phase. The former is dedicated to closure and a self-contained narration, while the latter will end with a riddle or an "open wound" (Taylor 15).[9] Closure is thus routinely denied to the audience, as the case of *Network* reflects: While one of Zuckerberg's attorneys hints at the fact that the trials will end with a settlement agreement and a non-disclosure condition, the last shot of him waiting for his ex-girlfriend Erica (Rooney Mara) to accept his friend request on Facebook suggests that his private life is a much more pressing issue at this moment. The end titles reveal how the company developed after the depositions ("Facebook has 500 million members in 207 countries. It's currently valued at 25 billion dollars"), but the viewer can only wonder if Zuckerberg will ever succeed in building sustainable human relationships.

The division of the biopic into different phases not only helps to classify contemporary examples, but also enables academic commentators to make informed conjectures about its typical occurrence and frequency. In view of the high popularity of the genre between the wars and the postwar era and its resurgence in the 2000s, Jan Romein's assessment of the biography as a "phenomenon of crisis" (28)[10] might provide a useful framework: Idealized and highly successful stories of Great Men (and, more rarely, Women) may offer a sense of stability and security in times of fast technological change and financial and ecological crises. The biopic comforts its audience that everything will inevitably turn out fine after all: Curie will discover radium, Lincoln will abolish slavery, and Jobs and Zuckerberg will become billionaires through their ingenious inventions. At the same time, it routinely confirms the status quo by reassuring the ordinary viewer that a normal, less agitated life may actually be preferable to the trials and tribulations of history's chosen few, who will invariably have to suffer for their unparalleled vision and genius. Accordingly, *Network* leaves us with a gloomy and worn-out Zuckerberg, all alone at night in a big corporate office building after the devastating deposition meeting against his former best friend Eduardo has ended. He stares into the laptop screen in front of him, again and again refreshing his Facebook page to check whether Erica has finally accepted his friend request. Evidently, this is the price Zuckerberg has to pay for becoming "the youngest billionaire in the world" (*Network*), as the end credits inform us. It is preferable, the movie seems to suggest, to enjoy one's fame-

---

ject, especially if it involves less celebratory figures such as outlaws or controversial politicians, often oscillates between attraction and repulsion, a tension which many films deliberately choose not to resolve (Neale, *Genre* 65). This point may even be enhanced by an unusual choice of actor, whose image clashes with the character he or she portrays. See also Landy (155-163) and this chapter's discussion of Jean-Louis Comolli's "The Body Too Much" (1978).

9 Original wording: "offene Wunde."

10 Original wording: "Krisenphänomen."

less existence and mediocre salary as long as one is surrounded by their loved ones. So why should the average American renounce all of this for a vague notion of money and fame, which clearly cannot buy happiness? Biopics tacitly confirm that this would be a more than unwise choice, especially in times of national crisis and turmoil.

This tendency suggests that the standard biopic production refrains from promoting subversive tendencies or providing a mouthpiece for dissident voices. While it may, through the story of its protagonist, be overtly dedicated to the Enlightenment ideal of progress and modernity, the biographical film "tends to be a conservative genre" (Anderson 334). Through its routine articulation of past ideologies and "a vanished world of values" (29), as Custen argues, the conventional biopic may be slow in adjusting its stance to a changed social order, be it in the postwar studio era or through present-day portrayals of gender, race, class, and sexuality. Especially in the American context, the triumph of the individual over social, racial, and other inequalities will be routinely confirmed through the evidence of 'actual' life stories, a tendency which becomes apparent in the two movies analyzed for this study. Both *Jobs* and *Network* present the protagonists' utmost devotion to their profession as the cause for their eventual success in the respective field. Zuckerberg is portrayed as utterly obsessed with coding and his wish to join one of Harvard's prestigious Final Clubs. Indeed, while he claims to be aware that "there's a difference between being obsessed and being motivated" (*Network*), the viewer immediately senses that he has long passed this line. Although *Network* remains, overall, skeptical toward the promises of fame and openly thematizes its downsides, it still presents Zuckerberg's aspirations as praiseworthy and understandable. Similarly, *Jobs* repeatedly excuses its protagonist's rude behavior, his outbursts of anger, and his volatile treatment of friends and employees with his complete dedication to creating the perfect product for the consumer. His obsessive attention to detail is represented as the underlying reason for his own and his company's success: Consequently, the impulsive decision to fire an employee for asserting that "typeface is not a pressing issue" (which anyone working on the construction of a computer's basic functions would probably subscribe to) is justified by Jobs's notorious perfectionism and unwavering devotion to improvement (*Jobs*). Through the movie's investment in a (not too distant) past, these traits become solidified as indispensable factors in Jobs's (and, to a lesser degree, Zuckerberg's) rise from his humble origins. Both examples show that contemporary Hollywood "simply reinforces commonly held notions of fame, agency, and history" (K. Doyle 401) through the use of the conventionalized biopic, which offers a perfect canvas for preserving, rather than subverting the dominant discourse.

Given its overall conservative orientation, the biopic must be understood, ultimately, as telling its audience more about the present than the past, whether distant or near. It is this tendency of historical drama and biographical filmmaking that

Landy seems to have in mind when she lauds the genre for "its ability to draw on the past in ways that are meaningful to the present" (190). Similarly, Colin McArthur's seminal *Television and History* (1980) revealed early on in the academic debate of period dramas that movies set in the past are, in effect, "providing for the ideological needs of the present" (40). Wozniak's argument that "nobody wants to buy a computer" (*Jobs*) would clearly have convinced any bystander in the late 1970s, especially in a time of civil unrest that faced greater problems than a battle against limited storage space. At the same time, one would have dismissed Jobs's attempts of persuading Woz of the opposite as delusional and naïve. Contemporary viewers, however, who know how the story is going to end both by their own first-hand experience of the world and the movie's introductory scene showcasing the triumphant iPod, will adamantly root for Jobs and admire his prescient vision.

Creating a past that sheds light on present conditions is a strategy that also functions for larger political concerns. With regard to the biographical films of the 1930s, mostly set in 'exotic' and presumably decadent European courts, Kevin Hagopian argues that these films taught "the lessons of the past within a conservative politics as a way of proposing a social analysis of the [...] present" (185). Current political, economic, and social conditions are thus coded as the direct results of clearly bounded actions set by key figures. Through the presentation of an individual's life story as 'inevitable' and largely 'guided by fate,' biopics tend to "draw on and reinforce narratives already circulating in American culture" (K. Doyle 385). Societal progress is not motivated by mass movements or upheaval, nor is there any potential for subversion of the present system. Indeed, any dissident elements in an individual's life story, such as Muhammad Ali's political affiliation with the Nation of Islam and his support of racial segregation and Black Nationalism, are sympathetically glanced over, as the movie *Ali* (2001) exemplarily reflects (K. Doyle 2006). In a similar vein, *Network* hardly thematizes Zuckerberg's Jewish origins, the underlying reason for his exclusion from the prestigious Final Clubs at Harvard, nor does *Jobs* portray Apple's blatant labor exploitation in Chinese factories.[11] Elements of an individual's life story that are publicly known, but still excluded from the movie version are hence as crucial as those elements that are included (K. Doyle

11 See Duhigg and Barboza (2012) and this chapter's section on "Cool Capitalism and the Scientist-Entrepreneur." According to the *New York Times*, Jobs vitally contributed to the appalling conditions at Foxconn and other manufacturing factories in China. Shortly before the launch of the iPhone, he ordered to exchange its plastic screens with scratch-proof glass, causing 12-hour overtime shifts for the Foxconn employees to deliver in time for the scheduled release (Duhigg and Bradsher 2012). It would be interesting to see whether the biopic could even manage to justify this outrageously exploitative behavior through references to Jobs's genius, his perfectionism, and unconventional management techniques.

385). Changes in society are typically not presented as being brought about by broader social, economic, and political forces, but can be attributed to key individuals. In the case of *Jobs*, this tendency becomes particularly blatant, as the protagonist is not only presented as the sole reason for a global company's unparalleled success, but also as propelling the whole of Western society into the information age almost single-handedly. The information revolution can thus be traced down to a few gifted individuals, all of them notably white, heterosexual males. Kegan Doyle's classification of the biopic as "dramas of co-opted dissent" (385) may hence be regarded as highly accurate, at least as far as mainstream Hollywood life writing is concerned.

The conservative orientation of the biopic is abetted by its highly formulaic structure and perpetuation of genre conventions in a historical setting. Since the genre can be characterized by a "weak narration" (Taylor 18),[12] it often relies on narrative templates, stereotypes, seriality, and elements of popular myth for compensation. As one of the earliest critiques of biographical filmmaking, Custen's seminal *Bio/Pics* (1992) listed its defining conventions. While his analysis explicitly focuses on Hollywood's studio era, most of the conventions he detects still apply to contemporary examples, as the films analyzed in this study also reflect. Frequently (and, considering its cross-generic approach, rather paradoxically), the biopic is indeed defined by its rather strict adherence to genre conventions. Custen argues that studio era biopics present us with a "construction of a highly conventionalized view of fame" (3), which functions through plot, characterization, imagery, mise-en-scène, and paratextual elements such as title cards and movie posters. Loosely defined as a film that is "minimally composed of the life, or the portion of a life, of a real person whose real name is used" (Custen 6),[13] the biopic, first and foremost, can be understood as a highly intertextual genre. Not only does it rely extensively on research and actual sources such as official (auto)biographies, interviews, and testimonials for an 'authentic' representation of events, but it will also contain direct references to newspaper headlines or famous TV appearances of its protagonist. In the studio era, the newsreel was a common technique to represent the rise to fame and/or notoriety in a compressed manner. Today, *Network* refers to several articles that appeared in the Harvard paper *The Crimson* and repeatedly uses images from the actual Facebook to illustrate plot lines. *Jobs* replicates its protagonist's

12 Original wording: "Schwache Narration."

13 This definition outrules classics like *Citizen Kane* (1941), which is based on the life story of publishing tycoon William Randolph Hearst, but does not employ his actual name. Similarly, *I'm Not There* (2007) might be treated as a borderline case, since the movie does not contain a character named 'Bob Dylan,' nor is the name referenced anywhere in the film, apart from a short caption in the opening credits. See Bingham, *Whose Lives* 50-71 and 377-403.

famous keynote speeches at the Apple Town Hall Meeting as well as the company's 1997 'Think Different' campaign. The use of direct citations or sound recordings is not unusual, so that the biopic is ultimately constituted of "bits of previous incarnations of already lived lives" (Custen 111), whether directly referenced or artistically recreated.

On the formal level, *Jobs* and *Network* employ several strategies that Custen has identified as defining the genre. Both movies start in medias res, the preferred expository strategy for movies that will highlight the nurture, rather than nature element of their character. This means that the audience is not presented with a linear exposition from the characters' birth or early childhood (or, as Laurence Sterne would have it, *ab ovo*), which would allow the movie to closely trace their formative years and the value system with which they were raised. Rather, the hero will already have reached (young) adulthood or indeed appear as his or her elder self, looking back on an eventful life. Accordingly, *Network* opens with a scene in a students' pub, where Zuckerberg's girlfriend Erica breaks up with him. The founding moment of Facebook happens minutes later, when Zuckerberg, out of hurt pride and grief, creates a sexist website that allows users to compare and rate female co-eds, among them his ex Erica. Throughout the whole movie, we never see Zuckerberg's parents or siblings, nor does he, in contrast to other characters such as Eduardo and the Winklevoss twins, ever reference them.[14] *Jobs* opens with the middle-aged Steve, presenting his first ingenious product, the iPod, to the Apple staff at the Town Hall Meeting. The whole movie is then narrated in the form of a flashback, starting with Jobs the college student. While the viewer learns that Jobs did not grow up with his biological mother, but his adoptive parents, the latter are hardly presented as having had a significant influence on his triumphant career. This mode of introduction allows us to see "the figure past the age where his or her values can be influenced by the family. Because of this structure, explanations of causality other than those involving the family come to dominate the lives of the famous" (Custen 149). In the case of the present examples, this becomes particularly noteworthy, since both Zuckerberg and Jobs are very young at the beginning of their respective biopics: The Facebook founder is in his sophomore year at Harvard, while Jobs has just dropped out of Reed College and still lives at his parents' place. Indeed, his parents' garage plays a vital role in the movie as the mythological, hippie-esque 'cradle' of the first Apple computer, but little credit is given to his parents for providing this space. Similarly, no importance is placed on his adoptive father's occupation as a mechanic; the latter's dealing with electronics such as radio and television and Jobs' access to the requisite equipment would in all likelihood have con-

14 Eduardo repeatedly fears disappointing his dad, while the Winklevosses are seen telephoning their father, head of a big consulting firm, for legal advice after Zuckerberg's betrayal.

tributed to his chosen career path. While the movie does hint at Jobs's adoption as the underlying reason for his often volatile, anti-social behavior and initial inability to assume parental responsibility for his child, it does not explore this issue in great depth. Both *Jobs* and *Network* strive to assert the protagonists as self-made men, who have risen by their own means, talent, and hard work, rather than through family ties or education.

Just as the biopic will typically not start with the subject's birth, it refrains from portraying his or her entire life. Most films focus on specific episodes in a person's lifetime, particularly those that have led to success and fame. Dramatizing a whole lifetime would considerably decelerate the narration, which could prove fatal for the production given the current cross-generic trend to fast tempo (Anderson and Lupo 92). Sometimes, movies will even focus solely on one specific episode or event in a person's life. Ultimately, this may affect genre designations: As Steve Neale points out, biopics "become more generically ambiguous when their span is more restricted" (*Genre* 62). Indeed, of the two movies in this study, *Network* can clearly be identified as the less conventional example, largely pertaining to the fact that it focuses on a very self-contained stage in Zuckerberg's life, namely the months leading up to Facebook's global success. Apart from the frame narrative of the deposition scenes, the whole action takes place in 2003, which is a rather short period of time for a genre that purports to depict a lifetime. Additionally, *Network* posits that the pivotal moments of a person's life need not necessarily be presented chronologically: The narration repeatedly alternates between the invention of Facebook in 2003 and the two depositions Zuckerberg faces in 2008. In a similar vein, *Jobs* begins with Apple's 2001 presentation of the iPod and then jumps back to the protagonist's time at college in the 1970s. This is a popular biopic trope, employed to create narrative tension through flashbacks and foreshadowing, to clarify a character's motivation, and to demonstrate the apparent inevitability of fate.

Typically of the genre, the protagonist is often given a mentor, friend, or sidekick, who serves as a foil to the hero. This can either be a real-life friend, such as Jobs's business partner Wozniak, or an invented (or composite) character, if there is no actual companion available to assume this role. Wozniak's characterization in *Jobs* follows the trajectory of "the collaborator with whom the protagonist later parts ways, usually to reconcile in the end" (*Whose Lives* 59), as Bingham shows. Indeed, Woz decides to leave the company after he feels that his partner no longer adheres to the principles that inaugurated the foundation of Apple back in the garage. This example already reflects that friends often act as conscience to the main characters, speaking out when the latter threaten to stray from their intended paths or forsake their formerly high morals. Through the figure of the close friend, the audience is thus given clear instructions of how to read the protagonists' impending corruption and their reaction to the internal and external struggles which they have to overcome on their way to the top, similar to how the Greek chorus summarized and interpreted the action for the spectators. At the same time, however, the close friend can also be used to "articulate the amazing qualities of the famous person, intoning what makes him or her great" (Custen 162). *Jobs*'s use of this trope seems particularly effective: Through the figure of the burly, slightly comical Wozniak, the viewers are repeatedly given to understand that Jobs is "smart," "quick-witted," and simply "the coolest" (*Jobs*). The praise for the protagonist emphasizes the friend's own shortcomings, so that the lead character emerges as even more visionary and exceptional in contrast. Indeed, *Jobs* needs to considerably downplay Wozniak's role as the actual inventor behind the early Apple products so that Jobs can emerge as the more central founding figure of the company: "In all fairness, the movie's called *Jobs*. And it's about Steve Jobs and the legacy of Steve Jobs, and so I think it focuses more on what his contribution to Apple was" (Kutcher, Interview II), as Ashton Kutcher attempted to justify the rationale behind the resulting fictionalization of Apple's genesis. Hence, Wozniak is presented as failing to realize the potential of a personal computer for home use, while his business partner appears to have had outstanding managerial skills and a clear vision for the company's route from the very start, two issues that have been met with severe criticism from fans and Wozniak himself (see Wozniak 2013, McCracken 2013). Within the genre of the biopic, however, these tropes vitally aid the construction of the protagonist as exceptionally visionary, talented, and cool.

The less conventional *Network* slightly deviates from the genre standard by portraying Zuckerberg's friend and business partner Eduardo as equally gifted in his own field of expertise. In fact, Eduardo can be read, at least initially, as even more visionary than his partner, because it is him who realizes the need to "monetize the site" (*Network*), while Zuckerberg still remains focused on programming. He is also depicted as the more ethically sound character of the two, and is hence given almost equal weight in the narration. This construction conforms to Taylor's diagnosis of recent biopics, which place the focus on two characters, the hero and 'the ordinary person': "In the center of narration are not one, but two personalities, one of whom has attained historical publicity or famousness, while the other one led a relatively 'ordinary' and largely unspectacular life. This opposition between the extraordinary and the every-day facilitates a more detailed perspective on the nature of fame" (43).[15] Finally, another crucial function of the sidekick is to provide comic relief in an otherwise intense drama, which actor Josh Gad, sporting a much burlier frame than the real Wozniak in the 1970s, frequently achieves through his mere physique:

15 Original wording: "Sie behandeln im Zentrum der Erzählung nicht nur eine, sondern zwei Persönlichkeiten, von denen die eine historische Bekanntheit oder Berühmtheit erlangte, die andere ein vergleichsweise 'gewöhnliches' und kaum aufsehenerregendes Leben führte. Diese Gegenüberstellung des Außergewöhnlichen und des Alltäglichen ermöglicht auch eine genauere Perspektive auf den Charakter des Ruhms."

The chubby Wozniak in a deckchair and with a tiny umbrella to shade himself from the Californian sun or donning a dazzling Hawaiian shirt easily defuses the tension induced by Jobs's repeated outbursts of anger.

The choice of actor for the portrayal of a role is not only crucial for generating humor or tragedy, but influences the tone of the biopic as a whole. In contrast to the creation of fictional characters, the embodiment of a real person (who is perhaps even still alive) can achieve a level of realism that remains unattainable in other genres. This largely pertains to the fact that the thrill induced by a well performed biopic

> "flows from the magic when characters 'come alive'—there is a material authenticity of referent, a potential embodiment in biographical performance that cannot exist in the creation of a fictional character, no matter how fully and imaginatively developed. Biopic audiences hope, even expect, to witness and enjoy such embodiments. It is a central appeal of the genre." (Anderson and Lupo 94)

Indeed, the choice and performance of the actor can figure as vital factors in the film's appeal. The moviegoer is attracted by both the star and the role he or she impersonates, ideally emphasizing specific aspects of the former. Richard Dyer (1979) has identified these dynamics as the "structured polysemy" (3) of the star image, suggesting that viewers not only wish to see the performance of a (potentially inspiring or tragic) life story, but also wish to see the star acting this role, which in turn reflects back on his or her own image. Accordingly, the success of many famous studio era biopics must, among other factors, also be attributed to strategic and well conducted casting: "[O]ne admires Queen Elizabeth for her statecraft, but also because she is Bette Davis" (34), as Custen argues with regard to *The Private Lives of Elizabeth and Essex* (1939) and *The Virgin Queen* (1955). The choice of Ashton Kutcher for the role of Steve Jobs illustrates this strategy: While his casting was initially identified as a "problematic fit" (Dyer 129) by fans and the media (see Dargis 2013), noting Kutcher's lack of experience with serious roles and his image as the dim-witted frat guy with boyish charm, latter accounts stressed his involvement in the IT industry and even referred to him as a "tech-oriented entrepreneur" (Caro n.p.), which clearly associates him with the Apple founder. Kutcher seemed ready to seize this opportunity for reshaping his star image, even channeling Jobs's iconic style of lecturing in his acceptance speech at the Teen Choice Awards in August 2013: "The sexiest thing in the entire world is being really smart" (Kutcher, "Acceptance Speech"), he professed to an audience of cheering adolescents. While Kutcher thereby seemed to invoke the Apple founder's famed 1996 Stanford Commencement Address, his nomination also already suggests that his casting for the role of Jobs may not exclusively be attributed to his little known ventures into the computer industry, but to his popularity as a teen idol. For an independent production with a restricted budget, a Hollywood star may prove an invaluable asset for drawing an audience to an otherwise modestly publicized film. The genre of the biopic can hence profit from the appeal of two stars: the one who performs and the one whose life is performed. In the ideal case, fans of Jobs *and* Kutcher will flock to the movie theater.

On the other hand, casting a well-known Hollywood star in the leading role of a biopic carries the danger of the actor overshadowing the protagonist whose life he or she is supposed to portray, whether by means of physical properties or character. In these cases, "the star image threatens to overwhelm the historical figure" (189), as Landy argues with regard to Marlene Dietrich's role of Catherine the Great in *The Scarlet Empress* (1934). The result may be a contradictory, and consequently less than convincing, portrayal of the character, whose persona clashes with the image of the actor.[16] To a certain degree, this applies to Kutcher's performance of Jobs, who despite an evident physical similarity failed to embody the Apple founder convincingly, as critics remarked (see O'Sullivan 2013). The audience seems to have struggled with reconciling a face well-known from tabloids and sitcoms with their image of Jobs, another well-documented and much mediated public persona.

In his article "Historical Fiction: A Body Too Much" (1978), Jean-Louis Comolli shows that in historical films based on real-life characters, there are, in effect, two bodies competing for the audience's attention: the actor's body and the body of the biopic's subject, whom the audience will most probably be acquainted with from photographs, film material, and other sources. This is one of the key qualities differentiating biographical films (or, more broadly speaking, historical fiction) from fictional genres, whose characters stem from the writer's imagination, rather than real life:

> "If the imaginary person, even in a historical fiction, has no other body than that of the actor playing him, the historical character, filmed, has at least two bodies, that of the imagery and that of the actor who represents him for us. There are at least two bodies in competition, one body too much." (Comolli 44)

The audience's focus will initially oscillate between both bodies, but it is vital for the effective suspension of disbelief that they eventually recognize the actor's body *as* the historical character and become accomplices in the fiction; otherwise, the

16 For completion's sake, it must be noted that a portrayal of the biopic subject as contradictory and multifaceted by virtue of 'problematic' casting may also be compelling for the audience, especially if it concerns controversial characters such as villains or statesmen. Indeed, upon closer inspection of the genre's extensive history, the protagonist can more often than not be exposed as harboring underlying contradictions. See Custen (175-176), Landy (155-163), as well as footnote 8 of this chapter.

pleasure one gains from the historical re-enactment is considerably lessened. Drawing on Comolli, Taylor has referred to this phenomenon as the "double biographical body" (46),[17] which will typically have been transformed from a real into a symbolic body by the end of the movie to underline the lasting iconic influence of the subject. While both Comolli and Taylor identify the discrepancy between the actor's and the subject's body as a defining condition affecting all biographical films, the movies in this study demonstrate that it must be understood as a matter of degree. Whereas the leading actor's body clearly impedes the easy suspension of disbelief in *Jobs*, the 'body too much' does not greatly affect the second film in this study, *Network*, whose actor Jesse Eisenberg was certainly less present in the public eye than Kutcher when the film was produced. Equally, Zuckerberg has had much fewer media appearances than the iconic Apple founder, so that many viewers would only be familiar with a photograph, and not necessarily his demeanor, way of talking, and tone of voice. *Network* hence benefits from the fact that both the real-life Zuckerberg and Eisenberg were, at the time of production, considerably less known than Jobs and his performer. This circumstance abets the viewer's immersion into the story, even though the possibility of such an unadulterated viewer-approach remains rare in a genre dedicated to depicting the most famous representatives of bygone ages.

Also related to the suspension of disbelief is another technique typically employed by the biopic, namely the use of signifiers of facticity such as title cards or captions that assert the 'authenticity' of the story on screen. In most cases, as in *Network* and *Jobs*, this will be the specification of date and place: "Harvard University Fall 2003" (*Network*) and "Apple Town Hall Staff Meeting 2008" (*Jobs*).[18] Additionally, *Jobs* displays photos of the real-life characters alongside their actors' during the movie's end credits and uses actual quotes, ad slogans, and parts of speeches given by the protagonist. These techniques establish requisite context and serve as a "reminder of a fact so obvious that we might overlook it: that most films made in Hollywood are not supposed to be taken true" (Custen 51). The claim to truth is indeed a distinct quality of the genre, and is, despite postmodern attempts to challenge it, still heavily employed in the industry.[19] In interviews, directors, screen

17 Original wording: "der doppelte biographische Körper."

18 As *Jobs* is episodic in character and spans a greater period of time than *Network*, title cards are used at various points of the story to establish time and place, such as "Reed College 1974" (*Jobs*). They thus not only act as signifiers of facticity, but are also essential in providing orientation for the viewer.

19 It should, however, also be noted that these signifiers of facticity may, in effect, achieve the opposite effect, i.e. emphasize the fictitiousness of the story by demonstrating that what the viewer is witnessing is precisely not the real story, but a temporarily and spatially removed re-enactment. Additionally, the use of actual quotes in *Jobs* must also be un-

writers, and performers will routinely stress their minute preparations and extensive research for the film. Accordingly, Kutcher elaborated on his compilation of an extensive sound file archive of recordings of Steve Jobs's voice, which he listened to every night before falling asleep (McCracken 2013). Similarly, *Network* can boast of being based on Ben Mezrich's bestselling *The Accidental Billionaires* (2009), for which Zuckerberg's former best friend and business partner Eduardo Saverin acted as consultant. Indeed, the use of consultants for veracity (rather than technical, scientific, or other expert knowledge) is another distinct feature of the genre: The testimony of people who directly witnessed this extraordinary life story should guarantee the most accurate version of it and simultaneously reassure audiences of its historical value. Consequently, *Jobs* enlisted Daniel Kottke, Jobs's housemate at college and his first employee at Apple Computer, to consult on the early stages of the script. The crew's attempt to have co-founder Wozniak consult on the final script were, however, rejected: According to his testimony in *The Verge*, he felt that the script was already too advanced at this stage to allow any major alterations, which would have been direly needed (Miller n.p.). In another interview, Kutcher responded to this criticism by implying that Wozniak's role as consultant on the second Jobs biopic on which Aaron Sorkin (based on Walter Isaacson's bestselling biography) is currently working might have significantly influenced Wozniak's judgment on Stern's movie (Kutcher, Interview II). What followed was a much publicized debate between Kutcher und Wozniak, with the latter accusing the movie of inaccuracy and unfair portrayal of both himself and his late partner. In response to this, Kutcher, Gad, and Stern repeatedly stressed the extensive amount of research that went into the project, as well as the accuracy of location and character portrayal.[20] More than anything, this impassioned exchange between the subjects and the creators of a biopic reflects that, despite frequent avowals that "the issue of truth or verisimilitude in representation in a narrow and litigious manner becomes irrelevant" (Landy 187), in recent years, its professed presence or absence has remained central for both the productive and the receptive side and continues to constitute a key distinctive quality of the genre.

Another main strategy for attaining the elusive 'truth value' of the biopic is the heavy use of montage, flashback, and prolepsis. These techniques present the protagonist's eventual rise to fame and power as inevitable, foreseeable, and ultimately positive. Landy defines the flashback as the most crucial part of the movie's "special historical reenactments" (20), hence classifying it as vital in the cinematic con-

---

derstood as a way of catering to the fans of the late Apple founder, who will know his famous speeches by heart.

20 *Jobs* was shot in the actual garage of his parents' house in Los Altos, California.

struction of the past.[21] One key function of flashbacks, in biopics as in other genres, is to merge a subjective, private vision of the past (through the eyes of the protagonist) with an 'official,' political version of it. The viewer is thus presented with a curious mix of both remembered and recorded past (Turim 2). At the same time, the flashback frames the significance of the presented life on screen (Custen 183) and, if originating from the vantage point of a particular character, showcases a particular point of view. Interestingly enough, *Jobs* is conceived as one big flashback from, one may assume, the subject's personal perspective;[22] the Town Hall Meeting acts as a frame for everything that is portrayed after the initial scene. Since the audience is thus already presented with the alleged moment of Jobs's ultimate success at the very beginning of the movie, all events that happen after it must inevitably lead up to this triumph. Everything we are presented with is safely relegated to the (near) past and simultaneously serves to shed light and meaning on the present.

While *Network* is low on flashbacks, it achieves a similar effect of inevitability through prolepsis, or foreshadowing: In the very first scene, Erica tells Mark that he is "probably going to be a very successful computer person" (*Network*) before she rushes out of the pub. Since the audience will already be familiar with the story of Facebook before they enter the movie theater, this does not come as a surprise. However, what the prolepsis achieves (besides introducing a faint element of humor in the rather tense opening sequence) is to place "emphasis on the *affect* accompanying the predicted outcome" (Landy 24, my emphasis). In the expository tradition of melodrama, it helps to suggest that Zuckerberg's path is largely predestined. Comparable scenes like his ingenious response during a lecture which "brighter men than [him] have tried and failed" (*Network*) or a fellow student's enthusiastic exclamation that "he [i.e. Gates] was looking at [Mark] when he said the next Bill Gates could be right in this room" (*Network*) further enhance the impression of a young man destined to great success. In a similar vein, angel investor Mike Markkula (Dermot Mulroney) acts as prophet when he tells Jobs and his team that they are "on to something big" (*Jobs*) upon first meeting them in their garage.

Montage scenes, present in both movies, fulfill a similar function.[23] Through the heavy intercutting of newspaper clippings, headlines, or quickly changing frames, the life of the protagonist seems to be guided by external forces of destiny, rather

---

21 According to Custen's qualitative study of studio era biopics, four out of five movies in his sample employed flashbacks, which attests to the strong entwinement of the technique with the genre (182).

22 The movie does not make the personal origin of the flashback clear. While the last frame before the prolepsis depicts Jobs on stage during the Apple Town Hall Meeting, typical conventions of the frame narrative, such as close-up or blurred lens, are notably absent.

23 According to Custen, over 80% of the movies in his sample contain montage sequences (184).

than coincidence or chance. *Network* displays such a scene early on in the narrative when Zuckerberg's first website facemash.com goes online and crashes the Harvard server at four o'clock in the morning. The montage shows various groups of male students clicking through the website and screaming 'left' or 'right' to rate the pictures. Female students look at the screen in disgust, but still contribute to the site's traffic. The whole montage is accompanied by a subdued, eerie tune, again foreshadowing the rather negative consequences awaiting the site's creator. The onlooker is left with the impression that the success of the website (together with all its legal and emotional consequences) was instant and unstoppable from the minute Zuckerberg put it online. Originally a technique employed by avant-garde filmmakers, the montage not only helps to move the plot forward quickly, but also acts as "a powerful marker of the teleology of fame, of its relentlessly forward march to a predetermined goal" (Custen 185). The rapid succession of shots creates an overwhelming impression of unalterable fate into which the protagonist is *nolens volens* propelled. Together with the voice-over, another central convention of the genre, the techniques of montage, flashback, and prolepsis aid the narration through condensing events and bridging gaps so that a whole human life (or at least key episodes of it) fits into two hours. Additionally, as Neale points out, "they help endow the life with a pattern" (*Genre* 63), i.e. create links, recurring motifs, and causal connections where there might not have been any. In both *Jobs* and *Network*, they contribute to establishing the protagonists' dominant traits of unparalleled ingenuity paired with utmost emotional and social incapability, which fast emerges as their personal nemesis.

While both films hence heavily rely on typical conventions of the genre, it should be noted that *Network* presents a more postmodern approach by relativizing its claims to truth through the use of two different frame narrations. After the first fifteen minutes of the movie, the viewers learn that what they have been watching so far is, in fact, an account given by Eduardo and the Winklevoss twins in their respective depositions with Zuckerberg. Consequently, the film suggests that all narrated events stem from these subjective memories, possibly tainted by resentment at Zuckerberg's betrayals, and may not be overwhelmingly accurate. This technique allowed screenwriter Sorkin to introduce multiple perspectives, instead of a unified point of view:

> "[T]wo lawsuits were brought against Facebook at the roughly same time, [...] and three different versions of the story were told. Instead of choosing one and deciding that's the truest one or choosing one and deciding that's the juiciest one, I decided to dramatize the idea that there were three different versions of the story being told." (Callaghan n.p.)

Since the main characters of the story had to sign nondisclosure contracts in real life, it was hard to determine which version might be 'more truthful' than the other.

Network thus introduces a level of self-reflexivity unusual for the conventional biopic by questioning what it depicts on screen. Given their role as prosecutors in a trial against their former best friend and/or business partner, Eduardo and the Winklevosses must certainly be classified as unreliable narrators when it comes to representing Zuckerberg's personality and his rise to fame. While this aspect distinguishes *Network* from the bulk of more orthodox examples, the film also relies, to a large extent, on the above explored conventions of the genre, similar to the more generic *Jobs*. Both movies hence reveal one of the central paradoxes of the biopic: "It tells the story of a glorious, unique life in a manner identical to the story of every other glorious unique life" (283), as K. Doyle observes. The following section will explore in greater detail the patterns of ingenuity, success, and upward mobility to which such 'unique lives' are subject.

TECHNOSCIENCE, ENTREPRENEURSHIP, AND THE AMERICAN DREAM: THE COOL CHARM OF THE SELF-MADE MAN

The above listed conventions of the biopic all contribute to the establishment of a classic narrative structure that is highly linear, formulaic, and predictable. "Hollywood modelled the lives it depicted according to dramatic and fictional formulae" (*Genre* 61), as Neale explicates its reliance on safe and proven templates. One of the most pertinent narrative models of the genre, the "Great Man variant" (Bingham, *Whose Lives* 7), displays a more or less clearly traceable rise-fall-rise pattern, with the male protagonist ultimately triumphing against all odds. The biopic thereby offers a highly compelling version of the American Dream, stressing the Puritan values of self-reliance, hard work, individualism, and innovation as the foundations of the modern self-made man. The two films explored in this study are no exception to the rule. Indeed, as I will argue in this section, *Jobs* and *Network* combine the commonly held notion of the American Dream with an anti-communitarian orientation, stark individualism, and emotional restraint, some of the defining characteristics in today's shared cultural image of the nerd. In combination with the narrative of the typical American self-made man and a rags-to-riches trajectory, however, these qualities become central in the construction of both protagonists as the leading tech icons of the American information society in the early twenty-first century. Rather than being depicted as either pathetic nerds or greedy businessmen, this strategy allows Zuckerberg and Jobs to emerge as cool innovators, who not only introduced a new paradigm in their respective fields of expertise, but indeed revolutionized the very foundations of Western society.

Before the following pages expound upon this argument with reference to the formula of biographical self-invention identified by Custen and others, it is requisite to note that these characteristics refer almost exclusively to biopics with a male, white American (or Western European) subject. The female biopic follows a starkly different narrative pattern and characterization, both in its studio era variant and in contemporary expressions. Bingham goes as far as differentiating the male and the female biopic as two separate genres, since their visual and narrative conventions are highly disparate (Bingham, *Whose Lives* 25-27). While the classic male biopic typically traces its protagonist's more or less linear rise to power and fame through self-invention, genius, and hard work, movies about exceptional women will focus on victimization, failure, and their private lives. The realms of professions portrayed will typically be restricted to movie actresses, singers, dancers, and royalty, with the occasional exception of the woman scientist (*Madame Curie*, 1943) or the politician (*The Iron Lady*, 2011). In these latter cases, however, the femininity of the protagonist will routinely have to be emphasized through romantic love, motherhood, or charity work (suggesting heightened empathy and compassion) to make the subject acceptable to mainstream audiences. In accordance with Bingham's differentiation between the male and the female biopic, the following two sections will consistently employ the masculine pronoun to clarify that all observations refer exclusively to the male subset of the genre.

Paralleling the relative paucity of female biopics, mainstream films depicting the lives of non-white characters are rare. In the cases when they do get made, radical differences are usually negated, appropriated, or cautiously glanced over. Controversial political and social points of view harbored by the protagonist are smoothed out and integrated into Hollywood's broadly 'liberal' orientation. Racial inequality will be transcended by the abilities of the exceptional individual, rather than through political action. In view of this tendency, Thomas Doherty has fittingly referred to the biopic as "the motion picture equivalent of the all-white neighborhood" ("Malcolm X" 37). The two movies explored in this study certainly conform to this trend: Not only are both protagonists white Anglo-Saxon heterosexual middle class males, but they also almost exclusively interact with others who fall into these categories. The blatant lack of positive female characters beyond nagging girlfriends or highly sexualized love interests and the complete absence of people of color are striking,[24] especially considering the films' settings in the near past. The auxiliary genre that both movies employ, the business film, reveals a similar distribution in terms of gender, race, and class (Pileggi et al. 217). Interestingly, any ref-

24 In fact, one of Zuckerberg's business opponents suing him for intellectual property theft together with the Winklevoss twins is of Indian descent in real life, but was cast with an actor from a completely different ethnic background.

erences to Zuckerberg's Jewishness or Jobs's mixed racial background (his biological father was Syrian) are studiously avoided.

These contemporary elaborations of the American Dream stand in marked contrast to its original formulations as articulated, among others, by Captain John Smith (1616) and J. Hector St. John de Crèvecœur (1792), which assumed a manifestly inclusive tone to strengthen and unite the new nation composed of various religious, ethnic, and class backgrounds. In these early accounts, America figures as a 'New Eden' and the land of never-ending opportunity which will reward hard work with personal and societal progress, wealth, and happiness.[25] This vision of self-creation as upheld in the earliest days of settlement is supported by the Puritan creed, which through its Calvinist doctrine of predestination regards material success as a sign of salvation. Evidently, such a belief system perfectly accords with a society that from its founding moments has emphasized individualism over community, industriousness over indolence, and the notion that prosperity and status must be earned, rather than inherited. While the American Dream may incorporate conflicting tendencies, such as a materialist myth of individual advancement versus a more egalitarian oriented narrative of brotherhood (see Fisher 1973, Winn 2003), it essentially always amounts to the same promise: The American citizen will reap the rewards of wealth, power, social status, and fame, if he only applies his "energies and talents to the fullest" (W. Fisher 160).[26] Accordingly, the inability to rise from one's modest means is not a matter of unfavorable social or political conditions, but a result of personal failure. In the idea of the American Dream and the self-made man, "achievement takes on a moral quality" (Edington 63) and is appreciated as a clear indication of a man's civil character and innate goodness. The frontier myth, which has already been explored in chapter 7 of this study, is closely related to these narratives, stressing the importance of perseverance, self-reliance, and individual courage.[27]

25 In his *Letters from an American Farmer* (1792), Crèvecœur advises the prospective American to "[g]o thou and work and till; thou shalt prosper, provided thou be just, grateful, and industrious" (32). More than a century and a half earlier, Smith's *Description of New England* (1616) already contains the confident promise that "every man may be master and owner of his owne [sic] labour and land; or the greatest part, in a small time. If hee [sic] have nothing but his hands, he may set up this trade and by industrie [sic] quickly grow rich; spending but halfe [sic] that time wel [sic] which in England we abuse in idlenes [sic], worse or as ill" (332).

26 Again, my formulation here consciously excludes the female pronoun, similar to how female settlers were largely excluded from the promises of the American Dream.

27 For more background information on the formation of the American Dream and the notion of the self-made man, as well as their perpetuation in popular culture, see Berk

Fittingly, both films in this analysis perpetuate an updated vision of westward expansion, as California and the Silicon Valley feature as the places where Jobs's and Zuckerberg's respective businesses can seriously start to prosper. Hence, the primary reason why Zuckerberg and his business partner Eduardo eventually decide to expand the Facebook network from Harvard to Stanford University is because "[i]t's time for them to see this in Palo Alto" (*Network*). Not surprisingly, the plan works, and soon Zuckerberg and his little crew push their personal frontier by renting a house in the heart of the IT industry, right across the street from Napster founder and billionaire Sean Parker (ironically played by pop singer Justin Timberlake), who will prove decisive in transforming Facebook from a nerdy college student's project into the multi-billion venture it is today. At the same time, the film offers a clear vision of its take on inherited, rather than earned affluence: Zuckerberg's main opponents in his rise to fame are the Winklevoss twins, who "came from a family of means" (*Network*) and repeatedly use (or abuse, as the movie suggests) their family ties to receive top-league legal advice any time they need it, gain access to one of the most prestigious Harvard Final Clubs, and even procure an appointment with the President of Harvard. The latter is not, however, amused by their apparent sense of entitlement and tells them that they are "not worthy of Harvard" and "don't get special treatment" (*Network*), which Zuckerberg's observation in the deposition scenes confirms: "The Winklevi [sic] aren't suing me for intellectual property theft. They're suing me because for the first time in their lives, things didn't work out the way they were supposed to for them" (*Network*). While he was the one who sat down and coded for nights on end to create a functioning website, the Winklevosses' superior notion of being able to 'hire' a genius like him for their own purposes appears conceited and insolent. In the end, the viewer will concur with Zuckerberg's assertion that "[i]f you guys were the inventors of Facebook, you'd have invented Facebook" (*Network*).

In accordance with the basic tenets of the American Dream as a rags-to-riches narrative, it is persistence and hard work that should and will be rewarded, rather than inherited entitlements. And fittingly enough, the Winklevosses' series of failures rapidly unfolding in the second half of the movie (they lose a terribly close rowing race on the Thames on the same day that they are confronted with news about Facebook's lucrative expansion to Europe) seems to confirm the Puritan notion of predestination: As the only actual antagonists of the movie, their failures are the surest sign of bad character and, in an overtly religious dictum, damnation. They are destined, as the film suggests, to moral and material failure, while their opponent Zuckerberg is cast as the upwardly mobile self-made man whose life story the biopic is dedicated to display.

(1977), Cawelti (1965), Horton (1982), Levy (1991), Manley (1990), Parshall (1991), Traube (1989), Kimmel (1996), and Winn (2007).

In a similar vein, Jobs is repeatedly faced with opponents who, in contrast to his modest origins, come from a great tradition of classic entrepreneurs. The managerial class, primarily represented by the Apple board and shareholders, consists of elderly men in gray suits with no practical knowledge of computer engineering or design. They are old-fashioned, fail to adapt to new conditions, and "don't see past their own shadows" (*Jobs*). They, as the movie overtly suggests, are responsible for Apple's slacking revenues in the 1970s and 1980s. It takes a former underling with adoptive parents, no college education, and a lower middle class background to change the rules of the game for good. Embodying the credo of the American Dream that self-creation and hard work trump inherited privileges, Jobs eventually rises to the top after firing Gil Amelio (Kevin Dunn) and Mike Markkula, the characters most overtly associated with 'old' capital. While the film does not establish Jobs's opponents as descendants of 'a family of means' as clearly as Network does, Markkula is still presented as the slick capitalist who drives up in a Porsche and wears a tailored suit, while Jobs tinkers barefoot in his parents' garage. While Jobs and his team spend their days (and nights) building the first Apple computer by hand, Markkula can effortlessly invest a major amount of capital in the budding company without doing any visible work throughout the film. In contrast to Jobs, Markkula is not portrayed as a self-made man who has distinguished himself by hard work and perseverance;[28] this plotline is exclusively reserved for the protagonist, and hence becomes all the more effective. Even when the movie can no longer deny Jobs's ascent as one of the wealthiest entrepreneurs in American history, his character still retains the valuable connotations of modesty, sobriety, and thrift so typical of the self-made man: In contrast to the others who dine with him in a fancy restaurant and readily adopt to the lifestyle of the rich and famous, Jobs sticks to his fruitarian diet and seems to remain utterly unimpressed by the splendor surrounding him. He thus also embodies "the 'character-building effects' of work itself" (276), as Elizabeth G. Traube argues with regard to cinematic images of the self-made man in the 1980s. Three decades later, the cinematic representation of Jobs demonstrates that the promises of the American Dream have lost nothing of their allure.

As a formulaic and essentially conservative genre, the biopic provides a particularly viable vehicle for tracing the story of America's great self-made men, whether they emerge from the realms of science, business, or the entertainment industry. The classic biographical film as conceived in the studio era is overtly devoted to presenting its audience with the awe-inspiring, admirable, yet relatable life story of the Great Man. This tendency is greatly abetted by the Americocentric orientation of the genre. Not only was the biopic in its present form conceived and cultivated

28 The real-life Markkula retired at the age of 32 by virtue of his wealth acquired from prudent stock options. Jobs lured him out of his retirement to invest in Apple (Markoff 1997).

by the American film industry and largely depicted the (distinctly American) lives of US citizens, especially so after the Second World War,[29] it has also been overtly dedicated to the values and morals of its native country. Studio era biopics "self-consciously glorified America by sustaining myths of inclusion" (190), as Custen observes, and a recent study by Anderson and Lupo (2002) demonstrates that this trend has only increased over the last decades.[30] This can be explained, among other political and socioeconomic factors, by the circumstance that the pervasive myth of the American Dream provides a particularly suitable template for the classic biopic narration, which will usually trace its protagonist's journey from humble beginnings to immeasurable wealth and fame. Whether entertainers, politicians, or inventors, almost every male biopic subject will portray distinctive features of the self-made man, the living embodiment of the most pertinent American virtues: individualism, a high work ethic, emotional toughness, personal initiative, upward mobility, and eventual success. Specific conventions of the genre, such as the already introduced tendency to set the beginning of the movie at a time when the subject is already past his childhood and thus a product of his own making rather than genetic heritage or family ties, support this representation.

If the chosen subject's life does not conform to the pattern predetermined by the American Dream myth, the creation of the life story for the conventional biopic format will unavoidably involve a considerable degree of fictionalization. This matter becomes particularly acute with regard to potentially problematic, controversial aspects of the subject's lifestyle: "No matter how alienating the character's actual life, it had to be told in terms congruent with audience's own experiences and expectations about how the famous conducted their lives or lived. A [...] famous person had to have clear motivation for the decisions that brought him or her to greatness" (Custen 18). Usually, this entails that the reason for an individual's success or failure is not portrayed as pertaining to larger socioeconomic or political conditions, but can be reduced to individual, clearly identifiable factors. Clear-cut causality hence emerges as another distinguishing quality of the biopic, which it shares with the genre of melodrama (Custen 165, Landy 152-156). Indeed, one of the main recurring accusations against the critically berated Jobs was that it fails to portray exactly how and why the protagonist, after all an adopted, lower-middle class college

29 The inter-war era displays a predilection for European subjects (including, among others, Marie Curie, Louis Pasteur, Benjamin Disraeli, and Julius Reuters), which Custen connects to the postwar shift of focus from old elites (statesmen, scientists, royalty) to the new upper crust of entertainers and Hollywood celebrities (119).

30 Anderson's and Lupo's study of sixty-one biographical films of the 1990s shows that recent biopics are even more dedicated to an Americocentric view of the world than earlier examples: 78% of the films in their sample depict US citizens (93), in contrast to 198 out of 300 studio era biopics that were set in the United States (Custen 91).

dropout with questionable ethical standards, could ascend to the position of CEO of the world's most profitable company (see, for instance, Gleiberman 2013 and O'Sullivan 2013). The causal connections in the life story of the country's currently most popular tech tycoon have not been made clear enough to the disappointed viewers. While I would agree with critics that, considering its Hollywood production background, the film portrays generally rather feeble character motivations, I would argue that it does offer its own tacit assumptions about the forces driving the rise of America's ultimate self-made man of the late twentieth century. It was not, as the movie's ostensible politics strive to clarify, money or profit which motivated Jobs, but a much more ephemeral currency: "You guys thought it was cool. I just wanted to be one of you guys. Of all the guys I knew, you were the coolest" (*Jobs*), as Woz intimates to his business partner the reasons why he left the safety of Hewlett-Packard and joined Apple in the first place. Apparently, Jobs had already envisioned the company's key role in the production of consumer cool from its beginnings in a garage, while the nerdy, plump Wozniak just wanted to be a part of the cool gang. From the midst of this gang, the barefooted, long-haired, and insanely confident Jobs soon emerged as the ultimate leader, whose apparently natural charisma and coolness even impressed someone who was five years his senior and an incomparably better computer engineer than he would ever be. Appropriately, Jobs's main ambition after returning to the company in the 1990s is "making Apple cool again" (*Jobs*). This is the underlying incentive, so direly missed by most critical reviews, yet tacitly informing the majority of Jobs's and his team's actions.

The enthusiastic critical reception of Network, on the other hand, must chiefly be attributed to the fact that it offers a very clear, almost clichéd motivation for Zuckerberg's journey. His invention of Facebook's predecessor, facemash.com, is directly caused by the emotional break-up with his girlfriend, the very first scene of the movie. One of the reasons for the break-up is that he seems "obsessed with finals [sic] clubs" (*Network*) and hardly pays any attention to what she says. The wish to join an exclusive club is of utmost importance to him, not necessarily to excel academically and professionally, but to increase his social status among peers:

> Erica Albright: Is it true that they send a bus around to pick up girls who want to party with the next Fed Chairman?
> Mark Zuckerberg: You can see why it's so important to get in. […] Because they're exclusive. And fun and they lead to a better life.
>
> (*Network*)

In short, the lifestyle Zuckerberg aspires to through admission to one of the exclusive clubs at Harvard is a distinctly cool one, which his implicit and explicit characterization as a "freakin' nerd" (*Network*) underlines. Hence, what causes Zuckerberg to succeed is the pervasive feeling of rejection in almost any area of life other than programming, and, as the movie rather emotively suggests, the wish to be accepted. Both movies hence portray the wish for a coolness that is, underneath its face of callous indifference, fuelled by notions of peer approval and acceptance: In his 1990 essay "Coolness," Daniel Harris defines this particular vision of cool as "an aesthetic [...] of the passwords and Masonic handshakes through which card-carrying initiates gain entrance to the clubhouse" (48). Coolness works intrinsically through mechanisms of inclusion and exclusion, and by conveying an impression of elitism, secrecy, and belonging. As a personal attitude reflecting social prestige, cool has repeatedly been understood as sharply distinguishing between those who are familiar with and those who will forever remain ignorant of the required gestures, props, and poses. De-regulating access to the exclusive clubhouse of cool would essentially amount to its abolishment: The cool heavily rely on the un-cool to assert their difference and dominance. Eventually, as *Network* bleakly suggests, Zuckerberg has gained the cool he has so devotedly worked for, though in a somewhat different manner than he would have envisioned: Coming full circle, the film ends with the already introduced final shot of Zuckerberg, all alone in an empty office building, waiting for Erica to accept his friend request on Facebook. Although the viewer is given to understand the significance of this action and intuitively senses the inner turmoil Zuckerberg must find himself in after two trials (one of them against his former best friend) and the loss of his girlfriend (whom he evidently still misses), his face is completely blank and displays no emotions whatsoever. The former nerd, this shot implies, has attained the absolute cool.

Network hence reflects a version of the American Dream that cannot be fulfilled without certain tolls; the protagonist has to pay a considerable price for his success, especially where his social and romantic relationships are concerned. The ascent to the top is paved with betrayal, greed, hardship, frustration, personal losses, and a challenge to one's former idealistic visions. The bleak ending to *Network* leaves the viewer assured of the old folk wisdom that money just cannot buy happiness and that the Dream has, in this case, gone awry. The protagonist in *Jobs*, on the other hand, first needs to enjoy some downtime and reconsider his priorities before he can excel as Apple's CEO, thus striking a more conciliatory tone. Only after he has been happily married, turned into a loving father and devoted gardener, as well as assumed responsibility for his first daughter whose fatherhood he previously denied is he allowed to reap professional happiness. Even though the love story between Jobs and his wife Laurene (Abby Brammell) is only briefly touched upon toward the end of the film, it provides a strong counterpoint to the otherwise strict focus on business operations. This conforms to the standard biopic plotline, in which a recurring motif is the redeeming force of heterosexual love, which cleanses the misguided genius from moral failures. As Glenn D. Smith (2007) suggests with regard to the 'celebrity biopics' of Ray Charles and Johnny Cash, romantic love is a frequent trope employed by the genre to provide "the proper context for the lead character's redemption" (224) in the classic Hollywood tradition. This redemption "comes in the form of [the protagonist's] psychological and moral repair and continued professional success" (G. Smith 224). Both dimensions are clearly portrayed in *Jobs*, whose hero transforms from an erratic tyrant with no ethical or moral boundaries into a doting father and visionary leader guiding Apple out of its crisis. The image of the self-made man thus portrayed is one of constant struggle and never-ending hardship. Only one of the two movies in this study rewards these exertions with eventual gratification.

Due to these divergent endings, the films under scrutiny can be understood as presenting two sides of the medal that is the American Dream. While *Jobs* eventually rewards its protagonist and presents a flawed, yet ultimately encouraging vision of individual achievement, *Network* adopts a considerably less cheerful tone and frames its lead character as an example of the dark sides of material aspirations. The biopics of this study hence portray two superficially contradictory narratives that articulate different aspects of the same myth. Pileggi et al. (2000) have identified these alternatives as the 'hopeful tale' and the 'cautionary tale' respectively. As their study shows, both trajectories are vital ingredients of the American Dream, which concomitantly serves as encouragement for success and warning against failure, depending on which aspect is emphasized. The hopeful tale, which *Jobs* must be characterized as, stresses upward mobility and builds "on the idea that, irrespective of an individual's socioeconomic background, dreams, ambitions, and goals can be actualized" (Pileggi et al. 210). The Puritan virtues of hard work and perseverance play a key role in this narrative.

In marked contrast, the cautionary tale presents the downsides of the wish for personal advancement: Greed, exploitation, and moral decay are among the vices to which the protagonists succumb. The inescapable consequences are personal tragedy, misery, and sweeping losses.[31] *Network* depicts a particularly viable subset of this tale, defined as the 'poor little rich person' narrative: "This latter narrative, in particular, paints the wealthy as unhappy individuals who are lacking the sincerity and warmth of family relations, romance, and friendship" (Pileggi et al. 211). The

31 While Pileggi et al. merely comment that in the analyzed films, "an individual's socioeconomic position has little bearing on achieving success and upward mobility" (224), several critics have remarked that the parameter of social class plays a vital role in the distribution of the hopeful vs. the cautionary tale. Winn (2003), for instance, shows that in Hollywood business films, working-class protagonists typically strive "to achieve upward mobility but eventually fail because of their greed" (308). This assessment remains true for the biopic, as Custen argues: Biopic characters other than the protagonist will either emphasize "drives that encourage individualism (stardom, fame, etc.)" or "drives that encourage community, and lack of individualism" (79), depending on their social class. See also Traube (1989) and G. Smith (2009).

representation of Zuckerberg largely conforms to this characterization. It is not only the desolate ending that conveys an impression of personal failure, but also the very structure of exposition: Zuckerberg's story is told through the perspective of his prosecutors in two distinct law suits, a circumstance that already suggests his deceptive character and the unfortunate denouement of his supposed success story.

The content analysis of 120 Hollywood films between 1927 and 1955 by Pileggi et al. demonstrates that specific socioeconomic factors influence whether a film will be a 'hopeful' or a 'cautionary' tale. 'Hopeful' tales appear to correlate with the economic climate, reinforcing social stability in times of economic instability. 'Cautionary' tales, on the other hand, take their cue from internal or external assault, when the social order is threatened or even indeed reconfigured, such as during the Civil Rights Movement of the 1960s. Although *Network* and *Jobs* are separated only by a few years (production for the movies began in 2009 and 2012 respectively), they could be taken to reflect this basic tendency: Zuckerberg's biopic may be read as portraying the aftermath of the war in Iraq and the political turnover heralded by President Obama's election, while *Jobs* is more dedicated to the consequences of the global financial crisis, whose effects became most palpable at the beginning of the new decade. Zuckerberg's tale encourages Americans to seek moral support in the private sphere and reminds them of the surpassing value of love and friendship, while Jobs's ultimate success story proves reassuring for a country experiencing economic depression.

Drawing on this basic opposition, I would like to extend the persuasive argument made by Pileggi et al. by suggesting that each tale not only portrays a certain dimension of the American Dream, but also draws on and activates specific aspects of cool. The hopeful tale can be understood as stressing the positive and productive connotations of cool as an attitude of heightened individualism and self-stylization. Cool as an all-encompassing mode of being places the self in the center of attention: Narcissism, hedonism, and ironic detachment have been defined as the "core personality traits" (Pountain and Robins 26) of the cool character. *Jobs* clearly portrays these qualities in the depiction of its protagonist, whose degree of self-absorption frequently verges on the obsessive and egotistical: When his girlfriend (Ahna O'Reilly) informs him of her pregnancy, he reacts coolly and lets her know that he has no use for a family at this pivotal time of his life. His dedication to the budding company, for which he has just attracted an angel investor, and his wish for self-fulfillment take precedence over the needs and feelings of others. Similarly, he has no scruples parking in the slot reserved for the handicapped right in front of the company building.[32] While these scenes paint a less than likable image of the lead character, the movie quickly redeems him through showcasing that the unapologetic

32 The scene also nicely suggests that Jobs might after all be entitled to the parking spot, for he may not be physically, but most probably emotionally handicapped.

pursuit of his vision is the underlying reason for his success, thus guaranteeing that his is ultimately a hopeful tale. He *has* to take the parking slot closest to the entrance, because doing otherwise would cost him precious time and energy that he has to dedicate to his work. "It's not my job to be nice to people, it's my job to be better!" (*Jobs*), he yells at his intimates, and the audience, one may assume, is supposed to nod in agreement: Jobs is venerated and has been graced with his own biopic not because of his altruism and good manners. It is the unfaltering belief in his vision and the devotion of all his energies and talents to the creation of the perfect consumer experience that have enabled him to come forward as the century's most admired entrepreneur. Jobs thus emerges as the archetype of the American self-made man in the information age, who "makes his own way in life on a path defined by constant self-betterment and self-control" (*Whose Lives* 105), as Bingham argues with reference to Michael Kimmel's *Manhood in America* (1996). Self-creation and self-assurance are requisite for his success, and he can only achieve progress for both himself and, ultimately, society, by adopting a stance of cool individualism.

In *The American Dream in Contemporary Hollywood Cinema* (2007), J. Emmett Winn has referred to this pervasive trope as 'romantic individualism,' which is predicated on the assumption that "character, virtue, romantic allure, and other bodily, psychological, and emotional states of the self are a remedy for an imbalanced social structure" (*American Dream* 42). With regard to Jobs, I would argue that the primary emotional state for concentrating individual drive and talent is cool, which stresses individuality and personal stylization above all other sensibilities. In the hopeful tale, cool comes to figure as an antidote to any challenges the protagonist might face; he will invariably overcome them as long as he adopts a stance of stylized comportment. Cool is the ultimate form of self-fashioning and involves a high valorization of "spontaneity, expressiveness, and stylishness" (Clarke par. 13). At the same time, cool endows the character with irresistible charm that may well be applied to personal gain.

Not coincidentally is Jobs as much remembered for his innovative products as for his irresistible charisma. In "A Sociology of Steve Jobs" (2011), Kieran Healy makes clear that "charismatic leaders, by the way, do not have to be kind, ethical, or pleasant people" (n.p.); rather, it is vital for them to appear internally driven and exceptionally focused, two qualities which the adoption of a cool stance crucially abets. The film goes to great lengths to emphasize these charismatic traits of the protagonist underneath his arrogance and egotism. Early on, Jobs is established as a natural leader and negotiator, whose charming ways easily seduce customers and investors alike. A later scene depicts the company founder underneath a tree on the Apple campus, giving an inspiring lecture to his employees from the Macintosh team, who are huddled in a half circle in the grass in front of him. The scenic arrangement echoes an earlier scene of Jobs's calligraphy class in the Reed College's campus yard, headed by a Buddhist monk. Now it is Jobs who has assumed the leadership position as the ultimate consumer product guru. He never tires of communicating his vision of infinite innovation ("We don't stop innovating!") and supplies his team with a seemingly endless collection of inspirational slogans. The cinematic representation of Jobs thereby portrays two vital ingredients of cool individualism as suggested by Nick Southgate (2003), namely authenticity and self-expression. Southgate argues that "[t]he drive for self-expression [...] is a natural part of what it is to be cool" (456), while the concept of authenticity, the truest hallmark of cool, is defined as "people's desire to have ownership and autonomy over their own identities" (456). Both observations clearly apply to the cinematic Jobs, whose foremost ways of authentic self-expression are the very products to which he dedicates his life. They are introduced as "insanely cool" and "limitless" (*Jobs*), and thus serve as the simplest and simultaneously most persuasive proof of his success. His journey from rags to riches, as the film overtly demonstrates, has been a paradigmatic example of victory and success.

In contrast to the hopeful tale pursued by *Jobs*, the cautionary tale more overtly explores the negative affective range of cool by portraying its lead character as cold, socially detached, and emotionally crippled. In *Network*, antisocial sentiment and emotional restraint indeed figure as dominant motifs throughout. The extensive making-of footage on the movie's DVD edition suggests that these off-putting qualities of the cool persona were conceived of as the defining features for the character of Zuckerberg by director David Fincher and screen writer Aaron Sorkin. A particularly apt illustration of this can be found in the scene outside the Caribbean-themed party at the Jewish fraternity, possibly the least cool event of the entire semester, when Zuckerberg pitches his idea of a new social network to Eduardo. While the latter, scantily dressed in cargo shorts and a Hawaiian shirt, is shivering from the cold January air, actor Jesse Eisenberg correctly assumes that his acting should be more subdued: "I don't mind the cold, right, because Mark Zuckerberg doesn't feel anything" ("Making Of"), an interpretation of the role which Fincher confirms. This stance is maintained throughout the movie: Eisenberg's acting seems deliberately subdued, oftentimes not even focused on the very conversation he engages in. His character repeatedly fails to understand subtle insinuation and word play by Erica or Sean Parker, indicative of his lack of emotional intelligence. Eisenberg's interpretation of Zuckerberg is of one who is completely detached from whatever is going on around him, as if he has retreated into his own little world. The implication, of course, is that Zuckerberg is always writing code in his ingenious brain and contemplating his next big idea, all the while remaining completely oblivious to basic social courtesy.

This dimension of cool as a form of utmost emotional restraint, indeed handicap, has its roots in its original tribal conceptualization as a form of affective shield against threatening circumstances for the body and the mind (Thompson 1979,

1984). In its extreme cases, this originally constructive attitude of "immaculate concentration" and "*transcendental balance*" (Thompson, "Aesthetic" 41, italics in the original) has devolved into a stance that also helps to suppress the positive affective spectrum. The destructive forces of cool have been particularly well documented among African-American males in the twentieth century (see Majors and Mancini Billson 1992, Connor 1995, hooks 2004). While the strategy has been valuable as a vital coping mechanism in a world of discrimination, harassment, and anomie, it also afflicts the private interactions with family members, friends, and lovers. As the often last and desperate attempt "to counter the ice ages of our existence with affirmative strategies" (Poschardt 10),[33] coolness as a modern survival strategy for the urban jungle usually comes at a high personal cost and will often be fuelled by trauma and loss.

Indeed, the portrayal of Zuckerberg hints at a highly vulnerable inner core and, as the above analysis has shown, one who is tacitly motivated by repeated experiences of rejection and frustration. Constant rebuffs by girls and the prestigious Final Clubs, as well as an apparent boredom with the academic subjects at Harvard (which he passes without apparent effort or intellectual stimulation) not only fuel his desire to excel through Facebook, but also his emotionally restrictive range of expression. Before the success of the website, Eduardo was "your only friend" (*Network*), as he tearfully decries during the deposition, and it seems that by the end of the movie, he has lost even this one ally due to his repugnant ways. Zuckerberg's character thus seems to be dedicated to the credo "that a cool way of life is required by coldness both in symbolic and literal terms. 'Coolness' enables humans to live with coldness instead of freezing in it" (11),[34] as Ulf Poschardt argues with regard to the word's etymological roots denoting low temperature. Through its representation of Zuckerberg's story as a cautionary, rather than hopeful tale, *Network*, however, makes it clear that the strategy of countering any personal or professional challenge with a stance of absolute emotional frost is ultimately doomed to failure.

33 Original wording: "der Versuch, den Kältepassagen der Existenz affirmative Strategien entgegenzusetzen."

34 Original wording: "daß eine 'coole' Lebenspraxis herausgefordert wird durch die Kälte im symbolischen wie im realen Sinn. 'Coolness' ermöglicht den Menschen mit der Kälte zu leben, statt in ihr zu erfrieren."

NO RESPECT FOR THE STATUS QUO: THE COOL REBELLION OF THE BIOPIC SUBJECT

> Here's to the crazy ones. The misfits. The rebels. The troublemakers. The round pegs in the square holes. The ones who see things differently. [...] While some may see them as the crazy ones, we see genius. Because the people who are crazy enough to think they can change the world, are the ones who do.
>
> FROM APPLE'S 'THINK DIFFERENT' CAMPAIGN

Whether through the hopeful or the cautionary variant, both movies are greatly invested in the basic narrative formula of the biopic that tacitly frames the protagonists as personifications of the American Dream. Again, this prescribed plot structure was first identified by Custen with regard to studio era biopics and refers almost exclusively to the male variant of the genre. First of all, the biopic formula of the Great Man requires the lead character to be initially depicted as entertaining "antagonistic relations with members of a given community" (71), as Custen defines the central conflict of the genre. The battle against the establishment is one of the most persistent tropes of the biographical movie and allows filmmakers to emphasize the protagonist's intuitive vision, fearlessness, and perseverance against all odds. Not surprisingly for a genre whose roots have been traced back to ancient legends about the lives of saints, the protagonists pursue their goal with almost religious zeal (Bingham, *Whose Lives* 3, Taylor 95). This utmost dedication to progress is, however, usually interpreted by their immediate environment as a pathological obsession. The subject is thus represented as a 'misunderstood genius,' who is not only "different from the prevailing norms" (Bingham, *Whose Lives* 36), but who tries to change these norms for the better. His ideas are deemed outlandish, crazy, and outrageously unorthodox by the established experts in the field, and a major part of the classic biopic will be dedicated to depicting his internal and external struggle against these erroneous assumptions. Only the audience, of course, who are already familiar with the subject's great contribution to society, will root for his unconventional ideas and laugh at the foolishness of those who originally dismissed his great vision.

This strategy is particularly well developed in *Jobs*, in which almost every character will at one point or the other oppose the protagonist. Jobs argues, in no particular order, with his partner Wozniak, their first-time customer at a local computer store, angel investor Mike Markkula, board members, shareholders, his chosen

CEO John Sculley (Matthew Modine), and even his employees, for none of them are "seeing the Big Picture" (*Jobs*) or share his grand vision of personal computing: "Over and over again, minor characters explain to him why something can't be done; Kutcher-as-Jobs smiles enigmatically and waves away their concerns" (Newton n.p.). He thus becomes what K. Doyle has described as "that most American of men: a defiant individual" (394), for whose life story the biopic displays a particular fondness. *Jobs*, however, takes the standard biopic convention to a new extreme, in so far as the protagonist feels misunderstood by almost everyone he encounters. At the same time, this recurrent trope is presented as the very cause for Apple's initial decline. Only toward the end of the movie will Jobs find people who share his vision, such as industrial designer Jonathan Ive (Giles Matthey), and the eventual success of Apple underlines his unfailing instinct. The (informational) world we know today, the film suggests, was only made possible by the prophetic vision of Steve Jobs, while everyone else, as Sculley professes, "got blinders on" (*Jobs*) throughout the digital revolution. This trajectory overtly concurs with what Custen identified as "the biopic credo of the person who has personal visions where others merely wear blinders" (191). Accordingly, the main drama of the movie (though some critics would argue that it lacks any) does not lie in the invention of the first Mac or the iPod, but in the visionary Jobs's struggle to convince the disbelievers that his projected path for Apple is the right one. Passing mentions of the big competitors in the field, especially IBM and Microsoft, help to construct Jobs and, by analogy, the company he has come to personify both in the movie and in real life, as the 'rebel' who dares to oppose the complacent and fraudulent establishment. Fittingly enough, *Jobs* ends with the protagonist recording Apple's 1997 'Think Different' campaign, which is included in this chapter as epitaph and invokes exactly these connotations of the biopic subject as an (initial) outsider, underdog, and dissident. Not by chance does Custen laconically note that in the biopic, "[i]t is not the fittest that survive; it is the misfits" (156). Incidentally, this observation seems to echo Apple's animated marketing jargon, which enthuses that the 'crazy ones,' the 'misfits,' and the 'rebels' will change the world, or, as one of Jobs's favorite expressions goes, "mak[e] a dent in the universe" (*Jobs*). Fittingly, it is these exceptional individuals that the biopic purports to immortalize in its productions.

The routine portrayal of protagonists as utterly 'different' from prevailing norms and sentiments already shows how the classic biopic may construct its subject as a distinctly cool hero, who comes to embody seemingly countercultural virtues of rebellion, revolution, and nonconformity. Even though subjects deemed worthy of their own cinematic life story will have risen in and become accepted by mainstream American society, the biopic will construct them as unruly, vaguely anarchist figures who 'have no respect for the status quo,' as Apple's campaign professes, at least in the early days of their struggle. Accordingly, *Jobs* devotes an extensive montage to showcasing the protagonist's journey through India, capitalizing on all the popular stereotypes of 1970s hippie culture, from long, unkempt hair to bare feet. Zuckerberg, on the other hand, is invariably dressed in flip-flops and a hooded sweater, whether walking through the snow-covered campus or testifying during the depositions.[35] His utter disrespect for institutions and basic conventions of human conduct is apparent. This gets to the essence of contemporary popular notions of cool, which Pountain's and Robins's seminal *Cool Rules (*2000) defines as a "*permanent* state of *private* rebellion" (19, emphasis in the original) and "a new mode of individualism" (13). As the previous chapters of this study have shown, the cool persona is identified through a stance of defiance and stylized contempt for whatever happens to represent the cultural mainstream. Not by chance did several youth sub cultures, from the beatniks of the 1950s to the punks of the 1980s, make cool an integral part of their philosophy, frequently re-baptizing it in the process. When one of the early icons of American cool and the defining image of 1950s youth, Marlon Brando in his role as Johnny Strabler, is asked what he is rebelling against, he nonchalantly quips: "What have you got?" (*The Wild One*, 1953).

Hence, what ultimately redeems Zuckerberg and Jobs, at least to the degree that the viewer does not feel repelled by their antisocial behavior, is the contradictory appeal of cool: While their restricted emotional range and obsessive egotism are repulsive, the two tech founders also exude an irresistible aura of nonconformity, displaying utter contempt for social conventions, rules, and institutions. These still accessible connotations of cool originate in the American counterculture, whose members cultivated cool as a way of setting themselves apart from the detested bourgeoisie. Cool became a "body armor" (Pountain and Robins 41) for the underprivileged and the dropouts, disillusioned by the continued condescension and neglect by the establishment. Particularly through its early association with jazz culture and black masculinity, the attitude quickly established itself as the dominant mode of existence for the rebels, crooks, and underdogs of American society. The biopic tacitly draws on these associations to create a dramatic conflict between the exceptional, yet misunderstood protagonist and the ordinary characters, who are seemingly lulled into conformity. In that sense, the genre must be understood as a

35 At the same time, this example also demonstrates the inherently democratic orientation of cool and its "physically forgiving" (Harris 46) nature. Zuckerberg's clothing style appears bizarre and unfashionable, which is, however, much more conducive to a cool attitude than the impeccable attire of the dressed-up Winklevoss twins: "Part of the success of coolness among self-conscious teenagers stems from the fact that it is [...] so inclusive, rejecting as it does the cookie-cutter aesthetic of 'normal' people, of Barbie and Ken, of the voluptuous blonde cheerleader and the all-American quarterback boyfriend" (46). The Winklevosses' muscular frame, their handsome features, and their membership in the Harvard varsity crew clearly disqualify them from cool.

perfect vehicle for presenting its subjects as cool by virtue of their requisite resistance to the establishment and the resulting stance of rebellious defiance.

The protagonist's journey to convince a given community of the significance and feasibility of his groundbreaking ideas will typically involve an actual or metaphorical trial scene (Custen 186-192). This narrative strategy enables the subject to defend his cause and present his vision to a larger audience, who stands in for the audience at home or in the movie theater. The trial scene has several benefits for the narration: It establishes clearly defined, opposing sides, articulates the hero's motivation, and lays bare the central thematic concern of the film. Within the trajectory of the protagonist's fight against the establishment, which derides his innovative ideas, the trial constitutes a form of climax after which the tide will turn for the hero. As such, it may often represent a turning point in the narration and prefigure the eventual dénouement. Even though many modern biopics do not actually contain trials, they will include a "trial-like setting" (Custen 186), such as when the lead character gives a speech, performs a show, or spontaneously addresses an agitated mob. While the protagonist thereby assumes the role of defendant and the film's main antagonist (or a community of conservative 'experts') will function as prosecutor, the role of the jury is metaphorically extended to the viewers. Ideally, they will be convinced of the hero's innovative vision through his persuasive performance, as Custen explicates: "[B]iopics with audience and jury metaphors are simulacra of shared judgment; we too, are won over by the virtue of a cause or the genius of an entertainer. Judgments of applause or jury verdicts are a conduit to our own assessment of the life and the aptness of its filmic version" (187). Whereas Custen seems to suggest that the public judgment thus induced will invariably be favorable, other critics have pointed out that biopics may equally function as satire, especially when notorious public figures such as tyrants, gangsters, and villains are portrayed. In these instances, the classic trial scene takes on a degree of debunking and marks the protagonist as clearly deviant and abnormal (Neale, *Genre* 63, Anderson 332).

To a certain degree, this is also the case in *Network*, which contains not one, but two trials (or rather depositions to avoid an actual trial). This structure allows the audience to assess the protagonist's actions simultaneously to their depiction on screen and further solidifies Zuckerberg's defiant, cocky, and almost disrespectful character. He appears utterly disinterested in the proceedings, distracted by the tiniest disturbances, and frequently refuses to give testimony, or even indeed to defend himself. In accordance with the notion of cool individualism, this rebellious stance makes him infinitely more relatable to the viewer than the condescending, uptight Winklevosses and their swarm of attorneys. Doodling away on a piece of paper, he appears utterly jaded when asked by this prosecutors' lawyer if he has his full attention: "You have part of my attention—you have the minimum amount. The rest of my attention is back at the offices of Facebook where my colleagues and I are doing things that no one in this room, including and especially your clients, are intellectually or creatively capable of doing" (*Network*). What the character of Zuckerberg hence portrays can be summed up as follows: Rather than *what* the story's hero testifies, it is far more consequential *how* he comports himself throughout the trial, and he evidently comports himself in the coolest possible manner. Custen's observation that the opinion of the biopic audience is altered "through rhetoric, or by the rhetoric of performance" (187) would accordingly have to be modified: The distribution of public favor will frequently be inspired by physical demonstration, rather than verbal persuasion.[36]

The second movie in this study, *Jobs*, supports this impression. Its most trial-like scene occurs, rather unusually, at the very beginning of the movie, when the lead character presents his triumphant invention, the iPod, to members of the staff in the Apple Town Hall. While Jobs gives an inspiring speech, dramatically calling the device a "tool for the heart" that "is gonna revolutionize an entire industry" (*Jobs*), before pulling it out of his jeans' pocket, it is the actual revelation of the gadget that draws gasps of wonder and amazement as well as thundering applause from his onlookers. Fittingly, Jobs's face is reflected in the gleaming surface of the iPod in a grandiose arc shot, so that his whole *raison d'être* seems to be embodied by the device. His accompanying words, "I like to introduce you to the iPod" (*Jobs*), support the personification of the gizmo as a subject with an agency that is bestowed on it by its creator. An orchestral score underlines the sublime scene. The subsequent unfolding of Jobs's journey from his humble beginnings as a college dropout supports this interpretation: His struggles, deprivations, and frustrated attempts at convincing the ignorant establishment culminate in this little box, which will dispel any doubts in the twinkling of an eye. The presentation of the iPod does not require verbose explanations or descriptions (the only thing that Jobs reveals about its functions is that it is a "music-playing device" and can store "a thousand songs in your pocket"), but instantly overwhelms the metaphorical jury by its mere presence. Paralleling Zuckerberg's succinct responses to the verbose accusations of his prosecutors, the Apple founder employs cool as an effective figure of speech as defined by art critic and cultural theorist Dave Hickey in his elucidative essay "American Cool" (1999):

"To employ the figure of cool, [...] [o]ne doesn't assert anything. One simply declares some truth to be self-evident on one's own embodied authority as a citizen, without deigning to invest it with fancy justifications, personal explanations, or expressive urgency. One simply paints the soup can, or mounts the board in the slough of the wave and proceeds, sans drama,

36 In Custen's defence, however, one could also read his definition of the 'rhetoric of performance' as including nonverbal elements such as physical demonstration. This view would align Custen's notion of rhetoric with the classical *officia oratore* in ancient Rome. See, for instance, Ueding (2000).

into the fray. Cool, then, is theater without drama, *demonstration without pleading*, distinction without status, and dissent without violence." (n.p., my emphasis).

Cool is thus understood as an essentially democratic virtue, emanating "generous equanimity and visible integrity" (Hickey n.p.) as exemplified by outstanding personalities of American public life like Andy Warhol, George Washington, and Henry James, an accomplished circle among whom the film clearly also ranks its own subject, the great American innovator Steve Jobs. By analogy, the iPod as the materialized essence of his lifelong oeuvre emerges as "the very emblem of cool, of that which need only be seen to be believed" (Hickey n.p.).[37] The enthusiastic reaction of the intradiegetic audience as well as, by analogy, the projected audience of the film supports this argumentation. Via a cool mode of expression through demonstration rather than articulation, the classic biopic trial scene thus gains additional momentum by allowing the audience to directly witness the material evidence which confirms that the hero's outlandish, unconventional, and eccentric ideas have indeed proven visionary.

If the protagonist thus manages to gain favorable public judgment in the trial-like setting, he will eventually succeed in enforcing his vision and not only become accepted, but indeed venerated by the establishment against which he originally rebelled. His contributions to society are so grand and influential that he has "reformulate[d] the boundaries of a given community," or "create[d] a Kuhnian paradigm in an already constituted field" (Custen 72) by the end of the movie, and, as the plotline usually implies, the end of his journey to fame. Thus, the classic biopic ends with a scene of vindication. The films of this study largely achieve this effect in their end credits, which inform us of the lasting success of its subjects: Through title cards, the viewer learns that Zuckerberg is "the world's youngest billionaire" (*Network*) and that Apple "became the most valuable company in the world" (*Jobs*) in 2012. Ultimately, it is the production of the biopic itself that acts as a form of vindication, especially for subjects who are still alive.[38] Apart from that, scenes depicting the failures of the protagonist's original opponents present the ultimate poet-

37 Hickey further argues that "the idea of cool as an incarnation of democratic, secular virtue is a peculiarly American one" (n.p.), as first testified by the Founding Fathers. Their joint oeuvre, the Declaration of Independence, is therefore understood as a prime exercise in the use of the rhetorical mode of cool, as it plainly and straightforwardly states, rather than verbosely pleads beliefs that are simply deemed 'self-evident.' According to Hickey, such a self-assured, unagitated assertion of authority, which the fictional Jobs emulates trough his dramatic demonstration of the iPod, is not only typically American, but ostensibly cool.

38 This not only applies to *Network*, but is partly also true for *Jobs*, since production for the latter started while the protagonist was still alive.

ic justice for the hero: His former enemies have to concede that he has always been a true visionary whom they fatally failed to recognize. Consequently, CEO Gil Amelio and original angel investor Mike Markkula meekly beg Jobs to return to Apple after the company finds itself in a downward spiral, seemingly caused by the absence of its visionary genius. Fittingly, he fires them both after being announced the new interim-CEO, taking late revenge for their 'betrayal' a decade earlier. Despite its markedly bleaker vision of fame and power, *Network* also depicts the public recognition Zuckerberg receives with his invention—and, as the plot implies, rightfully so, despite the two tiresome lawsuits he had to sit through. The biopic's version of the American rags-to-riches trajectory can thus be summed up as follows: "The standard biopic involves a maverick who is at first scorned by the establishment but who him/herself has become part of, or even a symbol of, that (chastened and modified) establishment by film's end" (K. Doyle 391). *Jobs* and *Network* largely conform to this formula: Zuckerberg's name is at the tip of everyone's tongue when social media are invoked, while Jobs has become virtually synonymous with his company's much celebrated product range. More broadly speaking, they represent two of the most widely known representatives of the information revolution and have become renowned as the country's foremost tech entrepreneurs. And for many of their fans, one may well assume, their stories also function as inspiring accounts of two former nerds who made it big.

In both *Jobs* and *Network*, the association of the standard biopic plotline with notions of a cool lifestyle thus becomes particularly acute through the close entwinement with the values of the information society and high technology on the one hand and the persistent appeal of the American Dream on the other. The biopic hence merely needs to extend these traits to the person we come to associate most closely with the company, its notorious founder. In this way, the genre constructs lives and behavioral patterns as much as it purports to portray them. Through their definite depictions of certain practices and areas of life, whether professional or private, biopics illustrate how one should behave in a given situation or profession to gain as much success as the protagonist did. With the exception of the depiction of 'dark celebrities' (Kitch 2007), the plot will ultimately redeem its subject, who still achieves, after the requisite tribulations, misfortunes, and dead ends, the professed goal. After all, he or she would hardly have been deemed a worthy subject for a feature film if it had happened otherwise. Hence, the biopic not only constructs a popular version of both political and private history for its audience, it also "virtually define[s] for uninitiated viewers entire realms of endeavor and ways of being" (Custen 16), especially *successful* being. In the case of *Jobs* and *Network*, this is the realm of information technology, computer science, and, even if the films do not openly purport to display it, corporate business. Particularly because movies about inventors in the digital, rather than industrial age have, so far, been rare,[39] the biopics of this study construct a clear vision of what a "famous computer person" (*Network*) should be like, how he should found and lead a company, which gender and ethnic background he preferably has, which social interactions he engages in, and which (negative) character traits and behaviorisms are apparently obligatory for the contemporary tech genius. Hence, Susan Drucker's and Robert Carthart's *American Heroes in a Media Age* (1994) and its conclusion that "computer programmers constitute a new kind of hero" (23) may well prove prophetic: If one subscribes to the basic assumption that the choice of national heroes is symptomatic of the values, desires, and needs of a society, the information age of the twenty-first century seems to be in dire need of a cool nerd, a former oxymoron that no longer seems to hold.

"How does somebody know what they want if they've never even seen it?" Cool Capitalism and the Scientist-Entrepreneur

> When you give everyone a voice and give people power, the system usually ends up in a really good place.
> MARK ZUCKERBERG/INTERVIEW ON ABC, 2010

While the previous chapters of this study explored novel roles adopted by the classic researcher as scientist-detective (*CSI*) or scientist-hero (disaster movies), the present analysis focuses on yet another popular hybrid which one may frequently encounter in contemporary American popular culture: the scientist-entrepreneur. The cinematic portrayals of Jobs and Zuckerberg are prime instances of the deeply American figure of the inventor (Haynes, *From Faust to Strangelove* 163), who has been updated for the demands of the modern information age. However, and quite similar to the historical forerunners they are frequently compared to, above all national heroes Thomas Edison and Alexander Graham Bell (both of whom were early enshrined in their own celebratory biopics), Jobs and Zuckerberg are venerated as successful entrepreneurs as much as they are praised for their ingenious inventions. Indeed, Jobs never assumed a crucial role in the actual manufacturing of his

39 Examples for biopics about inventors from the early twentieth century would be *Edison, The Man* (1940), *The Story of Alexander Graham Bell* (1939), *The Story of Louis Pasteur* (1936), and *Madame Curie* (1943). See Custen 253-254.

company's products, as critics protesting Wozniak's marginalized contribution have asserted (e.g. Ch. Caldwell 2013, McCracken 2013). Similarly, Zuckerberg only became a worthy biopic subject as soon as his website incorporated advertisements and hence moved from a lone nerd's unprofitable obsession to a multi-billion endeavor. Therefore, the field of expertise which both tech founders have contributed to through their prescient vision and unorthodox methods is as much located in the corporate business world as it is in computer science.

The movies under scrutiny do tacitly reflect this observation: The turf which the protagonists interact in is not restricted to, as one may assume, the IT industry, but is, in effect, much vaster, involving the world of global corporate business. The invention of Facebook or the iPod are merely the means by which the protagonists establish themselves as true American entrepreneurs, whose material success owes itself to their laudable characters and not merely to their inventions. This is already suggested by the main antagonists of each film, i.e. the Winklevosses and Apple CEO John Sculley. Neither of them has any background in programming or computer science: The twins major in economics and will soon inherit their father's consulting firm, while Sculley is the former president of soda manufacturer PepsiCo, another world wide exporter of American cool. Hence, the 'establishment' which Jobs and Zuckerberg are depicted as fighting can hardly be identified as the American IT industry, which has to be convinced of unconventional coding techniques or the eccentric alignment of diodes on a motherboard. Indeed, one may assume that such a battle of nerds, safely kept within the confines of Silicon Valley, would have hardly attracted the attention of filmmakers.[40] It is, in effect, not the inventions of Zuckerberg and Jobs per se, but rather the unparalleled success, public acclaim, and commercial returns of these inventions that render their life stories viable for a Hollywood biopic. This circumstance is also noted in several reviews of *Jobs*, which identify the focus on corporate business operations as the defining link between the two biopics: Owen Gleiberman argues that "[l]ike *The Social Network*, *Jobs* is a technology yarn that's really a high-wire business story" (n.p.), while Christopher Caldwell muses that "it is neither the inventor Edison nor the artist Disney to whom Jobs bears comparison, but the manager Ford" (n.p.). In that sense, both movies can as reasonably be treated as business films as they trace the lives of

40 The only instance that could be identified as a dispute between engineers, rather than businessmen, occurs in *Jobs*, when the protagonist is on the phone with Bill Gates, whom he accuses of having stolen his ideas for a new operating system. Again, however, both men interact in their roles as entrepreneurs in an issue of corporate law, rather than inventors with divergent visions of personal computing. Interestingly, the movie here subverts a plotline employed by *Network*, in which it is the lead character who is accused of intellectual property theft.

famed twenty-first century inventors in a biopic fashion. So why do both movies ostensibly whitewash Jobs's and Zuckerberg's roles as shrewd entrepreneurs?

One possible answer to this question can be found in the underlying politics of the Hollywood biopic. The cinematic representation of the protagonists' professions is largely in line with the overall orientation of the genre from its early beginnings. In comparison with epics about scientists, researchers, and inventors, biographical films about famed businessmen are rare. Custen goes as far as declaring that "one of the many absences in biopics is the late nineteenth and early twentieth century entrepreneur" (77), an assessment that can safely be extended to later decades. Iconic figures of early corporate America, including Henry Ford, Andrew Carnegie, and Henry Frick, seem to have been symbolically annihilated. Their absence is indeed significant, as all three of them are remembered as the first truly self-made billionaires and could thus serve as poster boys for the American Dream, both in their own time and in contemporary reconfigurations. While the biopic reveals a strong predilection for the classic rags-to-riches trajectory and preferably imposes it, as the previous section has shown, on almost every life story irrespective of profession and historical era, it has so far noticeably refrained from exploring the myth in the very habitat in which it originated. Custen's analysis offers a potential explanation for this tendency, suggesting that the absence of the world of business helps to construct a "sanitized view of history" (77) by eradicating potentially problematic, controversial areas of public life. The dubious and, for the average viewer, inscrutable world of corporate business could prove too alienating a subject to be explored in the mainstream Hollywood league.[41]

Another conceivable reason for the apparent lack of entrepreneurial biopics is the fact that the genre traditionally stresses the importance of innovation and novelty for societal progress. Bingham formulates the overt dedication of the genre to the idea of progress as follows: "The biopic usually concerns a subject who in some way brought modernity to society. Almost any subject worth making a biopic about, famous or not, produced something new and unique" (*Whose Lives* 133). A true biopic hero must have brought lasting benefit to the world, which includes the actual audience of the respective film. Ideally, they will rejoice in the legacy of the protagonist, a vital factor why they might have chosen to watch the movie in the first place. Accordingly, it is not enough for the Hollywood biopic to show *who* the famed person was (professionally or privately), but also *why* he or she is rightfully remembered as one of our society's Greats. It is thus easily conceivable why entre-

41 Notable exceptions are, for instance, the already mentioned *Wall Street* (1987) and *The Secret of My Success* (1987) as well as Martin Scorsese's recent critical success *The Wolf of Wall Street* (2013). The majority of films exploring businesspeople and entrepreneurs, however, reveal a pronounced focus on the protagonists' personal struggles and private lives, rather than their obscure business operations.

preneurs and businesspeople have traditionally been avoided as viable subjects: While their life stories will usually conform to the American Dream narrative well enough, their occupational field is known for draining, rather than enriching society. For filmmakers, it may simply appear too hard to conceive of any invention the capitalist has gifted the world as being worth celebrating; on the other hand, putting across a story about innovations in a field of expertise that is utterly alien and obscure to the majority of viewers would prove challenging. Therefore, movies will attempt to shift the focus to other realms of enquiry in order to weaken or even eradicate any entrepreneurial associations with the subject. Consequently, financier Jim Brady is presented as frustrated lover rather than crafty capitalist in *Diamond Jim* (1935). In a similar vein, "Edison and Bell are framed as inventors, donors of fabulous inventions to the public welfare, and thus are sanitized as capitalists" (Custen 77), a strategy that also seems to have been applied to the cinematic lives of Jobs and Zuckerberg.

Indeed, *Network* and *Jobs* seem overtly intent on marginalizing or even disguising their protagonists' roles as cunning entrepreneurs by constructing them as inventors and innovators instead. Ashton Kutcher's much publicized assertion that Steve Jobs was "the Leonardo da Vinci of our time" (Kutcher, Interview I) might sound pretentious for an entrepreneur selling high priced gizmos allegedly produced through labor exploitation, but certainly reinforces the line taken by the movie: Both Kutcher's and the filmmakers' understanding of Jobs as a Renaissance polymath, who seamlessly unites unsurpassed artistic and scientific qualities, subscribes to a framing of the protagonist as devoted to much nobler occupations than base commerce. Da Vinci pursued a great number of professions, but one of the few things he is surely not remembered as is a mercenary businessman solely intent on maximizing his profit. *Network* and *Jobs* seem determined to create a similar remembrance for their flawed, yet ultimately triumphant subjects.

These ostensible politics of representation concur with the larger ethico-aesthetic orientation of the genre. In contrast to businesspeople and entrepreneurs, researchers have been among the most prized biopic subjects from the very beginnings due to their conspicuous connection to those virtues to which the biographical film has traditionally been dedicated, above all innovation, nonconformity, and the opposition to existing (and allegedly flawed) paradigms. Hence, they constitute one of the largest occupational groups portrayed, second only to artists and entertainers, who display the desired traits through their innovative performances: "Non-entertainers, on the other hand, can be framed as radical because they deal not in art, but in the knowledge business. Thus, opposition to them is opposition to progress, and scores of biopic films tell us that science is nothing else if it is not progress" (Custen 210). Hence, the construction of a seemingly natural proximity of the protagonists of *Jobs* and *Network* to computer science equips the films with a welcome prop by which to implement a generic plotline, typology, and mise-en-scène. Additionally, their professions as innovative programmers and engineers, redefining not only the boundaries of their immediate community, but also of the greater American media landscape, can easily be depicted to reflect back on their mentality and character: Just as their chosen profession requires abstract thinking and bold decision-making, Jobs and Zuckerberg are shown to be emotionally deficit, volatile, obsessive, and radical in their interaction with partners and employees, "as if their field of endeavor virtually defines their personality" (Custen 211). Their framing as scientific explorers rather than entrepreneurs hence proves fertile for the genre in several crucial ways.

Granted, not even the conventional biopic can construct the college dropouts Jobs and Zuckerberg as 'scientists' in the strict sense. However, and even more apposite to its purposes, they are depicted as pursuing the most American of intellectual professions: In line with national heroes like Bell, Edison, Benjamin Franklin, and Nikola Tesla (who, as Custen points out, had to be equally sanitized from corporatist associations before being granted their own biopic), the tech gurus are presented as practical inventors solely dedicated to the progress of their respective fields. Being much more realistic and hands-on in orientation than the pure scientist, the figure of the inventor and/or engineer ultimately emerges as a deeply American one. Notions of practicality, resourcefulness, and ingenuity figure as distinctly American virtues and speak to a country of inveterate tinkerers, builders, and former frontiersmen, who pride themselves on their inventiveness. Extended displays of actual building, coding, and engineering, depicting the strenuous labor upon which digital innovation relies, allow both movies to showcase typical American virtues of industriousness, physical strength, and practical ability. These images are meant to speak to the average backyard tinkerer and amateur scientist, who may pride themselves on their inventiveness. The creation of state-of-the-art technology from whatever happens to be at hand in one's garage or college bedroom establishes the United States as a country of ingenious builders and inventors. The American nation, as both films seem to imply, is essentially built on the ingenuity and resourcefulness of audacious individuals like Jobs and Zuckerberg.

Roslynn Haynes explains this predilection for applied knowledge work as follows:

> "The inventor whose discoveries prove to be of surpassing benefit to mankind is a feature of American rather than of European literature […]. In the new world, which still espoused many frontier values, the inventor who typified the ideals of progress and adventure was perceived as a vital member of society, while the pure scientist was regarded as not only less useful but potentially less trustworthy, even sinister, tainted with the moral flaws of the alchemist." (*From Faust to Strangelove*, 163)

While this overall negative image of the scientist has, as this study also strives to reflect, considerably changed over the last decades, the decisively conservative genre of the biopic still seems to portray a marked disdain for the pure scientist. Indeed, in its long history since the 1920s, the genre has tended to portray scientists only if they were of European descent and hence sufficiently 'exotic' to the American audience, such as in the classics *Madame Curie* (1943) or *The Story of Louis Pasteur* (1936). Typically, these scientists were portrayed as *idiots savants*, i.e. failures in every area of their life other than their scientific work (Custen 164). National heroes such as the cult figure Edison were preserved for the role of the inventor, an inherently more positive, productive, and powerful individual who can actively shape the country's future, rather than merely ponder it.

The cinematic portrayals of Zuckerberg and Jobs support this observation. Among the scant praises the latter received was that it conveys the impression of a "DIY-spirited movie" (Neumaier n.p.) and "channels the early paradox of computer culture: that to us the gizmos seemed light, virtual, magical, but to the people who invented them, they were *built*, diode by diode" (Gleiberman n.p., emphasis in the original). The material dimension of Jobs's unsurpassed gift to the world, i.e. the actual building of the early Apple devices, is nicely visualized through a slow-motion montage of the company's original crew slaving away in the garage to a catchy 1970s tune. Indeed, the mythic location of Apple's origins perfectly underlines the American inventor as a prime biopic subject, who is both practical and highly individualistic: Jobs's and Wozniak's careers are depicted as having germinated with some random garage tinkering with whatever utensils were at hand, which could not more effectively perpetuate the deeply American virtues of resourcefulness and ingenuity. Epitomizing the ideal of the American garage inventor, Jobs and his team prioritize practical skill and craftsmanship over arcane research and place emphasis on the hands-on aspects of their work. Devoting extensive screen time to these activities not only vitally assists in advancing a cool aesthetic, but also considerably democratizes the notion of abstract science by emphasizing the cognitive and economic superiority of practical engineering.

Network contains similar scenes of obsessive, deeply focused 'manual' labor, such as when Zuckerberg or one of his interns hacks away on the keyboard and is completely oblivious to everything around him because he is "wired in" (*Network*), the hacker slang for working intensely on a code. The opposition between the scientist who merely thinks (or even worse, talks) and the inventor who actually does becomes most apparent in the lecture hall scene: Zuckerberg is attending the Operating Systems class, considered the hardest class in Harvard. After a girl has handed him an insulting note, Zuckerberg abruptly leaves the auditorium, which prompts the professor to assume that he is "our first surrender" (*Network*). On the stairs, Zuckerberg turns around and calls back the correct answer to the coding problem from the back of the lecture hall. Outside the hall, he is caught by the Winklevoss twins, who have already been waiting for him and will shortly initiate him into their idea of a university social network.

This scene exemplarily illustrates some of the film's most pertinent underlying politics as concerns its treatment of intellectual genius: It is not that the inventor is not bright enough for an academic career, but he consciously decides to apply his genius to a useful (and profitable) purpose rather than letting it go to waste in an ivory tower. Paradoxically for biopics that laud scientific invention and technological progress, (academic) education is framed as impedimental, rather than conducive to upward mobility. As the young Steve Jobs tells one of his professors at Reed College: Higher education "comes at the expense of experience" (*Jobs*). The film suggests that if he had spent his time in the classroom, rather than in India or on epiphanic LSD trips, he would not have gained the insights necessary to excel in his business as rapidly and forcefully as he did. Although set at Harvard, *Network* subscribes to the same line of argumentation: Zuckerberg appears to be naturally gifted with outstanding programming skills, so that he is usually bored in his classes and does not seem to suffer for a lack of essential knowledge once he has left the lecture hall (and the campus at large) for good. Ultimately, both Zuckerberg and Jobs choose practical careers over abstract higher education and drop out of college to dedicate themselves fully to their world-changing inventions. Literally and metaphorically, they both have to venture out of the stifling environment of academic knowledge production in order to gift the American people (and indeed the whole world) with their genius. Similar to the publicly venerated Edison, they emerge as the "practical, democratic individualist" (Wachhorst 131) at the end of the movie, an expression that has become virtually synonymous with the ideal American citizen. Their current status as icons of US computer industry and the global technoscientific revolution (especially so in the case of the late Jobs) speaks to the continuing societal trend of framing American knowledge workers as neither scientists nor entrepreneurs, but practical inventors.

Through this deliberate presentation of the protagonists as anything but the global capitalist players they are, the biopic in general and the two movies analyzed in this study in particular perpetuate a powerful vision of capitalist co-option. Zuckerberg's and Jobs's crude business tactics, their 'unorthodox' (i.e. merciless) managerial style, and their willingness to transgress ethical and moral boundaries in pursuit of their 'grand vision' are all legitimized through their framing as ingenious innovators, whose inventions will reformulate the boundaries of their disciplines and overthrow our one common enemy, the dull and greedy establishment. The film suggests that the world must bear their volatile tempers, erratic decisions, and emotional deficits because, rather than be exploited by them, it will ultimately benefit from their gifted minds. Bad business behavior associated with global capitalism such as labor exploitation and the questionable treatment of employees (*Jobs*) or intellectual property theft and corporate betrayal (*Network*) are coded as necessary preconditions for true genius: "[W]ith an unusual gift comes unusual suffering" (Custen 75), both for the protagonists and their immediate environment. Zuckerberg and Jobs cannot be the 'bad guys,' the movies seem intent on informing its audience, because they are too idealistic, hippie, countercultural, unassuming, rebellious, charismatic, unconventional, defiant, and smart to bear any relation to those indistinguishable, middle-aged men in gray suits who maximize their profit by luring us into mass consumerism. They are, in short, too cool for capitalism.

As several crucial studies of the last decades have shown, however, it is exactly this stance of countercultural rebellion represented *par excellence* by the cinematic portrayals of Jobs and Zuckerberg that contemporary global capitalism thrives on. In 1997, Thomas Frank's highly influential study *The Conquest of Cool* already demonstrated how American counterculture and rebellion sparked by various youth cultures was co-opted by the rising business forces of the 1960s, which employed those very ideas, slogans, and images to sell their products. Counterculture thus paradoxically became one of the driving forces for accelerating national and global consumption in the postwar era. Consumer products were presented as the very antidote to mass consumerism, literally capitalizing on notions of nonconformity, resistance, difference, deviance, and escape which brands like Pepsi, Volvo, or Camel promised to magically project on to their buyers. "Non-conformity was fast becoming the advertising style of the decade, from the office antics of the now unleashed creative workers, to the graphic style they favored, to the new consumer whose image they were crafting" (95), as Frank asserts. By self-reflexively mocking the standard ad strategies of the industry and ironically debunking its empty promises, consumption of mass-produced consumer goods was promoted as transgressive, rebellious, subversive, and the most effective (and effortless) way of accentuating one's individuality. Through these strategies, capitalism not only co-opted the cool stance that once belonged to society's underdogs and rebels, but became, in fact, cooler than any countercultural movement had ever been.

One does not need to venture very far to ascertain that this strategy has flourished since the 1960s: The notorious marketing campaigns of Apple, which is, as the movie reminds us, "the most valuable company in the world" (*Jobs*), have always striven to invoke exactly these connotations of dissidence, difference, and revolution. Apple is the brand for 'people [who] are crazy enough to think they can change the world' and will do so by virtue of their newly acquired Macintosh or iPod. The notorious 1984 Super Bowl commercial, which *Jobs* depicts almost in its entirety during the protagonist's keynote speech at a staff meeting, even references George Orwell's dystopian *1984*. A young blonde woman in tiny red shorts smashes a hammer at a big blue screen depicting a preaching dictator in tight close-up,[42] which a homogenous mass of apparently brainwashed citizens in indistinctive gray overalls stare at. The screen bursts into flames and the onlookers appear to be freed from their invisible shackles. The associations this ad aims for are clear: We are the good guys, the rebels who will battle against the establishment and the mass consumerism promoted by our competitors. Interestingly, Apple has managed to retain its precious image of the industry's underdog even after attaining the position of market leader and surpassing former rivals IBM and Microsoft. *Jobs* must be read as vitally contributing to this carefully crafted corporate image by advancing a vision of Apple and its notorious founder as bold, eccentric, and unconventional.

Facebook ascended through a comparable, yet similarly unique strategy, namely by focusing on the notion of exclusivity. Indeed, *Network* draws attention to a fact that might appear surprising to anyone using the website a decade later, namely the frequently neglected fact that Facebook started as the virtual equivalent of an exclusive university club. Only Harvard students and people directly invited by them were allowed to join: "It's like a final club except we're the president!" (*Network*), Mark enthusiastically exclaims when he first comes up with the invention that will finally allow him to create his own cool clique. While the seemingly high culture connotation of an elite club within an already elite university seems to be diametrically opposed to the underground associations which Apple strives to invoke, both companies' marketing strategies as presented in the films ultimately amount to the same outcome: An openly portrayed disdain for mass consumerism and, by analogy, the mainstream consumer (whom no one would identify as anyway) that helps to mark their products as different, individualistic, and cool. The movies tacitly acknowledge the appeal of the depicted products as pertaining less to quality or even indeed innovativeness, but social status and prestige: "Why do people buy an Apple?," Jobs asks in a staff meeting, and instantly gives the answer: "[B]ecause it's social status, [...] it's social currency" (*Jobs*). The elite connotations of Facebook work in a similar way: Registration originally required a harvard.edu email address, "[t]he most prestigious email address in the country" (*Network*), as the Winklevosses proudly assert. The implications of both quotes are clear: In a world where cool has become global currency, what could possibly be more prestigious, lucrative, and desirable than two of the coolest tech brands of the decade?

Even the production conditions of both movies seem hinged upon the enveloping appeal of a non-conformist advertising style as dissected by Frank (1998): The status of *Jobs* as an 'indie' production with a minor budget was laudably noted by critics, who complimented the filmmakers' courage to compete with the establish-

---

42 Hardly coincidentally, IBM, Apple's biggest competitor at the time, is nicknamed 'The Big Blue.' Fittingly, the Apple board in *Jobs* reacts furiously when the protagonist seems intent on "starting a war with IBM" (*Jobs*).

ment, i.e. big studio productions. Its official world premiere in January 2013 at the Sundance Film Festival, one of the largest independent film festivals worldwide, abet this claim to the image of the industry's underdog. It seems as if director Joshua Michael Stern and his crew ingeniously exploited their status as cinematic rebels by concentrating the promotional efforts on alternative, 'non-mainstream' channels: One of the movie's trailers premiered on the online photo-sharing platform Instagram, while a somewhat failed marketing campaign tried to gather technophiles at the movie theater as if they were awaiting the launch of a new Apple product. The movie's soundtrack also capitalized on countercultural associations, above all by including several songs by Jobs's favorite musician Bob Dylan. *Network*'s advertising strategies appear less blatant, but seemed equally intent on evoking associations of anarchy and rebellion: The official movie posters kept to subdued, cool colors and depicted an emotionally blank Zuckerberg staring at the onlooker in tight close-up, with the catch words "punk—genius—traitor—billionaire" written across in big white letters. More than half a century after the onset of cool business culture in the 1960s, the appeal of cool as a strategy for stimulating sale figures appears undiminished, even in an industry that has been an early adopter, to borrow a phrase from the world of personal technology.

Taking as his departure point Frank's 1997 study and his focus on the development of hip advertising, Jim McGuigan has extended the argument to describe the state of global capitalism in the 2000s. His insightful analysis *Cool Capitalism* (2009) asks the cardinal question of how capitalism is justified today. Early stages of capitalist activity, famously dissected by Max Weber (1920), are understood as being indebted to Protestant ideology (especially its Calvinist ethic of predestination) and an accompanying ascetic value system: It stressed hard work, rationality, and industriousness, while gratification was deferred to the afterlife. Today, these notions are no longer associated with the capitalist system, which has developed a distinctly hedonistic spirit (see Boltanski and Chiapello 1999). Following Frank's analysis of countercultural co-option, McGuigan mockingly refers to this version as 'cool capitalism,' which he defines as "the incorporation of disaffection into capitalism itself" (*Cool Capitalism* 1). Contemporary capitalism is thus understood as being legitimated by incorporating, and indeed neutralizing, anti-capitalist critique, rebellion, and dissent into its very theory and practice. Through this strategy, neoliberal capitalism constructs a highly resilient popular legitimacy that has thoroughly seeped "into the texture and common sense of everyday life in spite of severe and recurrent economic crisis" (McGuigan "Coolness," 431). Currently, the 2009 credit crunch stands as its most noted example. Cool capitalism helps to disguise the dark sides of globalization, such as labor exploitation in manufacturing countries like China, of which Apple in particular has been repeatedly accused (see

Duhigg and Barboza 2012).[43] Through a carefully cultivated rebel image, companies like Apple and Facebook not only epitomize consumer cool, but also partake in the ideology of contemporary cool capitalism to legitimate their capitalist ventures and ethically questionable practices.

The films of this study perfectly illustrate this argument. A recurring theme in *Network* is Zuckerberg's biggest nightmare of ruining the intrinsic coolness of his creation—which to him as the quintessential outsider appears highly worthy of protection—with the addition of ads. When Eduardo suggests that they need to "monetize the site" (*Network*), Zuckerberg vehemently declines:

> Mark Zuckerberg: 'Cause theFacebook [sic] is cool. If we start installing pop-ups for Mountain Dew it's not gonna…
> Eduardo Saverin: Well I wasn't thinking Mountain Dew but at some point—and I'm talking as the business end of the company—the site…
> Mark Zuckerberg: We don't even know what it is yet. We don't know what it is, we don't know what it can be, we don't know what it will be. We know that it's cool, that is a priceless asset I'm not giving it up.
>
> (*Network*)

Later on in the film, Napster founder Sean Parker reiterates this point, thus fuelling Zuckerberg's resistance to exploiting the website too early: "TheFacebook [sic] is cool, that's what it's got going for it. […] You don't want to ruin it with ads because ads aren't cool" (*Network*). Parker and Zuckerberg thereby perpetuate the original countercultural stance before the onslaught of cool capitalism: Corporate business is dull, bothersome, deadly for creativity, and the natural enemy to anything cool. As Frank and McGuigan show, however, this dichotomy between the cool counterculture and the square capitalist mainstream must be exposed as inherently flawed: Just as business culture has less co-opted rebellion and dissidence than expanded and re-invented it to a never before seen degree, contemporary cool is similarly hatched in the ad agencies and think tanks of big corporate enterprises rather than on the streets or in the dorm rooms. In the age of cool capitalism, big business has become the ultimate locus of cool. The movie's portrayal of youthful, 'accidental' billionaires in hoodies, slacks, and with purportedly idealistic visions of their creations underlines this view. The latter quality is, however, soon revealed as tainted: The only reason Parker supports Zuckerberg's disdain for ads is not because he is devoted to a naive view of Facebook free of capitalist profit, but because

43 See also the independent not-for-profit organization China Labor Watch and its report on the series of suicides committed in 2010 by employees of the Foxconn construction site in Shenzhen at <http://www.chinalaborwatch.org/pro/proshow-176.html>, accessed 13 Sept. 2013.

he is an experienced entrepreneur who recognizes the utmost importance of determining the right moment for starting a business. Hence, he advises Zuckerberg to act prudently in order not to forego the greatest possible revenue: "A million dollars isn't cool. You know what's cool? A *billion* dollars" (*Network*, emphasis in the original). In the logic of cool capitalism, what is vastly cooler than a website that remains free of ads and hence provides a potential platform for genuine public interaction is, of course, maximum profit.

With its vision of the teenaged nerd eventually surrendering to market forces, *Network* supports contemporary bleak outlooks on the state of coolness today. As a case in point, McGuigan not only asserts that cool is an essentially free-floating, meaning- and colorless empty signifier, but also that it has exchanged its former leftist (or, in the political dictum of the US, 'liberal') associations with a more right-wing affiliation; it has become "more a sign of compliance than of resistance" ("Coolness" 432), as he resignedly observes. There are a number of studies which support his negative assessment. Kalle Lasn's *Culture Jam: The Uncooling of America* (1999), for instance, argues that cool has devolved into a "heavily manipulative corporate ethos" (xiii-xiv). Pountain and Robins bemoan the fact that today, even the smallest ad agency employee knows how to "deploy Cool as a selling tool" (169), while Harris (1999) calls cool "the ultimate consumerist world view" (43). Ted Gioia's *The Birth and Death of the Cool* (2009) even suggests that the continuing capitalist exploitation of cool has been its downfall:

> "[N]o contemporary phenomenon is more reified than the cool. What originally was an attribute of individuals, almost a type of charisma or personal magnetism, has been ruthlessly exploited by profiteers and forced to pretend that it exists inside a pair of running shoes, a fragrance, a new style of jeans, a movie, a logo." (21)

As already outlined in the theoretical part of this study, I would vehemently contest the argument that cool is in decline as a dominant cultural sensibility—the preceding chapters should have provided ample evidence that it is alive and kicking, though in previously unsuspected terrains. Nonetheless, I do concur with Gioia's identification of manufactured goods as the ultimate bearers of an (allegedly) easily procurable coolness. The most evident sites of the contemporary corporate ethic are indeed cool consumer products, with sleek techno gadgets on the front line. Accordingly, McGuigan calls the smart phone (or the *all-purpose mobile communication device*) "the coolest of all commodities today" ("Coolness" 432). Both Facebook and Apple heavily depend on it for sales: The mobile version of Facebook continues to be one of the most popular applications on (and tacit selling propositions for) the iPhone, and initial difficulties with properly integrating ads presented one of the most difficult problems for Zuckerberg in recent years. Today, Facebook gains almost half of its revenue from the mobile app, and remains confident that it will soon surpass desktop profits (Goel 2013). The value of the iPhone for its inventor Apple needs no further explanation.

Similar to my own analysis, McGuigan connects the unsurpassed cool consumer appeal of the smart phone to today's heightened significance of neoliberal virtues like individuality and personalization, which such a device excellently embodies. With the smart phone, the users undergo a highly individualized experience (see the corporate catch word 'customization') and find themselves in a "consumerist cocoon" ("Cultural Populism" 15) safely removed from the communicative potential of the public sphere, from which civic dissent may arise.[44] In *Network*, Zuckerberg's first creation, facemash.com, already has his co-eds hooked to the screen, mostly alone or in pairs. The montage depicting the tidal wave of success primarily consists of shots that peer through a window or an open door into the interior of a tiny dorm room, where students huddle around the only source of light emanating from the computer screen. A feeling of claustrophobia afflicts the onlooker. The activity Zuckerberg has ingeniously wheeled into existence is clearly reserved for indoors and the private sphere. Despite Facebook's later motto that it 'helps you connect and share with people in your life,' the smooth user surfaces and glittering gizmos of the digital age emerge as the ultimate consumer items for the "cool, technologically savvy and neoliberal *self*" (McGuigan, "Mobile Privatisation" 75, my emphasis), rather than encouraging communitarian use. Thereby, both companies, bluntly put, create the ideal customer for their products *through* the very products: The smart phone and its sheer endless range of apps ingeniously conceal its mass-produced condition by presenting the user with a highly customized experience. Online profiles, personalized shopping recommendations, and even paraphernalia as trivial as iPhone cases allow an unprecedented degree of creation, representation, and celebration of the (virtual) self for the iGeneration.[45] As Frank's and McGuigan's perceptive studies show, cool thereby achieves the ultimate paradox: It is the magic potion luring users into buying a mass-produced communication device for accentuating their individuality and rebelling against a somewhat obscure estab-

44 While virtual activities like blogging or sharing videos online have been lauded for their subversive civic potential, especially so in military dictatorships, McGuigan maintains that these practices have to be classified as "producerly consumption" ("Cultural Populism" 16), which only creates the illusion of participation in the public sphere.

45 'iGeneration' is a term coined by popular sociology to describe contemporary adolescents. See, for instance, Larry D. Rosen's *Rewired: Understanding the iGeneration and the Way They Learn* (2010), in which he describes the iGeneration as the legitimate heirs of the preceding Net Generation (or, more commonly, Generation Y or Millennials). Rosen defines the members of the iGeneration as "multitaskers, social networkers, electronic communicators and the first to rush to any new technology" (*Rewired* 2). See also chapter 2 of this study.

lishment, all the while safely cocooned in the carefully bounded 'private' space it has created for them.

Interestingly, the genre of the biopic, to return to the starting point of this chapter, offers a prime vehicle for the exploration of these dynamics. Drawing on Leo Lowenthal's content analysis of "Biographies in Popular Magazines" (1944), Custen's categorization of biopic subjects differentiates between Idols of Production and Idols of Consumption. The former were particularly prevalent between the Great Depression and the Second World War, reflected the Great Man (and occasional Woman) who made things for society, and was meant to encourage viewers to similar industriousness. Typical subjects included statesmen, soldiers, and, above all, practical researchers and inventors. With the onset of postwar consumer culture, whose effects on the budding counterculture Frank has dissected, the focus moved towards a fascination with celebrity culture. Therefore entertainers, artists, and sportsmen now dominated the genre, "whose conspicuous consumption the spectator can vicariously enjoy but also see the limitations of" (Bingham, *Whose Lives* 6). They were, in effect, themselves sold as the ultimate consumer product, which especially the sexualization of female actors and singers in the industry reflects. This development can be read as marking a shift from those who built the world around us to those who benefit from it by fast, yet (in comparison) fleeting fame, or, as Neale phrases it: "[T]he pre-war biopic tends to address its spectators as citizens whereas the postwar biopic tends to address its spectators as consumers of popular culture" (*Genre* 61). Some critics, such as Bingham (*Whose Lives* 5-7), argue that the Idol of Production has returned in neoclassical form since the 2000s. However, clear categorization has become more difficult nowadays, as many contemporary producer-oriented biopics contain a marked "warts-and-all or […] investigatory tinge" (Bingham, *Whose Lives* 6), i.e. a mixture of both productive and consumptive elements.

In the age of global capitalism, the distinction between consumption and production thus seems to have been irrevocably blurred, and indeed one cannot easily ascertain which category Jobs and Zuckerberg would have to be assigned to. Judging by their profession, they could, of course, be assessed as Idols of Production, since both of them invented an ingenious product and thereby, as the movies do not fail to stress, made "a dent in the universe" (*Jobs*). Similarly, they are portrayed as highly idealistic and, as their careless way of dressing and almost ascetic lifestyles suggest, rather disinterested in material consumption: "Mark doesn't care about money and he needs to be protected" (*Network*), his business partner Eduardo naively asserts to his lawyers, and will later have to learn Zuckerberg's cunningness the hard way. On closer inspection, thus, Jobs's and Zuckerberg's classification as Idols of Production becomes troubled, especially with Frank's and McGuigan's analyses in mind: The comparison of their consumer commodities to the development of a vaccine (*The Story of Louis Pasteur*) or the discovery of a chemical element (*Madame Curie*) appears somewhat daunting. At the same time, the preceding pages have shown that both figures are portrayed as two of the foremost celebrities of the digital age, and the films considerably glamorize their life stories by presenting the protagonists as heroic, larger-than-life characters that offer a perfect, presumably easily imitated surface to be admired and even indeed aspired to (McGuigan, "Mobile Privatisation" 82-87). Similar to an entertainer or a movie star, their looks, mannerisms, clothing styles, and way of talking have become iconic and highly recognizable, especially so in the case of the Apple founder who had been a tabloid staple for decades before his premature death. While the films contribute to their image as countercultural rebels and 'atypical' scientist-entrepreneurs by portraying neither Jobs nor Zuckerberg as concerned about material wealth and ostentation, their foremost activity is still that of conspicuous consumption and the vicarious pleasure that comes with it. Especially Jobs was repeatedly described as "the ultimate consumer" (Elliot and Simon 21) by his biographers and contemporaries, and this verdict is far from surprising when we take a closer look at the films of this study. Indeed, both of them tacitly suggest that the protagonists first and foremost thought of themselves as consumers and invented the respective product for themselves. "We just wanted to create cool toys for people like us" (*Jobs*), Wozniak muses, and Zuckerberg's enthusiastic realization that he will not only be able to finally join an exclusive club via his website, but act as its very president, takes a similar line. Contrary to the scientists of the studio era biopics, Jobs and Zuckerberg do not create their inventions to benefit the human race or for mere curiosity's sake; their devices are by no means "tool[s] for the heart" (*Jobs*), even if Jobs enthusiastically claims so. In accordance with one of the foremost principles of successful entrepreneurship, the tech gurus first identify an existing market for the future product, which they primarily base on their own sentiments. The term 'Idol of Consumption' hence adopts a new meaning in the age of cool capitalism: Not only did Jobs and Zuckerberg facilitate an unprecedented height of neoliberal consumption in the twenty-first century by introducing an innovative range of products, but they also created a seemingly empowering vision of the technologically savvy and markedly cool consumer through their own willing participation in the consumption of their commodities.

A key scene crystallizing the above points can be found in *Jobs*, as the following and concluding close reading will reflect. Tellingly, it is the first (and only) scene that was officially leaked before the release of the movie, which suggests that the producers may well have considered it representative of the film's larger politics as well as suggestive of its portrayal of the main characters. The scene is taken from the first half of the movie, after Jobs has witnessed Wozniak's unique talent for building hardware. He intuitively recognizes the monetary potential of this handicraft and becomes obsessed with the wish to generate fast and massive profit. He now tries to convince his business partner of the viability of his latest invention, i.e.

connecting a computer's keyboard to the television monitor so that "you can see what you're working on while you're working on it" (*Jobs*). Woz has thereby created one of the earliest operating systems, but remains doubtful that this invention is ripe for mass production (and capitalist exploitation), because he cannot imagine an ordinary person interested in buying such a set for home use. In the dimly lit garage of Hewlett-Packard's headquarters, i.e. literally underground the industry's establishment, a highly agitated, visionary Jobs picks up Woz from his day job as an HP engineer and tries to persuade him of his creation's enormous potential, hence playing a sort of adman for his own vision. Initially, Wozniak is highly skeptical and almost embarrassedly mumbles that he "was just working on it for my own…" (*Jobs*). His business partner immediately seizes this thought and frantically exclaims: "Exactly, for your own! For you! It's what you wanted! It's what your gut, your instinct wanted. Your big, evolved brain wanted something that didn't exist and so you just wheeled it into existence!" (*Jobs*). He thereby invokes their position as the ultimate consumers, who do not produce for any potentially beneficial effect on society, but for their own vicarious pleasure and, ultimately, for profit. Their genius is not applied to noble causes, but to their own hedonism and conspicuous consumption, which, in accordance with the ideology of cool capitalism, is portrayed as the driving force for innovation, individuality, and rebellion against the status quo. In line with the history of the biopic genre, the consumer has moved to the center of attention, even if the depicted subject has become renowned for a particular form of production or an original product.

Woz is not yet thoroughly convinced and assumes that his partner is "overreacting" (*Jobs*), because even if they were "developing this for freaks like us" (*Jobs*), he is certain that no one would be interested in buying such a device. Quite self-consciously, he thus categorizes both of them as stereotypical computer nerds, whose obsessive interest in electronics and hardware does not even slightly concur with the pastimes of the popular crowd. While not even their fellow backyard tinkerers will wholeheartedly support such a bold idea, the cool kids would deem their efforts pathetic, bizarre and, as his quote implies, "freak[y]" (*Jobs*), a judgment that appears to be equally often applied to Woz himself. While Woz is thus framed as incapable of recognizing the potential of personal technology not only for serious business, but also for its fabrication of consumer cool, the young Steve Jobs also proves to be visionary in this regard. He goes on to employ the typical jargon of cool corporatist co-option insinuated by Frank, with a voice that almost cracks from excitement and enthusiasm: "This is freedom… this is freedom to create, and to do, and to build... as artists, as individuals!" (*Jobs*). Both the vocabulary he employs and the way of delivery construct the mass consumer product he tries to sell to Woz, i.e. the operating system, as the bearer of difference, individuality, creativity, and, most of all, freedom—a freedom that is implicitly framed as freedom from the dull, restraining products of the establishment, in whose massive garage they happen to meet. As Frank shows, the co-option of cool rebellion by the capitalist mainstream is all about "speaking *like* the rebel young [sic]" (121, emphasis in the original) rather than speaking to them. The cinematic Jobs thus seems to have intuitively grasped his future clientele.

When Woz, however, still voices doubts about the marketable value of a highly technologized invention in a world where everyone is still using typewriters, Jobs seems to realize that he is not addressing a potential costumer, but his partner in corporate crime. Hence, his speech shifts from the alluring ad slang of Madison Avenue to down-and-dirty business jargon: "How does somebody know what they want if they've never even seen it?" (*Jobs*), he aggressively asks. Judging by the movie's ostensible politics, this line was apparently included to emphasize Jobs's visionary conceptualization of a market that did not even exist at the time he was already concocting its revolution, but which would prove decisive a few decades later. From the viewpoint of capitalist critique, however, he is thereby unwittingly articulating the very tenets of commodity fetishism, the most powerful accomplice of capitalism today: The perfect capitalist product will not only appease a need before the customers even realize that they had it, but indeed create the very need to begin with. The self-proclaimed misfit, rebel, and bare-footed hippie Steve Jobs has thus formulated the prime method for corporate success in the neoliberal capitalist system.

A TOOL FOR THE HEART: THE LEGACY OF COOL CONSUMER TECHNOLOGY

Today, in an age which presumably enjoys the results of Jobs's and Zuckerberg's geniuses, it appears somewhat a tautology to assert that technology is the epitome of cool. The preceding analysis has, however, demonstrated that coolness is far from a God-given, natural asset of the gleaming high-tech gizmo. Rather, it is constructed through very specific, intricate strategies that help to endow both the producers and the consumers of high technology with an irresistible aura of individualism, rebellion, and non-conformity. Especially *Jobs* allows the tacit observation that in the early days of ICTs, those creating the big or small boxes exuding this coolness were far from the smooth magicians they emerge as today. Companies like Microsoft and IBM, two of Apple's biggest competitors from its humble beginnings in a garage, are framed as utterly conventional, sedate, and as far from a cool brand as possible. In that sense, the movie not only presents Jobs as the ultimate innovator who by virtue of his visionary genius rejuvenates an entire field of expertise, but also tacitly credits Apple with endowing ICTs with the social prestige and cool image which they enjoy today. Computer science and technology, two of the most profitable sectors of contemporary American industry, are thus paradoxically constructed as the ultimate arbiters of countercultural values. At the same time, those practicing technoscience, whether it be Zuckerberg with his unmatchable programming skills or Jobs and his unshakable belief in the creative potential of the home computer, are no longer presented as the geeks and nerds they used to be: The icons of the information age are youthful, rebellious, defiant, and seemingly unconcerned with material gain. The fact that they simultaneously figure as the wealthiest entrepreneurs of the twenty-first century and the prime arbiters of 'cool capitalism' as established by Frank and McGuigan is a circumstance which both films gloss over.

One element that plays a crucial role in the appeal of their companies and vitally contributes to the creation of an aura of cool capitalism that effectively legitimates exploitative neoliberal practices is the aesthetic dimension of personal technology, which is noticeably neglected in both Frank's and McGuigan's socio-economic analyses. The heightened significance of haptic aesthetics for the production and consumption of technological gizmos is again excellently epitomized by the all-purpose mobile communication device, to whose global success both companies have majorly contributed. Regarding the 2010 introduction of the iPad and its often parodied 'infomercial,' Florian Sedlmeier (2013) observes that the cool aura of the Apple device is produced by "a tactile erotic, expressive of a sensuous and ludic individual consumption, which in turn hinges upon the logic of complex simplicity" (275).[46] With reference to Bill Brown's thing theory (2001, 2003), he persuasively argues that this form of playful consumption crucially depends on the representation of the product as thing, rather than object. Since the focus in *Network* and *Jobs* is placed on the rise of the company by means of their ingenious founders, rather than their innovative product range, this aspect has only presented a minor concern for the present analysis. Further studies exploring the popular cultural representation of leading personalities in the tech industry may, however, benefit from more closely associating the characters with their respective products. Similarly, an exploration of the movies' visual constructions of cool by means of cinematography and mise-en-scène may yield interesting results as regards characterization and narrative sequencing. Both *Network* and *Jobs* contain, in the melodramatic tradition to which the classic biopic is indebted, a rather high frequency of close-ups, which are, however, employed to display that there are precisely *no* emotions, or at least

46 Kieran Healey's sociological analysis of Jobs's charismatic personality (2011) reveals that there is a dangerous paradox underneath the nowadays omnipresent imperative for 'complex simplicity': "[G]oods that are simple and elegant to *use* are often difficult and dangerous to *make*. A wide gap may open between the consumers and the producers of beautiful pieces of personal technology: it's an elegant, creative, meaningful future for me, but a lifetime toiling on a Foxconn production line for thee" (n.p., emphasis in the original).

no positive emotionality, to be observed. More often than not, the films' framing and visual arrangements thus help to present the lead characters as the living embodiments of the allegedly alienating devices for which their creators have become recognized and venerated. As the preceding analysis suggests, the "cool touch" (Sedlmeier 291) promoted by the smart phone and similarly sleek techno gadgets corresponds to the affective level displayed by *Jobs*'s protagonist: emotionally restricted, highly individualistic, and irresistibly cool. Zuckerberg, on the other hand, can be read as personifying his own creation, as he displays a "pragmatic way of making friends" ("Making Of") that is guided by matching mutual interests and potential benefits, rather than sympathy.

Appropriately, thus, the last chapter of this study joins myriad notions explored in the previous analyses. Without Apple's and Facebook's outstanding branding and marketing campaigns, directly referenced and expounded in the respective biopics, and the framing of their founders as countercultural icons in popular media, one can well assume that texts like *CSI* or *The Big Bang Theory* would not have flourished in the American televisual landscape as easily and successfully as they did. *Network* and *Jobs* can thus be understood as the logical culminations of the conceptualizations and frameworks explored in this study, exemplifying the notion of cool technoscience *par excellence*: By exploiting some of the most pertinent strategies of cool that originate in the concept's historical associations with subversion, anti-establishment, and the American counterculture, the films transport an imagery and discourse of computer science and technology that constructs mass-produced high-tech devices as markers of individuality and bearers of personality. It is for this remarkably paradoxical framing, as the preceding pages have shown, that the genre of the biopic seems to offer a particularly viable template for exploring the extraordinary life story of an unruly, rebellious, yet eventually triumphant individual. The myths of the American Dream and the self-made man abet this construction, so that Zuckerberg and Jobs may eventually emerge as the ultimate scientist-entrepreneurs, whose supreme vision, hard work, and unfailing belief in the rewarding act of self-creation trump inherited privileges. The movies can thus be treated as prime cinematic reflections of how "the celebration and, indeed, mystification of the entrepreneurial hero has been embodied perfectly in the figure of Jobs" (McGuigan, "Coolness" 434) and, one may safely add, his programming equivalent Zuckerberg. The apparently dissenting sensibility conveyed by their persona vitally contributes to a cool image of their company, while simultaneously legitimating controversial capitalist ventures. Both characters have not only, as their biopics expect us to believe, reformulated the boundaries of their respective field, but also discovered a very lucrative formula for cool.

Beyond the Formula

A Conclusion

> Whether we're talking about race or gender or class, popular culture is where the pedagogy is, it's where the learning is.
>
> BELL HOOKS /
> CULTURAL CRITICISM & TRANSFORMATION

This study began with a quote by George Lipsitz on how the sideshow of popular culture may sometimes emerge as the main event. The preceding pages have demonstrated that in the seemingly elitist realms of technoscience and contemporary information culture, this proposition indeed holds true in many instances. In *Time Passages: Collective Memory and American Popular Culture* (1990), Lipsitz further notes that "the most important facts about people and societies have always been encoded within the ordinary and the commonplace" (20). His study thereby stresses the centrality of popular culture for the formation, sustainment, and challenge of contemporary societal discourses, but also defines legitimation as a key function of such representations, especially in the audiovisual realm of film and television. Whether it concerns family values, work ethic, national sentiments, consumer choices, or global capitalism, mainstream American television formats and feature films showcase and depict, but simultaneously shape and transform dominant social forces of both containment and resistance. Through reference to the past in the widest possible sense, controversial transformations and trends in the present are justified, explained, and even indeed naturalized for the viewing audience. In the framework of the present study, this entails that popular cultural texts like *CSI*, *The Big Bang Theory*, *The Day After Tomorrow*, *The Social Network*, and *Jobs* seek legitimacy with their respective audiences by representing different aspects of technoscientific existence and contemporary life in the modern information society, which crucially depends on effective knowledge work. At the forefront of the knowledge workforce in the twenty-first century is the scientist, who, as this study contends, has finally hit upon the formula for cool.

The underlying premise upon which my analysis is based is resonant with Lipsitz's view of popular cultural production as a crystallization of currently pertinent issues and underlying contradictions, frequently via a detour to nostalgia and the past. Not by chance do "commercial motion pictures invariably resonate with the value crises of the times in which they appear" (Lipsitz 164), as he convincingly argues, and the same pertains to the realm of television. In accordance with this assessment, the analyses in this study are connected to the legitimation crisis that has affected scientific practice and representation since the postwar era. As a defining feature of postmodern society in which metanarrative explanations no longer hold (Lyotard 1979), legitimation has emerged as a distinctive problem for contemporary science work, further augmented by the growing penetration of technoscientific innovation into nearly all aspects of contemporary life. The texts analyzed in this study were hence understood as a viable popular cultural response (and perhaps challenge) to this enduring moment of crisis: Similar to how the credibility and commercial success of television soap operas and game shows have been shown to depend on their ability to turn into play the daily chores of commercial purchase and the nurturing of the family (see Modleski 1986), cultural formations can be understood as gaining credibility by depicting the everyday work and personal lives of scientists, researchers, and, more generally speaking, knowledge workers, after all a growing socio-economic class in contemporary American information society and beyond. The scrutinized representations resort to one of the historically most popular and culturally 'lowest' affective attitudes, i.e. the quintessentially American sensibility of cool, and thereby offer a novel possibility for justifying contemporary science's cultural prestige, epistemological authority, and exploitation of financial and other resources. In that sense, cool is read as a popular cultural replacement of former, nowadays dysfunctional legitimatory discourses, which have traditionally displayed much more elitist, 'high-brow' connotations. In the analyzed examples, cool serves as an effective aesthetic and affective source of justification for scientific concerns, thereby challenging cognitive and/or ethical discourses. Popular culture can therefore be identified as a key locus of scientific legitimation: Productions of commercial mass culture such as the ones analyzed in the present study tacitly assume a legitimatory function in a society saturated with technoscientific imagery.

With reference to Habermas's *Legitimation Crisis*, however, one must concede, as Lipsitz does, that popular cultural texts may, in actual fact, achieve the opposite effect, which also holds true for the representations of 'cool technoscience' scrutinized in this study: Portrayals of technoscientific discourses in contemporary popular culture may legitimize the very forces that work to undermine these exact discourses (Lipsitz 72). Bluntly put, the celebrated center-stage depictions of nerdy characters, gleaming high-tech equipment, and scientific plotlines can, according to this premise, help to disguise and simultaneously counteract the fact that the claims of Western industrial science are today, in the best postmodern manner, thoroughly questioned, relativized, and fundamentally weakened. As the initial theoretical explorations of this study have also demonstrated, legitimation processes are essentially characterized by their instability and ambiguity and can hence easily work both ways. In unison with Stuart Hall's influential 1977 essay "Culture, Media, and the 'Ideological' Effect," representations in mass media must be understood as directing the popular focus from production to consumption, dividing audiences into isolated consumers. Applied to the thematic concerns of the present study, this contention would entail that active, direct, and in-depth scientific engagement is, in its last consequence, discouraged at the level of popular cultural exploration. Indeed, as the preceding analyses in this study have also demonstrated, social groups—'the scientists'—and experiences—'research,' 'the lab'—are either misrepresented as collectives or fragmented into individual elements, so that arising tensions, contradictions, and alliances do not have to be thematized beyond the occasional, sentimental side glance. In that respect, any critique of Western scientific practice in the realms and via the modes of popular culture, including eco-critical, feminist, and postcolonial voices, is quickly silenced or even indeed prevented from rising in the first place. Problematic cases like the gender dynamics in *The Big Bang Theory* or the capitalist underpinnings in *Jobs* and, to a lesser degree, *The Social Network*, blatantly showcase this problematic tendency. At the same time, the depiction of 'cool science' suggests that the elitist appeal and cultural prestige of Western technoscience might be considerably weakened by its unholy marriage with cool, a sensibility that has historically been at the very core of mass cultural production and commercial exploitation. Perhaps paradoxically, the analyses of this study may thus support contradictory conclusions: Depending on one's preferred perspective, the global authority and institutional practice of industrial science in twenty-first-century American culture are either fully legitimized or further destabilized by the scrutinized texts. Where does this, at the end of this lengthy exploration, leave us if we concur with bell hook's epigraphic notion of popular culture as the place of pedagogy and learning?

Let us begin the answer to this question and the concluding remarks to this study by recapitulating what each of the individual analyses has suggested so far in terms of the representation of technoscience and its closer than ever entwinement with notions of cool. With reference to the key research questions that have originally guided my analysis, this may be broadly grouped into two steps: firstly, the image(s) of science shaped and circulated by the respective text, and secondly, the entwinement of these images with popular cultural notions of cool, which the particular text is based on and informed by. The diverse reifications of cool found in each production were either freshly mobilized and reinforced or challenged and subverted, as the preceding pages have demonstrated. These analytic considerations were rounded off in each individual chapter by the final close reading of a particular scene or episode, which helped to crystallize the key argument and apply it to a pertinent and coherent example from the text. The two focal points of enquiry were thus combined, contrasted, and consolidated. Given the structural and thematic diversity of the five selected, highly multifarious primary texts, each analysis gave rise to varied and at times even indeed contradictory results, which shall be briefly reviewed and critically assessed in the following paragraphs. As a second step, I shall draw on these individual insights as an informed basis for more wide-ranging, generalized conclusions on the production and circulation of cool in cultural significations of technoscience and its wider implications for the study of American popular culture. This necessarily includes the final evaluation of my initial research aims, determining whether the growing emphasis of popular scientific imagery on cool may indeed be regarded as a response to the obsolescence of former sources of legitimation.

The first text under scrutiny, *CSI: Crime Scene Investigation* and its spin-offs, has suggested that science fulfills a vital aesthetic function, in addition to providing comfort, stability, and temporary closure in a world shaken by terrorism and organized crime. The specific technoscientific imagery upon which *CSI* capitalizes, from the chrome-and-steel laboratory to larger-than-life computer simulations, vitally contributes to the aesthetic stylization of the show and demonstrates how images of science are noticeably informed by dominant notions of coolness. It is only due to its combination with gleaming machinery that science can be meaningfully commodified and transformed into a visual spectacle, as the iconography provided by technology allows the show to focus on the aesthetic, rather than the cognitive or ethical dimension of science. Accordingly, sleek techno gadgets, flashing multimedia appliances and shiny electronic devices are seldom shown in use, but seem to be staged for ornamental reasons, guaranteeing visual entertainment and providing compelling illustrations when tedious procedures or obscure theories are explained. This is enhanced by *CSI*'s markedly cinematic camerawork, which typically directs the viewer's gaze onto the metallic, literally 'cool' texture of electronic instruments. It is exactly the overt emphasis on science as a key narrative and visual element that contributes vitally to the program's signature cool aesthetic. Coolness in *CSI* was thus understood as a direct consequence of, rather than antonym to, its scientific storylines, rational and detached worldview, and sleek high-tech surfaces. The analysis has also shown that the same verdict pertains to the protagonists of the show, who convey cool through their combination of an extreme, negative affective range (including, above all, anger and vengeance) and a cool exterior that is portrayed as camouflaging their actual state of mind.

The second cultural formation under scrutiny revealed a comprehensively different image of science, especially as concerns its visual-aesthetic dimension and narrative functions. While *The Big Bang Theory* can be regarded as the text that most explicitly foregrounds traditional notions of science and research as one of its central themes, one must concede that these notions typically assume the role of an insider joke within the narrative of a single episode. This overt stylization of science as one of the show's central topoi notably contrasts with the fact that the audience hardly ever gets to see actual scientific practice, which clearly differentiates *The Big Bang Theory* from *CSI*, yet aligns it with the remaining texts in this study. When the viewers do get to see science in *The Big Bang Theory*, it is almost exclusively for a comic effect. This effect is mostly achieved by emphasizing the comic contrast between two highly disparate life worlds, a common trope in the classic domestic sitcom. Additionally, the comic effect contrived by science is enhanced through teaming it with repeated references to geek culture and fandom. In analyzing the particular notions of 'geek cool,' Ilan Dar-Nimrod's study "Coolness" (2012) and his differentiation between cachet and contrarian coolness proved highly useful. The analysis has demonstrated that alternative notions of cool are directly linked to the protagonists' identity as scientists and geeks. Similar to how technoscientific expertise in *CSI* allows the scientist-detectives to emerge as superior and cool, it is *because of*, rather than despite the association with science that *The Big Bang Theory* characters display typical contrarian traits of coolness. In most instances, these traits are tacitly ascribed to the protagonists' rational mind-set and stout adherence to scientific principles, which the show presents as an inevitable consequence of their purely intellectual occupation as physicists. As a result, science comes to assume the role of what postcolonial theory has classified as 'the Other,' since it repeatedly serves to provoke comic, yet highly consequential deviations from the norm.

The third text chosen for analysis in this study, *The Day After Tomorrow*, revealed strong parallels with *CSI* as concerns the narrative function of science and technology for the story's dénouement and final sense of closure. With reference to a key critical text in the analysis of cinematic disaster and its representations of science, Susan Sontag's "The Imagination of Disaster" (1965), science was read as a great unifying force in an overtly utopian sense. While unchecked scientific progress is frequently condemned as unnatural, sinister, or immoral in the disaster genre, my analysis has shown that in this specific instance, science emerges as an idealized, omnipotent, and almost magical force, providing an antidote to human imperfection and the adversity of nature. At the same time, it is science in the shape of techno-industrial progress that caused the disaster in the first place, thus tacitly endowing the plotline with a particularly salient form of cool ambivalence. Finally, science is repeatedly attributed religious insignia and presented via a biblical rhetoric: The rescuer appears like an angel from up above when the movie's hero has to descend a snowy slope to retrieve a group of survivors. Through the personification via its central practitioner, science brings salvation in the very last second, when all hope, it appears, has been lost. Additionally, the cognitive value of science was read as a vital factor in making science and its allies appear cool, which aligns the disaster movie with the biopics *Jobs* and *The Social Network*. It is only by virtue of technoscientific knowledge *about* the disaster that the scientific characters are endowed with a higher chance of survival and may face the crisis in seemingly cool blood. Technoscience hence directly accounts for the scientist-hero's baffling cool pose during the global disaster. On the formal level, this is abetted by the movie's restricted use of facial close-ups and a high degree of digitally produced special effects, which replace the human agent as the site of affective audience involvement and equally depend on a high degree of technoscientific innovation. As with *The Big Bang Theory*, however, the movie does not depict any scenes of actual scientific practice.

Finally, the last analytical chapter examined the representation of computer sciences and the scientist-entrepreneur in two recent biographical films, *Jobs* and *The Social Network,* which mobilize notions of a countercultural cool hinged on the social and aesthetic value of modern ICTs. Both texts perpetuate images which may be considered the farthest removed from the traditional core values of Western science, which pertains to the necessary foregrounding of information technology and commercial culture in the fictional lives of the protagonists. Through its promotion of self-creation and individualism, the genre of the biopic has always offered a perfect vehicle for perpetuating the myth of the American Dream and the self-made man, which both *Jobs* and *The Social Network* connect to technological innovation. Most notably, the analysis has demonstrated that the biopics present their lead characters as scientific innovators and corporate rebels, rather than shrewd entrepreneurs, by projecting the dissenting aura of cool on the popular electronic brands they embody. The invocation of science and its cultural and epistemological authority is thus intrinsic to the portrait of both protagonists as visionaries with noble causes that reach beyond the mere acquisition of material wealth or worldly fame. The heavy use of this representational strategy aligns the biopics with the narrative approach employed in *The Day After Tomorrow*. The underlying capitalist ambitions of both companies, whose rise to power the films *nolens volens* trace through the life story of their respective founders, is deliberately downplayed by emphasizing the societal progress achieved through the technoscientific inventions of Apple and Facebook as the rationale for their success. The final close reading demonstrated how the cinematic representations of Jobs and Zuckerberg perpetuate the pervasive ideology of contemporary 'cool capitalism,' a term borrowed from Jim McGuigan (2009), to legitimate corporate ventures and technoscientific supremacy. The chosen scene from *Jobs*—the first and only clip published before the movie's official release—exemplifies the power of personal technology to fabricate consumer cool and position itself as the bearer of social change, individuality, and creative freedom, just like protagonists Jobs and Zuckerberg themselves.

The analyses of the individual primary texts suggest that the key hypothesis of this study has been verified: As the selected examples demonstrate, contemporary popular representations of technoscience circulated in American society are increasingly informed by notions of cool. Generally speaking, the close reading of the chosen cultural significations in terms of a cool representation of science has yielded strong concurrences between the individual texts. This also suggests that contemporary images of 'cool technoscience' are largely congruent and revolve around a set of key parameters, despite the divergent and highly multifarious notions of 'science' and 'technology' perpetuated in today's popular culture. These key parameters involve, among others, the visual use of technoscience and according equipment as a prerequisite for cool aesthetics (*CSI*, *Jobs*, *The Social Network*), the cognitive-ethical value of Western science and technology as inducive to the portrayal of emotionally 'cool,' courageous, and even indeed heroic characters (*The Day After Tomorrow*, *Jobs*, *The Social Network*), as well as the (at times comic) portrayals of so-called nerds and geeks as a form of non-hegemonic, contrarian, countercultural coolness opposed to dominant notions of beauty and success (*The Big Bang Theory*, *The Social Network*). Additionally, all texts in my sample reflect, to varying degrees, the co-option of historically countercultural cool by the capitalist mainstream, be it in the form of modern ICTs or the commodification of 'geek chic.'

In summary, the selected portrayals strongly suggest that cool not only pervades American society in the fields of advertisement, fashion, and celebrity culture, as a range of studies from various academic angles (e.g. Frank 1998, Pountain and Robins 2000, Southgate 2003) have persuasively demonstrated, but also operates as an underlying principle for popular representations of scientific practice in the early twenty-first century. As the specularization and aestheticization of science and its tight amalgamation with high-end technology in all of the scrutinized cultural formations has shown, cool can serve as an effective aesthetic and affective source of scientific legitimation, challenging and potentially replacing traditional cognitive and/or ethical justifications. In other words, the rationale for science in the twenty-first-century knowledge culture of the United States, including the entirety of its beneficial and detrimental effects on human life, can well be sought within the popular domain. It hence appears plausible to regard the growing emphasis of scientific imagery on cool in the selected sample of American popular cultural significations as a response to, or even indeed a substitution of, former and now invalid sources of legitimation, as originally hypothesized at the outset of this study.

CONTEXT AND DESIDERATA

At this point, it seems pertinent to situate the preceding discussion in the broader context of the academic study of popular culture and examine the inferences that might be drawn from my analysis. The wider contribution to knowledge made by the present study may be located in the ongoing debate between academic commentators on popular culture, which in the German-speaking world was prominently ignited by the influential writings of Theodor W. Adorno and Max Horkheimer, above all in their seminal *Dialectic of Enlightenment* (1944). As key proponents of Critical Theory in the tradition of the Frankfurt School, Horkheimer and Adorno differentiated between so-called 'authentic culture,' which is seen as intellectually and technically challenging, autonomous, characterized by its intrinsic critical value and inherent potential for transformation and progress, and 'art as commodity' or the so-called 'culture industry,' which is by definition superficial, trivial, and reduced to a mere product manipulating the masses into passivity. The term 'culture industry' is chosen deliberately, as it is supposedly guided by elites and undemocratic tendencies and thus opposed to a genuine mass culture for the people. In this view, individuals are reduced to mere consumers and subversion is no longer possible. The culture industry only serves global capitalism and is epitomized in modern mass media, including television and tabloids. In the age of standardization and sameness, genuine individualism is no longer possible, an argument that tacitly aligns with Walter Benjamin's notion of the 'aura' of an original artwork in his essay "The Work of Art in the Age of Mechanical Reproduction" (1936). In contrast to these genuine artistic expressions, easily reproducible cultural products under late capitalism are regarded as intellectually debased, formulaic, uncritical, and conformist, leading to mental stagnation and docility in times of crises.

This argumentation also has clear political implications, as the culture industry is understood as assisting the legitimization of existing hegemonic structures, rather than subverting them. Power is thus imposed from above and simultaneously naturalized via the circulation of commercial cultural products, which have more in common with factory goods than 'authentic' art. In essence, *Dialectic of Enlightenment* thereby extends Karl Marx's notion of commodity fetishism[1] and applies it to the realms of commercial mass culture: In contrast to high culture, which provides for 'true' human needs like creativity, emancipation, and elevation, the culture industry creates false needs that can only (seemingly) be satisfied by the purchase of capitalist products. Horkheimer and Adorno refer to this as 'Verblendungszusammenhang,' or context of delusion:

---

1 Additionally, Horkheimer and Adorno draw on Max Weber's theory of rationalization. See also Jameson (1979).

"The culture industry endlessly cheats its consumers out of what it endlessly promises. [...] This principle requires that while all needs should be presented to individuals as capable of fulfillment by the culture industry, they should be so set up in advance that individuals experience themselves through their needs only as eternal consumers, as the culture industry's object. Not only does it persuade them that its fraud is satisfaction; it also gives them to understand that they must make do with what is offered, whatever it may be." (113)

It goes without saying that all of the scrutinized texts of the present study would fall into this category. The 'sideshow' provided by contemporary popular cultural expressions such as the analyzed television formats and feature films would be regarded as distracting from the essential; the focus is now placed on the secondary, the mundane, the superficial, which Horkheimer and Adorno deem reprehensible, uncreative, and debased.

While I will refrain, at this point, from weighing into this intense academic dispute that has engaged cultural studies scholars from the early stages of the discipline,[2] it nonetheless seems requisite to position the present study in relation to the above argument. What my analysis may contribute to the debate is that all of the examined texts offer a fusion of 'high' and 'low' (or 'authentic' and 'commercial') cultural elements in their use of cool for the representation of modern-day techno-science, both on the visual and the narrative level. The above summaries have shown that the interests and legitimatory needs of Western industrial science, one of the most elitist realms of human activity for the last few centuries,[3] are abetted by a popular cultural sensibility which, given its historical trajectory, would easily have to be placed among the intellectually lowest and most commercialized forms of cultural expression and aesthetic stylization. If one follows the underlying premise of the present study, the mere differentiation between authentic and popular as two clearly bounded realms of human expression must be revealed as flawed and premature. Given the demonstrated capitalist co-option of cool's formerly counter-cultural values, I would not go as far as ascribing subversive potential to any of the scrutinized texts; they are, after all, closely entwined with the logic of capitalist consumption, which is indeed more often than not legitimized through an association with seemingly contrarian coolness, as texts like *Jobs* and *The Social Network* have revealed. Nonetheless, it appears shortsighted and reductive to dismiss any of the analyzed significations and, for that matter, popular culture at large, as a trivial

2 See, for instance, Gans (1974), Hall (1980), Tagg (1982), Ross (1989), Lipsitz (1990), and Cook (1996).

3 This classification must be assessed as historically contingent, because in past centuries, scientists did not necessarily enjoy a high social status nor was their activity regarded part of 'high' culture. Steven Shapin (1991), for instance, shows how the roles of gentleman and scientist were regarded as mutually exclusive in early modern England.

sideshow: If anything, the present study reflects how "[f]or all of their triviality and frivolity, the messages of popular culture circulate in a network of production and reception that is quite serious" (20), to return to Lipsitz as one of popular culture's chief advocates.

Accordingly, I would argue that the combination of what Horkheimer and Adorno classify as 'high' and 'low' forms of cultural expression (in this case, visual-aesthetic values of mass cultural cool buttressing the cultural prestige of elitist technoscience) is not, perhaps unfortunately for this very study, as unique and uncommon as one might be prone to think, considering the postmodern processes of parody and pastiche (see Hutcheon 1989, Jameson 1991) frequently encountered in the realm of popular culture. This line of argumentation is taken, among others, by the Birmingham School of Contemporary Cultural Studies and their chief representative Stuart Hall, who defines cultural forms as "deeply contradictory" ("Deconstructing" 448) and neither "wholly corrupt or wholly authentic" ("Deconstructing" 448):

> "[T]hey play on contradictions, especially when they function in the domain of the 'popular'. [...] For, from period to period, the *contents* of each category change. Popular forms become enhanced in cultural value, go up the cultural escalator—and find themselves on the opposite side. Other things cease to have high cultural value, and are appropriated into the popular, becoming transformed in the process." ("Deconstructing" 448, emphasis in the original)

In a similar vein, Fredric Jameson, another key critic of postmodernity, detects "fragmentary and as yet undeveloped tendencies in recent art production" ("Reification" 133), which leads him to conclude that there is "an increasing interpenetration of high and mass cultures" ("Reification" 133) in postmodern society and no eternal standard of high or low brow cultural production. Culture is not static, as Hall and Jameson remind us, and their argumentation holds very true for this study: Through its heightened exploitation by popular cultural formations, the realm of science may well be understood to be weakening its elitist associations, while simultaneously, appropriations of the alleged high culture contents (or rather higher education)[4] that

4 The role of education as a bastion of high culture is particularly noteworthy, as Hall also remarks that "a whole set of institutions and institutional processes are required to sustain each [category of culture]—and to continually mark the differences between them. The school and the education system is one such institution—distinguishing the valued part of the culture, the cultural heritage, the history to be transmitted, from the 'valueless' part. The literally and scholarly apparatus is another—marking off certain kinds of valued knowledge from others" ("Deconstructing" 448-449). Science as that core academic practice must thus be understood as undergoing a comprehensive transformation of its cultural capital when employed as popular cultural contents.

science represents in Western information societies such as the United States appear to have become more acceptable and even indeed desirable in the popular cultural arena. The texts scrutinized in this study reflect this tendency: Far from off-putting or tiresome, references to scientific formulas, lab work, and the academic world seem to be treasured by the respective audiences of *CSI* and *The Big Bang Theory*, which in all probability do not consist of forensic scientists or physicists alone. Indeed, as the preceding analyses should have amply demonstrated, it is *because of*, rather than despite the incorporation of scientific storylines and paraphernalia that the shows' characters, cinematography, and visual aesthetics may emerge as cool.

The ramifications of Horkheimer's and Adorno's controversial contribution to the high vs. low debate and the lines of critique inspired by it, however, extend even further. On the one hand, the use of the term 'culture industry' instead of mass culture has helped to demonstrate "the unexpected and imperceptible introduction of commodity structure into the very form and content of the work of art itself" ("Reification" 132), as Fredric Jameson contests in one of his earliest publications, the 1979 essay "Reification and Utopia in Mass Culture." Art has ceased to be a end in itself and is now experiencing an increasing instrumentalization by the market as it becomes subject to demands of practicality and performativity. In the realm of science and academic knowledge production, a similar argument was put forth by Jean-François Lyotard (1984), as the theoretical part of this work has shown: Due to a growing insistence on the eventual applicability of scientific research, knowledge is no longer produced for knowledge's sake, but must answer to specific needs. Idealist and humanist narratives of legitimation are thus replaced with those of performativity, efficiency, and usefulness; as Lyotard bluntly puts it: "[S]ince performativity increases the ability to produce proof, it also increases the ability to be right" (46). Consequently, scientific practice is reduced to the optimization of relationship between input and output (Lyotard 11). Adorno and Horkheimer also seemed to be aware of this tendency, as their account includes examinations of how, under the logic of late capitalism, academia has to produce knowledge that conforms to a basic benefit-cost ratio:

> "The place of science in the social division of labor is readily identified. Its task is to accumulate facts and functional connections between facts in the largest possible quantities. The storage system must be easily surveyed. It must enable individual industries to locate the desired intellectual commodity in the required variety. Already the compilation is largely made with an eye for certain industrial contracts." (201)

While this tendency has not been the focus of the present study, it seems pertinent to note that the analyzed primary material may also support such a line of argumentation, as the preceding survey of Lyotard's work has shown. Science has become ultimately useful in the information society through the application and appropriation of its aesthetics and paraphernalia in popular culture for the purpose of generating revenue, be it in the shape of hit sitcoms or blockbuster movies. In Adorno's and Horkheimer's dictum, these expressions must be classified as a form of commodification of science, in a similar way to which scientific innovation serves the military or the industry.

Conversely, however, the argument of the culture industry commodifying both art and science also works in the opposite direction, as Jameson is acutely aware. His analysis of the Frankfurt School perceptibly dissects the ramifications of the high vs. low culture split on aesthetics, arguing that this account of commodification necessarily "implies that everything in consumer society has taken on an aesthetic dimension" (Jameson, "Reification" 132). This contention can easily be extended to the realm of science, for which the present study has provided ample proof. As the preceding examination of five selected primary texts demonstrates, the sources of justification for technoscientific practice, social prestige, and epistemological authority do indeed appear to have shifted from former cognitive or ethical models to aesthetic and affective ones, of which cool stands as a prime example. This proposition, of course, corresponds to the characterization of postmodernity as the era that confuses and/or merges the previously discrete realms of the cognitive, the aesthetic, and the ethic (Waugh 1992). Cool thus acts a viable means for readjusting the borderlines of traditional science by allowing a heightened degree of fusion with the modes, aesthetics, and techniques of contemporary popular culture. To refer to Horkheimer and Adorno's analysis of the culture industry complex, this development would not only entail a commodification, but also indeed an *aestheticization* of science via popular cultural expressions. The rise of the modern information society after the Second World War has hugely abetted these processes, as the socio-political accounts by Bell (1973), Hughes (1986), and others show. In the words of Ulrike Felt, we are facing comprehensive "interdependencies and border readjustments" ("Wechselwirkungen" 47)[5] between formerly discrete territories, prioritizing neither the one nor the other. Accordingly, it appears reasonable to seek legitimation for cognitive activity within the aesthetic domain.

Following these considerations in the wake of Horkheimer's and Adorno's influential argument and the wider scholarly debate on the study of popular culture, the significance of the present study for the larger academic context must be located in its scrutiny of cool aesthetics beyond the usual realms of enquiry, including advertisement, counterculture and resistance, and global business culture. In contrast to established sociological (quantitative) studies of scientific representation and science images, the focus of this analysis was placed on popular cultural production as the locus of contemporary discourses of scientific legitimation. The methodological toolbox provided by cultural studies as a discipline that remains acutely aware of its

5 Original wording: "Wechselwirkungen und Grenzverschiebungen."

own epistemic engagement with the objects of study proved to be well suited for these purposes. Consequently, the uses of 'cool science' analyzed in this study can be considered a case in point for the increasing legitimatory demands of science in post-industrial American society and the concomitant outsourcing of these demands to novel, unusual channels of cultural expression. My study thereby responds to existing research gaps in American studies scholarship on scientific representation, which have only more recently been addressed by the heterogeneous and interdisciplinary cultural studies of science. Key proponents of the field, including Donna Haraway (1989, 1991, 1994), Bruno Latour (1987, 1993), and Joseph Rouse (1987, 1991, 1993), have early demonstrated how vitally the fringe areas, or 'sideshows,' of cultural production matter to processes of scientific representation, legitimation, and critique. In exploring its own sample of popular cultural formations, from the crime drama to the sitcom, from the disaster movie to the biopic, the present study was intended to shed light on current processes of interaction between science and popular culture, both of which constitute pivotal sources for change in twenty-first-century American culture.

Despite its varied perspective, my analysis is neither free of necessary omissions nor can it end without listing key desiderata. The identification and investigation of such a recent development in the socio-cultural fabric of the United States necessarily has to be conceived of as an investigation in process rather than a task destined for quick completion. The fields of activity from which cool originates, including youth and commercial cultures, are dynamic and fast changing. Given cool's ontological elusiveness and semantic mutability, its currently 'defining' characteristics will very well undergo further significant adaptations and transformations, which are worth monitoring from a cultural studies perspective. As regards the actual objects of study, future enquiries would benefit from a more in-depth scrutiny of specific genres, which could not be realized at this stage as such an approach would have exceeded the scope of the present study. The range of primary material chosen for analysis was deliberately varied and diverse; apart from the structural homogeneity of audiovisual productions, the analyzed texts are characterized by strong particularities in terms of narrative, mise-en-scène, character drawing, and generic development. Further analyses may wish to take these discrepancies into consideration. A particularly valuable area of study for the representation of science and its entwinement with notions of cool is the virtual world, including web 2.0 and social media. Practices like the public appearance of international 'Big Science' projects on Twitter and Facebook or individual scientists hosting their own blog for a more light-hearted, 'popularized' account of their everyday work could only be mentioned in passing, but would prove to be fertile areas of enquiry. One may also wish to support these findings with quantitative data and hence introduce a more empirical outlook on the popular cultural exploitation of scientific story lines; this was elided in the present study for the already mentioned methodological and disciplinary considerations. Finally, a closer focus on individual disciplines could also yield some interesting results. As my analysis of the selected examples has implicitly shown, representations of physicists may differ greatly from representations of computer scientists, and the aesthetic stylization of a fictional crime lab is notably different from that of a sitcom's living room arrangement. Such differentiation may allow a more detailed assessment and help to circumvent generalizations, which have, time and again, been unavoidable in the present study.

"The most profound intellectual questions," Lipsitz observes by way of conclusion, "emerge out of what seem to be ordinary and commonplace objects of study" (20). Popular culture, to resort to the original proposition with which this study started, is far from the insignificant sideshow it is still frequently made out be. Indeed, I hope that my study has provided ample proof for the appraisal of popular culture as a central source of both containment and change in the information age of the twenty-first century. The selected popular cultural representations analyzed for this purpose demonstrate that 'cool science' abounds in the age of globalized knowledge work. Far from trivial or arbitrary, cool has proven a viable and rewarding object of study. From the individual to the collective, the social to the economic, the mainstream to the countercultural, the radical to the conformist—cool encapsulates it all, and science would be foolish not to utilize this helpful accomplice. Whether the present study has indeed been able to crack the formula for cool remains to be seen; what is certain, however, is that popular culture has, once again, proven its viability as America's main event.

References

2012. Dir. Roland Emmerich. Columbia, 2009. Film.

"A Bullet Runs Through It." Part 1. *CSI: Crime Scene Investigation*. CBS. Dir. Danny Cannon. Writ. Richard Catalani and Carol Mendelsohn. 10 Nov. 2001. Television.

"A Special Advertising Section: The Phenomenon of Cool." *Wired* (November 2001): ix-xv.

Adorno, Theodor, and Max Horkheimer. *Dialectic of Enlightenment: Philosophical Fragments*. 1944. Ed. Gunzelin Schmid Noerr. Trans. Edmund Jephcott. Stanford: Stanford UP, 2002.

Airport 1975. Dir. John Smight. Universal, 1974. Film.

Airport. Dir. George Seaton. Universal, 1970. Film.

Ali. Dir. Michael Mann. Columbia, 2001. Film.

Allen, Michael, ed. *Reading CSI: Crime TV Under the Microscope*. London: I.B.Tauris, 2007.

—. "Introduction: This Much I Know…" *Reading CSI: Crime TV Under the Microscope*. Ed. Michael Allen. London: I.B.Tauris, 2007. 3-14.

Altmann, Rick. *Film/Genre*. London: British Film Institute, 1999.

Amend, Christoph, Florian Illies, and Tanja Stelzer. "Die nuller Jahre." *Zeit-Magazin* 11 Dec. 2008: n.p. Accessed 15 Jan. 2012. <http://www.zeit.de/2008/51/Inhalt-51>.

American Pie. Dir. Paul Weitz. Universal, 1999. Film.

"And They're Offed." *CSI: Miami*. CBS. Dir. Matt Earl Beesley. Writ. Barry O'Brien. 19 Jan. 2009. Television.

Anderson, Carolyn. "Biographical Film." *A Handbook of American Film Genres.* Ed. Wes D. Gehring. Westport, CT: Greenwood, 1988. 331-351.

—, and Jon Lupo. "Hollywood Lives: The State of the Biopic at the Turn of the Century." *Genre and Contemporary Hollywood.* Ed. Steve Neale. London: British Film Institute, 2002. 91-104.

Armageddon. Dir. Michael Bay. Touchstone, 1998. Film.

Arthur, Charles. "Teenagers would rather lose TV than internet or mobile—survey" *Guardian* 25 Oct. 2011: n.p. Accessed 16 January 2012. <http://www.guardian.co.uk/technology/2011/oct/25/teenagers-lose-tv-internet-mobile>.

Ash, Mitchell. "Wissenschaft, Politik, und Öffentlichkeit—Zur Einführung." *Wissenschaft, Politik, und Öffentlichkeit: Von der Wiener Moderne bis zur Gegenwart.* Ed. Mitchell Ash and Christian Stifter. Wien: Facultas, 2002. 19-43.

—, and Christian Stifter, ed. *Wissenschaft, Politik, und Öffentlichkeit: Von der Wiener Moderne bis zur Gegenwart.* Wien: Facultas, 2002.

—. "Wissenschaftswandlungen in politischen Umbruchzeiten—1933, 1945 und 1990 im Vergleich." *Acta Historica Leopoldina* 39 (2004): 75-95.

—. "Wissenschaft(en) und Öffentlichkeit(en) als Ressourcen füreinander: Weiterführende Bemerkungen zur Beziehungsgeschichte." *Wissenschaft und Öffentlichkeit als Ressourcen füreinander: Studien zur Wissenschaftsgeschichte im 20. Jahrhundert.* Ed. Sybilla Nikolow and Arne Schirrmacher. Frankfurt: Campus, 2007. 349-362.

—. "Wissenschaft und Verantwortung: Zur Historisierung einer diskursiven Formation." *Universität im öffentlichen Raum.* Ed. Rainer Christoph Schwinges. Basel: Schwabe, 2008. 311-344.

Ashcroft, Bill, Gareth Griffiths, and Helen Tiffin. *Post-colonial Studies: The Key Concepts.* London: Routledge, 1998.

Austin, John L. *How To Do Things With Words.* 1962. 2nd ed. Harvard: Harvard UP, 1975.

Back to the Future. Dir. Robert Zemeckis. Universal, 1985. Film.

Back to the Future II. Dir. Robert Zemeckis. Universal, 1989. Film.

Back to the Future III. Dir. Robert Zemeckis. Universal, 1990. Film.

Barnes, Barry. *Scientific Knowledge and Sociological Theory.* London: Routledge & Kegan Paul, 1974.

Barthes, Roland. "Death of the Author." Trans. Richard Howard. *Aspen* 5-6.3 (1968): n.p.

—. "Réponses." *Tel Quel* 47 (Autumn 1971): 89-107.

—. *Mythologies.* Trans. Annette Lavers. New York: Farrar, Straus and Giroux, 1972.

—. *S/Z: An essay.* Trans. Richard Miller. New York: Hill and Wang, 1975.

Basalla, George. "Pop Science: The Depiction of Science in Popular Culture." *Science and Its Public: The Changing Relationship.* Ed. Gerald Holton and William Blanpied. Dordrecht: Reidel, 1976. 261-278.

Bates, J. Leonard. *The United States, 1898-1928: Progressivism and a Society in Transition*. New York: McCraw-Hill, 1976.

Baudrillard, Jean. *Simulacra and Simulation*. 1981. Trans. Sheila Faria Glaser. Ann Arbor: U of Michigan P, 1995.

—. *Symbolic Exchange and Death*. Trans. Iain Hamilton Grant. London: Sage, 1993.

Beck, Ulrich. *Risikogesellschaft: Auf dem Weg in eine andere Moderne*. Frankfurt am Main: Suhrkamp, 1986.

Becker, Markus. “Klimaforscher streiten übers Emmerichs Eismaschine.” *Spiegel Online* 13 May 2004: n.p. Accessed 2 Feb. 2013. <http://www.spiegel.de/wissenschaft/natur/the-day-after-tomorrow-klimaforscher-streiten-ueber-emmerichs-eismaschine-a-299652.html>.

Bell, Art, and Whitley Strieber. *The Coming Global Superstorm*. New York: Pocket, 1999.

Bell, Daniel. *The Coming of Post-Industrial Society*. New York: Basic, 1973.

Belmont, Cynthia. “Ecofeminism and the Natural Disaster Heroine.” *Women’s Studies: An Inter-Disciplinary Journal* 36 (2007): 349-372.

Benjamin, Walter. “The Work of Art in the Age of Mechanical Reproduction.” 1936. *Essays and Reflections*. Ed. Hannah Arendt. Trans. Harcourt Brace Jovanovich. New York: Schocken, 1969. 217-252.

Berk, Lynn M. “The Great Middle American Dream Machine.” *Journal of Communication* 27.3 (1977): 27-31.

Beverley Hills, 90210. Prod. Aaron Spelling. Fox. 1990. Television.

Biagioli, Mario, ed. *The Science Studies Reader*. New York: Routledge, 1999.

Bianculli, David. “The CSI Phenomenon.” *Reading CSI: Crime TV Under the Microscope*. Ed. Michael Allen. London: I.B.Tauris, 2007. 222-226.

Bignell, Jonathan. “Seeing and Knowing: Reflexivity and Quality.” *Quality TV: Contemporary American Television and Beyond*. Ed. Janet McCabe and Kim Akass. London: I.B. Tauris, 2007. 158-170.

Billen, Andrew. “Cops Without Frontiers.” *New Statesman* 18 June 2001: n.p. Accessed 22 July 2012. <http://www.newstatesman.com/node/140564>.

Billig, Michael. *Laughter and Ridicule: Towards a Social Critique of Humour*. London: Sage, 2005.

Bingham, Dennis. *Whose Lives Are They Anyway: The Biopic as Contemporary Film Genre*. New Brunswick: Rutgers UP, 2010.

—. “Living Stories: Performance in the Contemporary Biopic.” *Genre and Performance: Film and Television*. Ed. Christine Cornea. Manchester: Manchester UP, 2010. 76-95.

Böhme, Gernot, and Nico Stehr, ed. *The Knowledge Society: The Growing Impact of Scientific Knowledge on Social Relations*. Dordrecht: Kluwer, 1986.

Boltanski, Luc, and Eve Chiapello. 1999. *The New Spirit of Capitalism*. London: Verso, 2005.

Bones. Prod. Kathy Reichs, Emily Deschanel, and David Boreanaz. Fox. 2005. Television.

Bowles, Simon. "'The Day After Tomorrow' heats up a political debate." *USA Today* 26 May 2004: n.p. Accessed 12 Dec 2013. <http://usatoday30.usatoday.com/life/movies/news/2004-05-26-weather-main_x.htm>.

Boyd, Brian. "Laughter and Literature: A Play Theory of Humor." *Philosophy and Literature* 28.1 (April 2004): 1-22.

Brandt, Christina. *Metapher und Experiment: Von der Virusforschung zum genetischen Code*. Göttingen: Wallstein, 2004.

—, and Anja Casser. "Populärkultur und Wissenschaft: Science-Fiction und populäres Bild als Medien der Wissenschaftskommunikation." *Wissenschaft und Öffentlichkeit als Ressourcen füreinander: Studien zur Wissenschaftsgeschichte im 20. Jahrhundert.* Ed. Sybilla Nikolow and Arne Schirrmacher. Frankfurt: Campus, 2007. 165-176.

Brecht, Bertolt. "In the Jungle of Cities." 1927. *Collected Plays: One*. Ed. John Willett and Ralph Manheim. Trans. Gerhard Nellhaus. London: Methuen, 1970. 117-178.

—. *The Threepenny Opera*. 1928. Trans. Desmond Vesey and Erich Bentley. New York: Grove, 1964.

Brown, Bill. "Thing Theory." *Critical Inquiry* 28.1 (Fall 2001): 1-22.

—. *A Sense of Things: The Object Matter of American Literature*. Chicago and London: U of Chicago P, 2003.

"Bully for You." *CSI: Crime Scene Investigation*. CBS. Dir. Thomas J. Wright. Writ. Ann Donahue. 18 Oct. 2001. Television.

Bürger, Peter. *Kino der Angst: Terror, Krieg und Staatskunst aus Hollywood.* Stuttgart: Schmetterling, 2007.

Burke, Edmund. *A Philosophical Enquiry into the Origin of our Ideas of the Sublime and the Beautiful*. 1759. Ed. Adam Phillips. Oxford: OUP, 1990.

Burke, Peter. *A Social History of Knowledge: From Gutenberg to Diderot*. Cambridge: Polity, 2000.

Burlingham, Bo, and George Gendron. "The Entrepreneur of the Decade." *Inc. Magazine* 1 April 1989: n.p. Accessed 13 Sept. 2013. <http://www.inc.com/magazine/19890401/5602_pagen_5.html/6>.

Bush, Vannevar. *Science—The Endless Frontier: A Report to the President on a Program for Postwar Scientific Research.* Washington: Government Printing Office, 1945.

Butler, Judith. "Performative Acts and Gender Constitution: An Essay in Phenomenology and Feminist Theory." *Theatre Journal* 40.4 (December 1988): 519-531.

—. *Gender Trouble*. 1990. New York: Routledge, 2006.

—. *Undoing Gender*. New York: Routledge, 2004.

—. *Precarious Life: The Powers of Mourning and Violence*. New York: Verso, 2004.

Butsch, Richard. "A Half Century of Class and Gender in American TV Domestic Sitcoms." *Cercles* 8 (2003): 16-34.

Butterflies. Prod. Gareth Gwenlan and Sydney Lotterby. BBC. 1978. Television.

Caldwell, Christopher. "Why Steve Jobs worked as a manager but not in a movie." *Financial Times* 23 Aug. 2013: n.p. Accessed 24 Aug. 2013. <http://www.ft.com/intl/cms/s/0/499cee1a-0a86-11e3-9cec-00144feabdc0.html#axzz2dwjBnXvc>.

Caldwell, John. *Televisuality: Style, Crisis, and Authority in American Television*. New Brunswick: Rutgers UP, 1995.

Callaghan, Dylan. "Face Value: Aaron Sorkin finds tragedy in the creation of Facebook and makes The Social Network an era-defining film." *Writers Guild of America* 1 Oct. 2010: n.p. Accessed 14 Sept. 2013. <http://www.wga.org/content/default.aspx?id=4348>.

Calsamiglia, Helena. "Popularization discourse." *Discourse Studies* 5.2 (2003): 139-146.

Caro, Mark. "Channeling 'Jobs' calls for occasional bouts of jerkiness." *Chicago Tribune* 9 Aug. 2013: n.p. Accessed 14 Sept. 2013. <http://articles.chicagotribune.com/2013-08-09/entertainment/ct-ae-0811-jobs-kutcher-interview-20130809_1_matt-whiteley-ashton-kutcher-joshua-michael-stern/2>.

Carter, Bill. "*Big Bang Theory* Is Now Bigger Than *American Idol*." Media Decoder. *New York Times* 27 January 2012: n.p. Accessed 21 Sept. 2012. <http://mediadecoder.blogs.nytimes.com/2012/01/27/big-bang-theory-is-now-bigger-than-american-idol>.

Castells, Manuel. *The Rise of the Network Society. The Information Age: Economy, Society and Culture*. 1996. 2nd ed. Malden: Blackwell, 2000.

Cavender, Gray, and Sarah K. Deutsch. "*CSI* and Moral Authority: The Police and Science." *Crime Media Culture* 3 (2007): 67-81.

Cawelti, John. *Apostles of the Self-Made Man*. Chicago: U of Chicago P, 1965.

—. *Adventure, Mystery, and Romance*. Chicago: Chicago UP, 1976.

Chalmers, Alan F. *What Is This Thing Called Science?* 3rd ed. Maidenhead: Open UP, 1999.

Chisum, W. Jerry, and Brent E. Turvey. "Evidence Dynamics: Locard's Exchange Principle and Crime Reconstruction." *Journal of Behavioral Profiling* 1.1 (January 2000): 1-15.

Citizen Kane. Dir. Orson Welles. RKO Radio Pictures, 1941. Film.

Clarke, George Elliott. "Cool Politics: Styles of Honour in Malcolm X and Miles Davis." *Jouvert: A Journal of Postcolonial Studies* 2.1 (1998): n.p. Accessed 12 January 2012. <http://english.chass.ncsu.edu/jouvert/v2i1/CLARKE.HTM>.

Cleopatra. Dir. Joseph L. Mankiewicz. Twentieth Century Fox, 1963. Film.

Cold Case. Prod. Jerry Bruckheimer. CBS. 2003. Television.

Cold Case Files. Prod. Mary Frances O'Conner. Arts and Entertainment Network. 1999. Television.

Collins, Harry M. “Public Experiments and Displays of Virtuosity: The Core-Set Revisited.” *Social Studies of Science* 18.4 (November 1988): 725-748.

Collins, Paul. “Must-Geek TV: Is the world ready for an Asperger’s sitcom?” *Slate Magazine.* 6 February 2009: n.p. Accessed 3 August 2012. <http://www.slate.com/id/2210635/pagenum/all/#p2>.

Comolli, Jean-Louis. “Historical Fiction: A Body Too Much.” *Screen* 19-2 (1978): 41-54.

Conley, Lucas. “A Craving for Cool.” *Fast Company.* 1 July 2006: n.p. Accessed 12 January 2012. <http://www.fastcompany.com/magazine/107/next-essay.html>.

Connor, Marlene Kim. *What is Cool? Understanding Black Manhood.* New York: Crown, 1995.

Cook, Deborah. *The Culture Industry Revisited.* Lanham: Rowman & Littlefield, 1996.

“Cool.” *The Oxford English Dictionary.* Online edition. Oxford: OUP. <www.oed.com>.

“Cool.” *Urban Dictionary.* Accessed 13 January 2012. <http://www.urbandictionary.com/define.php?term=cool>.

Cooter, Roger, and Stephen Pumfrey. “Separate Spheres and Public Places: Reflections on the History of Science Popularization and Science in Popular Culture.” *History of Science* 32 (1994): 237-267.

Costanza, Robert, ed. *Ecological Economics: The Science and Management of Sustainability.* New York: Columbia UP, 1991.

Cox, Brian, et al. “How Science Became Cool.” *Guardian* 13 April 2010: n.p. Accessed 3 Jan 2012. <http://www.theguardian.com/science/2010/apr/13/science-cool>.

Craven, Paul, and Barry Wellman. “The Network City.” *Sociological Inquiry* 43 (1973): 57-88.

Crawford, T. Hugh. “Book Review: Bruno Latour, We Have Never Been Modern.” *Configurations* 2.3 (1994): 578-580.

“Creatures of the Night.” *CSI: NY.* CBS. Dir. Tim Hunter. Writ. Pam Veasey. 29 Sept. 2004. Television.

Crèvecœur, J. Hector St. John de. “Letter III by an American Farmer.” 1782. *The American Studies Anthology.* Ed. Richard P. Horwitz. Oxford: Rowman & Littlefield, 2001. 23-32.

Criminal Minds. Prod. The Mark Gordon Company, CBS Television Studios, and ABC Studios. CBS. 2005. Television.

“Cross Jurisdictions.” *CSI: Crime Scene Investigation / CSI: Miami* (Crossover Episode). CBS. Dir. Danny Cannon. Writ. Ann Donahue, Carol Mendelsohn, and Anthony E. Zuiker. 9 May 2002. Television.

Crossing Jordan. Prod. Damon Lindelof. NBC. 2001. Television.

CSI: Crime Scene Investigation. Prod. Jerry Bruckheimer. CBS. 2000. Television.

CSI: Miami. Prod. Jerry Bruckheimer. CBS. 2002. Television.

CSI: NY. Prod. Jerry Bruckheimer. CBS. 2004. Television.

Custen, George F. *Bio/Pic: How Hollywood Constructed Public History*. New Brunswick: Rutgers UP, 1992.

Daiches, David. *Critical Approaches to Literature*. Englewoods Cliffs: Prentice-Hall, 1956.

Danesi, Marcel. *Cool: The Signs and Meanings of Adolescence*. Toronto: U of Toronto P, 1994.

Danton, Eric R. "Freddie Mercury Biopic Casts Lead Role." *Rolling Stone* 10 Dec. 2013: n.p. Accessed 11 Jan. 2014. <http://www.rollingstone.com/movies/news/freddie-mercury-biopic-casts-lead-role-20131210>.

Dargis, Manohla. "Portrait of the Artist Behind Apple: Ashton Kutcher in a Biopic About Steve Jobs." *Times* 16 Aug. 2013: C4.

Dar-Nimrod, Ilan, et al. "Coolness: An Empirical Investigation." *Journal of Individual Differences* 33.3 (2012): 175-185.

Daston, Lorraine, and Peter Galison. *Objectivity*. Boston: Zone, 2007.

Daumann, Rudolf. *Protuberanzen: Ein utopischer Roman*. Berlin: Schützen, 1940.

Dawson, Michael, and John Bellamy Foster. "Virtual Capitalism." *Capitalism and the Information Age*. Ed. Robert W. McChesney, Ellen Meiksins Wood, and John Bellamy Foster. New York: Monthly Review, 1998. 51-67.

Dean-Ruzicka, Rachel. "Vengeance, Healing and Justice: Post 9/11 Culture Through the Lens of *CSI*." *Quarterly Review of Film and Video* 26 (2009): 118-130.

Deep Blue Sea. Dir. Renny Harlin. Village Roadshow, 1999. Film.

Deep Impact. Dir. Mimi Leder. Paramount, 1998. Film.

Derrida, Jacques. "Structure, Sign and Play in the Discourse of the Human Sciences." 1966. *Twentieth Century Literary Theory: A Reader*. Ed. K. M. Newton. 2nd ed. New York: St. Martin's, 1997. 115-119.

—. *Of Grammatology*. Trans. Gayatri Chakravorty Spivak. London: Johns Hopkins UP, 1976.

Dharma and Greg. Prod. Chuck Lorre. ABC. 1997. Television.

Diamond Jim. Dir. A. Edward Sutherland. Universal, 1935. Film.

Diana. Dir. Oliver Hirschbiegel. Entertainment One, 2013. Film.

Dietrichsen, Dietrich. "Coolness: Souveränität und Delegation." *Person/Schauplatz*. Ed. Jörg Huber. Wien: Edition Voldemeer / Springer, 2003. 243-254.

Dinerstein, Joel. *The Origins of Cool: Jazz, Film Noir, and Existentialism in Postwar America*. Chicago: U of Chicago P. (forthcoming)

Disraeli. Dir. Alfred E. Green. Warner Bros., 1929. Film.

Dixon, Wheeler Winston. *Vision of the Apocalypse: Spectacles of Destruction in American Cinema*. London: Wallflower, 2003.

Doane, Mary Ann. "Information, Crisis, Catastrophe." *Logics of Television: Essays in Cultural Criticism.* Ed. Patricia Mellencamp. Bloomington: Indiana UP, 1990. 222-239.

Dobson, Nichola. "Generic Difference and Innovation in *CSI: Crime Scene Investigation.*" *The CSI Effect: Television, Crime, and Governance*. Ed. Michele Byers and Val Marie Johnson. Plymouth: Lexington, 2006. 75-89.

"Does Obama make it 'cool' to be American?" *Fox News* 9 November 2008. Accessed 14 January 2012. <http://video.foxnews.com/v/3917474/americas-coolagain>.

Doherty, Thomas. "Malcolm X: In Print, On Screen." *Biography* 23.1 (Winter 2000): 29-48.

—. "Cultural Studies and 'Forensic Noir.'" *Chronicle of Higher Education* 50.9 (2003): 15-16.

Dove, George N. *Suspense in the Formula Story*. Bowling Green: Bowling Green State U Popular P, 1989.

Dovey, Jon. "Reality TV." *The Television Genre Book.* Ed. Glen Greeber. London: British Film Institute, 2001. 134-137.

Doyle, Arthur Conan. *The Poison Belt.* 1913. New York: Adamant Media, 2000.

Doyle, Kegan. "Muhammad Goes to Hollywood: Michael Mann's *Ali* as Biopic." *The Journal of Popular Culture* 39.3 (2006): 383-406.

"Dr. Urkel and Mr. Cool." *Family Matters.* ABC. Dir. John Tracy. Writ. Jim Geoghan. 12. Nov. 1993. Television.

Dragnet. Prod. Jack Webb. NBC. 1951. Television.

Drucker, Susan J., and Robert S. Carthart. *American Heroes in a Media Age.* Cresskill: Hampton P, 1994.

Dubeck, Leroy W., Suzanne Mosher, and Judith E. Boss. *Science in Cinema: Teaching Science Fact Through Science Fiction Films.* New York: Teacher's College, 1988.

Duhigg, Charles, and David Barboza. "In China, Human Costs are Built Into an iPad." *The New York Times* 26 January 2012: A1.

Duhigg, Charles, and Kevin Bradsher. "How the U.S. Lost Out on iPhone Work." *New York Times* 22 Jan. 2012: A1.

Dyer, Richard. *Stars.* 1979. London: British Film Institute, 1998.

Dyson, Esther, George Gilder, George Keyworth, and Alvin Toffler. "Cyberspace and the American Dream: A Magna Carta for the Knowledge Age." *Future Insight* 1.2 (August 1994): n.p. Accessed 2 February 2012. <http://www.alamut.com/subj/ideologies/manifestos/magnaCarta.html>.

Eagles of Death Metal. "Anything 'Cept the Truth." *Heart On*. Downtown, 2008. CD.

Earthquake. Dir. Mark Robson. Universal, 1974. Film.

Ebert, Teresa. "The Crisis of Representation in Cultural Studies: Reading Postmodern Texts." *American Quarterly* 38.5 (Winter 1986): 894-902.

Eco, Umberto. *A Theory of Semiotics*. Bloomington: Indiana UP, 1976.

Edelman, Lee. *No Future: Queer Theory and the Death Drive*. Durham: Duke UP, 2004.

Eden, Jenny. "Caruso Control." *Radio Times* 5-11 August 2006: 10-12.

Edge, David. "Reinventing the Wheel." *Handbook of Science and Technology Studies*. Ed. Sheila Jasanoff, Gerald E. Markle, James C. Petersen, and Trevor Pinch. Thousand Oaks: Sage, 1995. 3-27.

Edgerton, David. "The Linear Model Did Not Exist: Reflections on the History and Historiography of Science and Research in Industry in the Twentieth Century." *The Science-Industry Nexus: History, Policy, Implications*. Ed. Karl Grandin and Nina Wormbs. New York: Watson, 2004. 31-57.

Edington, K. "The Hollywood Novel: American Dream, Apocalyptic Vision." *Literature/Film Quarterly* 23.1 (1995): 63-67.

Edison, The Man. Dir. Clarence Brown. Metro-Goldwyn-Mayer, 1940. Film.

Elliot, Jay, and William L. Simon. *The Steve Jobs Way: iLeadership for a New Generation*. Philadelphia: Perseus, 2011.

Elsaesser, Thomas. "Film History as Social History: The Dieterle/Warner Brothers Bio-pic." *Wide Angle* (Spring 1986): 15-31.

Emmerich, Roland, dir. *The Day After Tomorrow*. Interview. Fox Home Entertainment, 2004. DVD.

Erin Brockovich. Dir. Steven Soderbergh. Universal, 2000. Film.

Family Matters. Prod. Thomas L. Miller, Robert L. Boyett, and William Bickley. ABC. 1989. Television.

Family Ties. Prod. Ubu Productions and Paramount Television. NBC. 1982. Television.

Fellner, Astrid M., Susanne Hamscha, Jennifer Moos, and Klaus Heissenberger. *Is It 'Cause It's Cool? Affective Encounters with American Culture*. Vienna: LIT, 2014.

Felt, Ulrike, Helga Nowotny, and Klaus Taschwer. *Wissenschaftsforschung: Eine Einführung*. Frankfurt: Campus, 1995.

Felt, Ulrike. "Science and its Public: Popularization of Science in Vienna 1900-1938." Paper presented at the International Symposium When Science Becomes Culture in Montreal, April 1994. Accessed 2 March 2012. <http://www.cirst.uqam.ca/pcst3/PDF/Communications/FELT.PDF>.

—. "Wissenschaft, Politik, und Öffentlichkeit—Wechselwirkungen und Grenzverschiebungen." *Wissenschaft, Politik, und Öffentlichkeit: Von der Wiener Moderne bis zur Gegenwart*. Ed. Mitchell Ash and Christian Stifter. Wien: Facultas, 2002. 47-72.

Feuer, Jane. "The Unruly Woman Sitcom (I Love Lucy, Roseanne, Absolutely Fabulous)." *The Television Genre Book*. Ed. Glen Greeber. London: British Film Institute, 2001. 68-69.

—. “Situation Comedy, Part 2.” *The Television Genre Book*. Ed. Glen Greeber. London: British Film Institute, 2001. 67-70.

Feyerabend, Paul. *Against Method: Outline of an Anarchistic Theory of Knowledge*. 1970. Rev. ed. London: Verso, 1991.

Fisher, Donald. “Boundary Work and Science: The Relation between Power and Knowledge.” *Theories of Science in Society*. Ed. Susan E. Cozzens and Thomas F. Gieryn. Bloomington: Indiana UP, 1990. 98-119.

Fisher, Walter R. “Reaffirmation and Subversion of the American Dream.” *Quarterly Journal of Speech* 59 (1973): 160-167.

Fleck, Ludwig. *Entstehung und Entwicklung einer wissenschaftlichen Tatsache*. 1935. Frankfurt am Main: Suhrkamp, 1980.

Forman, Paul. “The Primacy of Science in Modernity, of Technology in Post-modernity, and of Ideology in the History of Technology.” *History and Technology* 23.1 (2007): 1-52.

Foucault, Michel. *The Archaeology of Knowledge*. New York: Pantheon, 1972.

—. *The Birth of the Clinic: An Archaeology of Medical Perception*. 1961. Trans. Alan Sheridan. New York: Pantheon, 1973.

—. *Discipline and Punish: The Birth of the Prison*. 1975. Trans. Alan Sheridan. New York: Pantheon, 1977.

—. *Power/Knowledge: Selected Interviews and Other Writings 1972-1977*. Ed. Colin Gordon. Brighton: Harvester, 1980.

—. “Space, Knowledge, and Power.” *The Foucault Reader: An Introduction to Foucault's Thought*. Ed. Paul Rabinow. London: Penguin, 1991. 239-256.

Frank, Thomas. *The Conquest of Cool: Business Culture, Counterculture, and the Rise of Hip Consumerism*. Chicago: U of Chicago P, 1998.

Freud, Sigmund. *Jokes and their Relation to the Unconscious*. Trans. James Strachey. London: Penguin, 1907.

Fricke, Harald. “Mobilisierung für die Massen: Amerikanische Untergangsängste, Mythen und Science-fiction: *Armageddon* von Michael Bay.” *Nach dem Ende: Auflösung und Untergänge im Kino an der Jahrtausendwende*. Ed. Margit Frölich, Reinhard Middel, and Karsten Visarius. Marburg: Schüren, 2001. 117-126.

Friends. Prod. Kevin S. Bright, Marta Kauffman, and David Crane. NBC. 1994. Television.

“Friends and Lovers.” *CSI: NY*. CBS. Dir. Lou Antonio. Writ. Andrew Lipsitz. 3 Nov. 2000. Television.

Frölich, Margit, Reinhard Middel, and Karsten Visarius, ed. *Nach dem Ende: Auflösung und Untergänge im Kino an der Jahrtausendwende*. Marburg: Schüren, 2001.

Fuchs, Christian. *Internet and Society: Social Theory in the Information Age*. New York: Routledge, 2008.

Funtowicz, Silvio O., and Jerome R. Ravetz. "A New Scientific Methodology for Global Environmental Issues." *Ecological Economics: The Science and Management of Sustainability.* Ed. Robert Costanza. New York: Columbia UP, 1991. 137-152.

—. "Post-Normal Science." *Internet Encyclopedia of Ecological Economics* (February 2003): 1-10. Accessed 30 January 2012. <http://www.ecoeco.org/pdf/pstnormsc.pdf>.

Galison, Peter, and Bruce Hevly. *Big Science: The Growth of Large-Scale Research.* Stanford: Stanford UP, 1992.

Galison, Peter, and David J. Stump, ed. *The Disunity of Science: Boundaries, Contexts, and Power.* Stanford: Stanford UP, 1996.

Gans, Herbert J. *Popular Culture and High Culture: An Analysis and Evaluation of Taste.* New York: Basic Books, 1974.

Gardner Conklin, Barbara, Robert Gardner, and Dennis Shortelle, ed. *Encyclopedia of Forensic Science.* Westport: Oryx, 2002.

Garrad, Greg. "Apocalypse." *Ecocriticism.* By Garrad. London: Routledge, 2004. 85-107.

"Geek." *The Oxford English Dictionary.* Online edition. Oxford: OUP. <www.oed.com>.

Geertz, Clifford. *The Interpretation of Culture.* New York: Basic Books, 1973.

Geiger, Annette, Gerald Schröder, and Änne Söll, ed. *Coolness: Zur Ästhetik einer kulturellen Strategie und Attitüde.* Bielefeld: transcript, 2010.

Geiger, Annette. "Cool ist out: Warum James Bond heute weinen muss und die Avantgarde immer weniger Gefühle zeigt." *Coolness: Zur Ästhetik einer kulturellen Strategie und Attitüde.* Ed. Annette Geiger, Gerald Schröder, and Änne Söll. Bielefeld: transcript, 2010. 85-101.

Gerbner, George, Larry Gross, Michael Morgan, and Nancy Signorelli. "Health and Medicine on Television." *New England Journal of Medicine* 305.15 (October 1981): 901-904.

Gere, Charlie. "Reading the Traces." *Reading CSI: Crime TV Under the Microscope.* Ed. Michael Allen. London: I.B.Tauris, 2007. 129-139.

Gever, Martha. "The Spectacle of Crime, Digitized. *CSI: Crime Scene Investigation* and Social Anatomy." *European Journal of Cultural Studies* 8.4 (2005): 445-463.

Gibbons, Michael, Helga Nowotny, Camille Limoges, et al. *The New Production of Knowledge: The Dynamics of Science and Research in Contemporary Societies.* London: Sage, 1994.

Gieryn, Thomas F. "Boundary-Work and the Demarcation of Science from Non-Science: Strains and Interests in Professional Ideologies of Scientists." *American Sociological Review* 48 (1983): 781-795.

—. "Science and Coca-Cola." *Science & Technology Studies* 5.1 (Spring 1987): 12-21+31.

—. "Boundaries of Science." *Handbook of Science and Technology Studies*. Ed. Sheila Jasanoff, Gerald E. Markle, James C. Petersen, and Trevor Pinch. Thousand Oaks: Sage, 1995. 393-443.

—. *Cultural Boundaries of Science: Credibility on the Line*. Chicago: U of Chicago P, 1999.

Gilchrist, Todd. "*The Day After Tomorrow*: An Interview with Roland Emmerich." *Blackfilm.* Accessed 2 Feb. 2013. <http://www.blackfilm.com/20040528/features/rolandemmerich.shtml>.

Gioia, Ted. "Cool Jazz and West Coast Jazz". *The Oxford Companion to Jazz*. Ed. Bill Kirchner. London: OUP, 2000. 332-342.

—. *The Birth and Death of the Cool*. Golden: Speck Press, 2009.

Gladwell, Malcolm. "The Coolhunt." *The New Yorker* 17 March 1997: n.p. Accessed 12 January 2012. <http://www.gladwell.com/1997/1997_03_17_a_cool.htm>.

Gleiberman, Owen. "Movie Review." *Entertainment Weekly* 19 Aug. 2013: n.p. Accessed 27. Aug. 2013. <http://www.ew.com/ew/article/0,,20483133_20725984,00.html>.

Goel, Vindu. "Facebook Is Erasing Doubts on Mobile." *New York Times* 25 July 2013: B1.

Godzilla. Dir. Roland Emmerich. Sony, 1998. Film.

Gold, Herbert. *Bohemia: Digging the Roots of Cool*. New York: Simon and Schuster, 1993.

Goldner, Orville, and George Turner. *The Making of King Kong*. New York: Ballantine, 1975.

Goode, Ian. "*CSI: Crime Scene Investigation*. Quality, the Fifth Channel and 'America's Finest.'" *Quality TV: Contemporary American Television and Beyond*. Ed. Janet McCabe and Kim Akass. London: I.B. Tauris, 2007. 118-128.

"Grand Murder at Central Station." *CSI: NY*. CBS. Dir. Scott Lautanen. Writ. Zachary Reiter. 5. Oct. 2005. Television.

Greeber, Glen. *The Television Genre Book*. London: British Film Institute, 2001.

Gregory, Jane, and Steve Miller. *Science in Public: Communication, Culture, and Credibility*. New York: Plenum, 1998.

Grossman, Lev. "And The Geek Shall Inherit the Earth." *Time Magazine* 25. Sept. 2005: n.p. Accessed 16 Dec. 2013. <http://content.time.com/time/magazine/article/0,9171,1109317,00.html>.

Grunig, James E. "Communication of Scientific Information to Non-Scientists." *Progress in Communication Studies*. Vol. 1. Ed. Brenda Dervin and M. J. Voight. Norwood: Ablex, 1980. 167-214.

Habermas, Jürgen. *Legitimation Crisis*. 1973. Trans. Thomas McCarthy. London: Heinemann, 1976.

—. "Modernity versus Postmodernity." Trans. Seyla Ben-Habib. *New German Critique* 22 (Winter, 1981): 3-14.

Hagendijk, Rob, and Jan Meeus. "Blind Faith: Fact, Fiction and Fraud in Public Controversy over Science." *Public Understanding of Science* 2 (1993): 391-415.

Hagopian, Kevin Jack. "Hollywood Restoration: Genre and the American Film Industry in the Culture of American Affirmation, 1936-1945." Unpublished PhD thesis. U of Wisconsin, Madison, 2006.

Hall, Stuart. "Culture, Media, and the 'Ideological' Effect." Ed. James Curran, Michael Gurevitch, and Janet Woollacott. *Mass Communication and Society*. London: Edward Arnold, 1977. 315-348.

—. "Cultural Studies: Two Paradigms." *Media, Culture and Society* 2 (1980): 57-72.

—. "Notes on Deconstructing 'the Popular.'" 1981. *Cultural Theory and Popular Culture: A Reader*. Ed. John Storey. Pearson: Prentice Hall, 1998. 442-453.

—. *Representation: Cultural Representations and Signifying Practices*. London: Sage, 1997.

Hamamoto, Darrell. *Nervous Laughter: Television Situation Comedy and Liberal Democratic Ideology*. New York: Praeger, 1989.

Hamit, Francis. "The CSI Shot." *Emmy Magazine* 24.3 (June 2002): 101.

Haraway, Donna. *Primate Visions: Gender, Race, and Nature in the World of Modern Science*. New York: Routledge, 1989.

—. "A Cyborg Manifesto: Science, Technology, and Socialist-Feminism in the Late Twentieth Century." *Simians, Cyborgs and Women: The Reinvention of Nature*. By Haraway. New York: Routledge, 1991. 149-181.

—. "A Game of Cat's Cradle: Science Studies, Feminist Theory, Cultural Studies." *Configurations* 2.1 (1994): 59-71.

—. *Modest_Witness@Second_Millenium. FemaleMan©_Meets_OncoMouse™: Feminism and Technoscience*. New York: Routledge, 1997.

Harding, Sandra. *Whose Science? Whose Knowledge? Thinking from Women's Lives*. Ithaca: Cornell UP, 1992.

—. "Is Science Multicultural? Challenges, Resources, Opportunities, Uncertainties." *Configurations* 2.2 (1994): 301-330.

Hardt, Michael, and Antonio Negri. *Empire*. Cambridge: Harvard UP, 2000.

Harrington, Ellen Burton. "Nation, Identity and the Fascination with Forensic Science in Sherlock Holmes and *CSI*." *International Journal of Cultural Studies* 10.3 (2007): 365-382.

Harris, Daniel. "Coolness." *The American Scholar* 68.4 (Autumn 1999): 39-49.

Hartley, John. "Situation Comedy, Part 1." *The Television Genre Book*. Ed. Glen Greeber. London: British Film Institute, 2001. 65-67.

Haselstein, Ulla, Irmela Hijiya-Kirschnereit, Catrin Gersdorf, and Elena Giannoulis. *The Cultural Career of Coolness: Discourses and Practices of Affect Control in European Antiquity, the United States, and Japan*. Lanham et al.: Lexington, 2013.

Haug, Wolfgang Fritz. *High-Tech-Kapitalismus*. Hamburg: Argument, 2003.

Haynes, Roslynn. "The Scientist in Literature: Images and Stereotypes—Their Importance." *Interdisciplinary Science Reviews* 14.4 (1989): 384-398.

—. *From Faust to Doctor Strangelove: Representations of the Scientist in Western Literature*. Baltimore: John Hopkins UP, 1994.

—. "From Alchemy to Artificial Intelligence: Stereotypes of the Scientist in Western Literature." *Public Understanding of Science* 12 (2003): 243-253.

Healy, Kieran. "A Sociology of Steve Jobs." 10 Oct. 2011: n.p. Accessed 10 Sept. 2013. <http://kieranhealy.org/blog/archives/2011/10/10/a-sociology-of-steve-jobs/>.

Heidegger, Martin. *Die Technik und die Kehre*. Stuttgart: Klett Cotta, 1962.

Heisenberg, Werner. *Physics and Beyond: Encounters and Conversations*. 1969. Trans. Arnold J. Pomerans. New York: Harper, 1971.

Hickey, Dave. "American Cool." *Art Issues* (January-February 1999): 10-13.

"Hide Sight." *CSI: NY*. CBS. Dir. Alex Zakrzewski. Writ. Bill Haynes. 5 Nov. 2010.

Hilgartner. Stephen. "The Dominant View of Popularization: Conceptual Problems, Political Uses." *Social Studies of Science* 20.3 (August 1990): 519-539.

Hiltz, Starr Roxanne, and Murray Turoff. *The Network Nation*. 1978. Cambridge: MIT Press, 1993.

Hodgson, Simon. "A Wild Ride." *Magazine of the IEMA* (August 2004): 14-15.

Hoff, Al. "*The Day After Tomorrow*: Chilly Forecast." *Pittsburgh City Paper* 3 June 2004: n.p. Accessed 2 Feb. 2013. <http://www.pghcitypaper.com/pittsburgh/the-day-after-tomorrow/Content?oid=1336354>.

Hofmann, Franck. "Nach der Coolness. Lob der Kälte—Lob des Gartens: Bertolt Brechts Erneuerung der Literatur." *Coolness: Zur Ästhetik einer kulturellen Strategie und Attitüde*. Ed. Annette Geiger, Gerald Schröder, and Änne Söll. Bielefeld: transcript, 2010. 127-147.

Holton, Gerald, and William Blanpied. *Science and Its Public: The Changing Relationship*. Dordrecht: Reidel, 1976.

hooks, bell. Interview. *Cultural Criticism & Transformation*. Dir. Sut Jhally. Media Education Foundation, 1997. Film.

—. *We real cool: Black Men and Masculinity*. New York: Taylor & Francis, 2004.

Hooper, Rowan. "Television Scrambles the Evidence." *New Scientist* 12 (Fall 2005): 12-13.

Horton, John. "Class Struggle and the American Dream: A Marxist Analysis of Communication." *Studies in Communication* 2 (1982): 111-141.

Hottois, Gilbert. "Technoscience: Nihilistic Power versus a New Ethical Consciousness." *Technology and Responsibility*. Ed. Paul T. Durbin. Dordrecht: Reidel, 1987. 69-84.

House, M.D. Prod. Paul Attanasio et al. Fox. 2004. Television.

How I Met Your Mother. Prod. Craig Thomas, Carter Bays, et al. CBS. 2005. Television.

Hughes, Thomas P. "The Seamless Web: Technology, Science, Etcetera, Etcetera." *Social Studies of Science* 16 (1986): 281-292.

Hüppauf, Bernd, and Peter Weingart, ed. *Science Images and Popular Images of the Sciences*. New York: Routledge, 2008.

Hutcheon, Linda. *The Politics of Postmodernism*. New York: Routledge, 1989.

Huxley, Thomas H. *Man's Place in Nature and Other Anthropological Essays*. London: Macmillan, 1897.

Hyde, William. "My Day After Tomorrow Review(ish) thing." *Mindstalk*. Blog. Accessed 4 Feb. 2013. <http://mindstalk.net/hyde>.

Ihde, Don, and Evan Selinger. *Chasing Technoscience: Matrix for Materiality*. Bloomington: Indiana UP, 2003.

I.Q. Dir. Fred Schepisi. Paramount, 1994. Film.

I'm Not There. Dir. Todd Haynes. The Weinstein Company, 2007. Film.

Independence Day. Dir. Roland Emmerich. 1996. Film.

J. Edgar. Dir. Clint Eastwood. Warner Bros., 2011. Film.

Jameson, Fredric. "Reification and Utopia in Mass Culture." *Social Text* 1 (Winter 1979): 130-148.

—. "Foreword." *The Postmodern Condition: A Report on Knowledge*. By Lyotard. Trans. Geoff Bennington and Brian Massumi. Manchester: Manchester UP, 1984. vii-xxv.

—. *Postmodernism, or, the Cultural Logic of Late Capitalism*. London, New York: Verso, 1991.

Jasanoff, Sheila. *The Fifth Branch: Science Advisers as Policymakers*. Cambridge: Harvard UP, 1990.

—, Gerald E. Markle, James C. Petersen, and Trevor Pinch, ed. *Handbook of Science and Technology Studies*. Thousand Oaks: Sage, 1995.

Jay, Martin. *The Dialectical Imagination: A History of the Frankfurt School and the Institute for Social Research 1923-1950*. Berkeley: U of California P, 1996.

Jensen, K. Thor. "11 Reasons Geeks Hate *The Big Bang Theory*." *UGO Entertainment*. 12 Jan. 2012: n.p. Accessed 15 Sept. 2012. <http://www.ugo.com/tv/11-reasons-geeks-hate-the-big-bang-theory>.

Jermyn, Deborah. "Body Matters: Realism, Spectacle and the Corpse in *CSI*." *Reading CSI: Crime TV Under the Microscope*. Ed. Michael Allen. London: I.B.Tauris, 2007. 79-89.

Jobs. Dir. Joshua Michael Stern. Writ. Matt Whiteley. Open Road Films, 2013.

Jurassic Park. Dir. Steven Spielberg. Universal, 1993. Film.

"Justice Is Served." *CSI: Crime Scene Investigation*. CBS. Dir. Thomas J. Wright. Writ. Jerry Stahl. 26 April 2001. Television.

K19: The Widowmaker. Dir. Kathryn Bigelow. Paramount, 2002. Film.

Kakoudaki, Despina. "Spectacles of History: Race Relations, Melodrama, and the Science Fiction/Disaster Film." *Camera Obscura* 50 17.2 (2002): 1-153.

Kant, Immanuel. *Critique of Judgment*. 1790. Trans. John Henry Bernard. London: Macmillan, 1931.

Kath and Kim. Prod. Jane Turner and Gina Riley. ABC. 2002. Television.

Keane, Stephen. *Disaster Movies: The Cinema of Catastrophe*. London: Wallflower, 2001.

Keller, Evelyn Fox. *Reflections on Gender and Science*. New Haven: Yale UP, 1985.

—. *Secrets of Life—Secrets of Death: Essays on Language, Gender, and Science*. New York: Routledge, 1992.

—. *Refiguring Life: Metaphors of Twentieth-Century Biology*. New York: Columbia UP, 1995.

"Killer Date." *CSI: Miami*. CBS. Dir. Karen Gaviola. Writ. Elizabeth Devine and John Haynes. 18 April 2005. Television.

Kimmel, Michael. *Manhood in America: A Cultural History*. New York: Free Press, 1996.

King, Geoff. *Spectacular Narratives: Hollywood in the Age of the Blockbuster*. London: I.B. Tauris, 2000.

Kirby, David A. "Science Consultants, Fictional Films, and Scientific Practice." *Social Studies of Science* 33.2 (April 2003): 231-268.

—. *Lab Coats in Hollywood*. Cambridge: MIT Press, 2011.

Kirchner, Bill. *The Oxford Companion to Jazz*. London: OUP, 2000.

Kitch, Carolyn. "'It Takes a Sinner to Appreciate the Blinding Glare of Grace': Rebellion and Redemption in the Life Story of the 'Dark' Celebrity." *Popular Communication* 5.1 (2007): 37-56.

Klaver, Elizabeth. *Images of the Corpse: From the Renaissance to Cyberspace*. Madison: U of Wisconsin P, 2004.

Knorr Cetina, Karin. *The Manufacture of Knowledge: An Essay on the Constructivist and Conceptual Nature of Science*. Oxford: Pergamon, 1981.

—. "The Reappropriation of Technology." *Alteration in Food Production*. Ed. Dietrich Knorr and Tom R. Watkins. London: Van Nostrand Reinhold, 1984. 223-237.

—. "Laboratory Studies: The Cultural Approach to the Study of Science." *Handbook of Science and Technology Studies*. Ed. Sheila Jasanoff, Gerald E. Markle, James C. Petersen, and Trevor Pinch. Thousand Oaks: Sage, 1995. 140-165.

Knowing. Dir. Alex Proyas. Summit, 2009. Film.

Knox, Simone. "Five's Finest: The Import of *CSI* to British Terrestrial Television." *Reading CSI: Crime TV Under the Microscope*. Ed. Michael Allen. London: I.B.Tauris, 2007. 183-197.

Kohlenberger, Judith. "C.ool S.exy I.ntelligent: Cool Forensics and the Spectacle of Technoscience in *CSI*." *Is It 'Cause It's Cool? Affective Encounters with Amer-*

ican Culture. Ed. Astrid M. Fellner, Jennifer Moos, Susanne Hamscha, and Klaus Heissenberger. Vienna: LIT, 2013. 117-139.

König, Anna. "Who Are They? Style Codes of the *CSI* Investigators." *Reading CSI: Crime TV Under the Microscope.* Ed. Michael Allen. London: I.B.Tauris, 2007. 103-106.

Kuhn, Thomas. *The Structure of Scientific Revolutions.* 1962. 3rd ed. Chicago: U of Chicago P, 1996.

Kutcher, Ashton. "Acceptance Speech." *Teen Choice Awards.* Gibson Amphitheater, Los Angeles, CA. 11 Aug. 2013.

—. Interview I. *The Tonight Show with Jay Leno.* NBC. 24 July 2013.

—. Interview II. *The Hollywood Reporter* 8 Aug. 2013: n.p. Accessed 14 Sept. 2013. <http://www.hollywoodreporter.com/news/jobs-ashton-kutcher-responds-steve-602679>.

Kuznick, Peter J. *Beyond the Laboratory: Scientists as Political Activists in 1930s America.* Chicago: U of Chicago P, 1987.

Kruse, Corinna. "Producing Absolute Truth: *CSI* Science as Wishful Thinking." *American Anthropologist* 112.1 (March 2010): 79-91.

Labov, Teresa. "Social Language Boundaries among Adolescents." *American Speech* 67: 339-366.

Lacan, Jacques. *Écrits.* 1936-1966. Transl. Bruce Fink. New York: W.W. Norton & Company, 2007.

—. *The Seminar, Book XI: The Four Fundamental Concepts of Psychoanalysis.* 1964. Ed. Jacques-Alain Miller. Trans. Alan Sheridan. New York: W.W. Norton & Company, 1977.

LaFollette, Marcel C. *Making Science Our Own: Public Images of Science, 1910-1955.* Chicago: U of Chicago P, 1990.

Landy, Marcia. *Cinematic Uses of the Past.* Minneapolis: U of Minnesota P, 1996.

Lane, Robert E. "The Decline of Politics and Ideology in a Knowledgeable Society." *American Sociological Review* 31 (1966): 649-662.

Lashinsky, Adam. "The Decade of Steve: How Apple's imperious, brilliant CEO transformed American business." Fortune Magazine 5 Nov. 2009: n.p. Accessed 13 Sept. 2013. <http://archive.is/20120604101213/http://money.cnn.com/2009/11/04/technology/steve_jobs_ceo_decade.fortune/index.htm>.

Lasn, Kalle. *Culture Jam: The Uncooling of America.* New York: Eagle, 1999.

Last Night. Dir. Don McKellar. Lionsgate, 1998. Film.

Latour, Bruno. "Give Me a Laboratory and I Will Raise the World." 1983. *The Science Studies Reader.* Ed. Mario Biagioli. New York: Routledge, 1999. 258-275.

—. *Science in Action: How to Follow Scientists and Engineers through Society.* Milton Keynes: Open UP, 1987.

—. *We Have Never Been Modern.* Transl. Catherine Porter. Cambridge: Harvard UP, 1993.

Latour, Bruno, and Steve Woolgar. *Laboratory Life: The Social Construction of Scientific Facts*. 1979. Princeton: Princeton UP, 1986.

Latta, Robert L. *The Basic Humor Process: A Cognitive-Shift Theory and the Case against Incongruity*. Humor Research 5. New York: Mouton de Gruyter, 1999.

Lavigne, Carlen. "Death Wore Black Chiffon: Sex and Gender in *CSI*." *Canadian Review of American Studies* 39.4 (2009): 383-398.

Law & Order: Criminal Intent. Prod. Dick Wolf et al. NBC. 2001. Television.

Leiserowitz, Anthony. "Before and after *The Day After Tomorrow*: A U.S. study of climate change risk perception." *Environment* 46.9 (2004): 22-37. Accessed 12 Dec. 2013. <http://environment.yale.edu/leiserowitz/climatechange/TDAT_assets/LeiserowitzTDAT2.pdf>.

Levy, Emanuel. "The American Dream of Family in Film: From Decline to Comeback." *Journal of Comparative Family Studies* 22 (1991): 187-204.

Lewenstein, Bruce. "The Meaning of 'Public Understanding of Science' in the United States after World War II." *Public Understanding of Science* 1 (1992): 45-68.

—. "From Fax to Facts: Communication in the Cold Fusion Saga." *Social Studies of Science* 25 (1995): 403-436.

Leydesdorff, Loet, and Henry Etzkowitz. "Emergence of a Triple Helix of University-Industry-Government Relations." *Science and Public Policy* 23 (1996): 279-86.

Lincoln. Dir. Steven Spielberg. Twentieth Century Fox, 2012. Film.

Lingis, Alphonso. "Some questions about Lyotard's postmodern legitimation narrative." *Philosophy & Social Criticism* 20 (January 1994): 1-12.

Lipsitz, George. *Time Passages: Collective Memory and American Popular Culture*. Minneapolis: U of Minnesota P, 1990.

Liu, Alan. *The Laws of Cool: Knowledge Work and the Culture of Information*. Chicago: U of Chicago P, 2004.

Logan, Robert A. "Popularization vs. Secularization: Media Coverage of Health." *Risky Business: Communicating Issues of Science*. Ed. Lee Wilkins and Philip Patterson. New York: Greenwood, 1991. 43-59.

Loretan, Matthias, and Charles Martig. "Weltuntergang im Film zwischen Spektakel und Vision. Sinn und Gefahren apokalyptischer Rede. Kritische Auseinandersetzung mit ethischen, ästhetischen und theologischen Aspekten zeitgenössischer Filme." *Internationale Zeitschrift für Kommunikation in Religion, Kirche und Gesellschaft* 32 (1999): 115-148.

Lowenthal, Leo. "Biographies in Popular Magazines." *Radio Research: 1942-1943*. Ed. Paul Lazarsfeld and Frank Stanton. New York: Duell, Sloan and Pearce, 1944.

Luhmann, Niklas. "Ökologie des Nichtwissens." *Beobachtungen der Moderne*. By Luhmann. Opladen: Westdeutscher Verlag, 1992. 149-220.

Lury, Karen. *Interpreting Television*. London: Hodder Arnold, 2005.

Lynch, Michael. "Representation is Overrated: Some Critical Remarks about the Use of the Concept of Representation in Science Studies." *Configurations* 2.1 (1994): 137-149.

Lyons, Margaret. "The Theorists: Big Bang Theory's big Belarusian rip-off." *Popwatch*. 9 Feb. 2010. Accessed 21 Sept. 2012. <http://popwatch.ew.com/2010/02/09/big-bang-theory-belarus>.

Lyotard, Jean-François. *The Postmodern Condition: A Report on Knowledge*. 1979. Trans. Geoff Bennington and Brian Massumi. Manchester: Manchester UP, 1984.

MacAdams, Lewis. *Birth of the Cool: Beat, Bebop, and the American Avant-garde*. New York: Free Press, 2001.

Machlup, Fritz. *The Production and Distribution of Knowledge in the United States.* Princeton: Princeton UP, 1962.

Madame Curie. Dir. Mervyn LeRoy. Metro-Goldwyn-Mayer, 1943. Film.

Madrigal, Alexis C. "Sorting the Real Sandy From the Fakes." *The Atlantic* 29 Oct. 2012: n.p. Accessed 12 Feb. 2013. <http://www.theatlantic.com/technology/archive/2012/10/instasnopes-sorting-the-real-sandy-photos-from-the-fakes/264243/>.

Mailer, Norman. "The White Negro." *Advertisements for Myself.* By Mailer. Cambridge: Harvard UP, 1992. 337-358.

Majors, Richard, and Janet Mancini Billson. *Cool Pose: The Dilemmas of Black Manhood in America.* New York: Touchstone, 1992.

"Making Of: How Did They Ever Make a Movie of Facebook? Feature Length Documentary." *The Social Network.* Dir. David Fincher. Writ. Aaron Sorkin. 2010. Sony Home Entertainment, 2011. DVD.

Manley, John F. "American Liberalism and the Democratic Dream: Transcending the American Dream." *Policy Studies Review* 10.1 (1990): 89-102.

Marc, David. *Comic Visions: Television Comedy and American Culture*. Boston: Unwin Hyman, 1989.

Marcus, George E., ed. *Technoscientific Imaginaries: Conversations, Profiles, and Memoirs*. Chicago: U of Chicago P, 1995.

Marcus, Greil. *Dead Elvis.* Cambridge: Harvard UP, 1991.

Markoff, John. "An 'Unknown' Co-Founder Leaves After 20 Years of Glory and Turmoil." *New York Times* 1 Sept. 1997: n.p. Accessed 18 Sept. 2013. <http://www.nytimes.com/1997/09/01/business/an-unknown-co-founder-leaves-after-20-years-of-glory-and-turmoil.html?pagewanted=all&src=pm>.

Martig, Charles. "Filmische Apokalypsen: Wie das Kino dem Ende der Zeit entgegenfieberte." *Herder Korrespondenzen* 54 (2000): 32-38.

—. "Viel Vernügen beim Weltuntergang: Filmästhetik zwischen Spektakel und Vision." *Nach dem Ende: Auflösung und Untergänge im Kino an der Jahrtausendwende.* Ed. Margit Frölich, Reinhard Middel, and Karsten Visarius. Marburg: Schüren, 2001. 49-69.

Martin, James. *The Wired Society*. New Jersey: Prentice-Hall, 1978.

Maude. Prod. Norman Lear. CBS. 1972. Television.

McArthur, Colin. *Television and History*. London: British Film Institute, 1980.

McCabe, Janet. "Mac's Melancholia: Scripting Trauma, 9/11 and Bodily Absence in *CSI: NY*." *Reading CSI: Crime TV Under the Microscope*. Ed. Michael Allen. London: I.B.Tauris, 2007. 167-180.

—, and Kim Akass, ed. *Quality TV: Contemporary American Television and Beyond*. London: I.B. Tauris, 2007.

McCracken, Henry. "An Apple History Nerd's Take on the Jobs Movie." *Time Magazine Online*. 6 Aug. 2013. Accessed 24 Aug. 2013: n.p. <http://techland.time.com/2013/08/06/an-apple-history-nerds-take-on-the-jobs-movie/>.

McGuigan, Jim. *Cool Capitalism*. London: Pluto, 2009.

—. "From Cultural Populism to Cool Capitalism." *Art & The Public Sphere* 1.1 (2011): 7-18.

—. "The Coolness of Capitalism Today." *Communication, Capitalism & Critique* 10.2 (2012): 425-438.

—. "Mobile Privatisation and the Neoliberal Self." *Key Words: a Journal of Cultural Materialism* 11 (2013): 75-89.

McKee, Robert. *Story: Style, Structure, Substance, and the Principles of Screenwriting*. New York: itbooks, 1997.

McLuhan, Marshall. *Understanding Media: The Extensions of Man*. New York: McGraw-Hill, 1964.

Mead, Margaret, and Rhoda Métraux. "Image of the Scientist among High-School Students." *Science* 126 (1957): 384-390.

Medical Investigation. Prod. Paramount Network Television and Universal Television Studio. NBC. 2004. Television.

Melancholia. Dir. Lars von Trier. Concorde, 2011. Film.

Mellencamp, Patricia. "Situation Comedy, Feminism, and Freud: Discourses of Gracie and Lucy." 1986. *Feminist Television Criticism: A Reader*. Ed. Charlotte Brunsdon. Oxford: OUP, 1997. 60-73.

Mentges, Gabriele. "Coolness—Zur Karriere eines Begriffs. Versuch einer historischen und analytischen Annäherung." *Coolness: Zur Ästhetik einer kulturellen Strategie und Attitüde*. Ed. Annette Geiger, Gerald Schröder, and Änne Söll. Bielefeld: transcript, 2010. 17-35.

Merton, Robert. "The Normative Structure of Science." 1942. *The Sociology of Science: Theoretical and Empirical Investigations*. By Merton. Ed. Norman W. Storer. Chicago: Chicago UP, 1973. 267-278.

Messerschmidt, James W. *Masculinities and Crime: Critique and Reconceptualization of Theory*. Lanham: Rowan & Littlefield, 1993.

Mezrich, Ben. *The Accidental Billionaires: The Founding of Facebook: A Tale of Sex, Money, Genius and Betrayal*. New York: Doubleday, 2009.

Miller, Ross. "Steve Wozniak turned down offer to work on Jobs after reading early script." *The Verge* 26 Jan. 2013: n.p. Accessed 14 Sept. 2013. <http://www.theverge.com/2013/1/26/3919472/steve-wozniak-turned-down-offer-to-work-on-jobs-after-reading-early-script>.

Mills, Brett. "Studying Comedy." *The Television Genre Book.* Ed. Glen Greeber. London: British Film Institute, 2001. 61-62.

—. "Comedy Verité: Contemporary Sitcom Form." *Screen* 45.1 (Spring 2004): 63-78.

—. *The Sitcom.* Edinburgh: Edinburgh UP, 2009.

Missner, Marshall. "Why Einstein Became Famous in America." *Social Studies of Science* 15 (1985): 267-291.

Mitchell, Charles P. *A Guide to Apocalyptic Cinema.* Westport: Greenwood, 2001.

Mitchell, W.J.T. *Picture Theory.* Chicago: U of Chicago P, 1994.

—. "Representation." *Critical Terms for Literary Study.* Ed. Frank Lentricchia and Thomas McLaughlin. Chicago: U of Chicago P, 1990. 11-22.

Modleski, Tania. "Femininity as Mas(s)querade: A Feminist Approach to Mass Culture." *High Theory/Low Culture: Analysing Popular Television and Film.* Ed. Colin MacCabe. New York: St. Martin's, 1986. 37-52.

Moirand, Sophie. "Communicative and cognitive dimensions of discourse on science in the French mass media." *Discourse Studies* 5.2 (2003): 175-206.

Monbiot, George. "A hard rain's a-gonna fall." *Guardian* 14 May 2004: n.p. Accessed 2 Feb. 2013. <http://www.guardian.co.uk/film/2004/may/14/climatechange>.

Montgomery, Kathryn C. *Generation Digital: Politics, Commerce, and Childhood in the Age of the Interne*t. Cambridge: MIT Press, 2007.

Mopas, Michael. "Examining the 'CSI Effect' through an ANT lens." *Crime, Media, Culture* 3.1 (2007): 110-117.

Moraes, Lisa de. "What were the Top 10 most-watched shows this season?" *Washington Post* 23 May 2012: n.p. Accessed 3 August 2012. <http://www.washingtonpost.com/blogs/tv-column/post/what-were-the-top-10-most-watched-shows-this-season/2012/05/23/gJQANudXlU_blog.html>.

Morreale, Joanne, ed. *Critiquing the Sitcom: A Reader.* Syracuse: Syracuse UP, 2003.

Morreall, John. *Taking Laughter Seriously.* Albany: SUNY Press, 1983.

—. *The Philosophy of Laughter and Humor.* Albany: SUNY Press, 1987.

Morris-Suzuki, Tessa. "Capitalism in the Computer Age." *Cutting Edge: Technology, Society and Social Revolution.* Ed. Jim Davis, Thomas Hirschl, and Michael Stack. London: Verso, 1997. 13-27.

"Murder Sings the Blues." *CSI: NY.* CBS. Dir. Oz Scott. Writ. Sam Humphrey. 1 Nov. 2006. Television.

My Week with Marilyn. Dir. Simon Curtis. The Weinstein Company, 2011. Film.

Myers, Greg. "Discourse Studies of Scientific Popularization: Questioning the Boundaries." *Discourse Studies* 5.2 (2003): 265-279.

Neale, Steve. *Genre and Hollywood*. London: Routledge, 2000.

—. "Sketch Comedy." *The Television Genre Book*. Ed. Glen Greeber. London: British Film Institute, 2001. 62-65.

Nelkin, Dorothy. "Science Controversies: The Dynamics of Public Disputes in the United States." *Handbook of Science and Technology Studies*. Ed. Sheila Jasanoff, Gerald E. Markle, James C. Petersen, and Trevor Pinch. Thousand Oaks: Sage, 1995. 444-456.

Nessel, Sabine. "Zurschaustellung und Performanz: Visual Effects als Stars der Katastrophenfilme der 90er Jahre." *Nach dem Film* 1 (Dec. 1999): n.p. Accessed 20 Jan. 2013. <http://nachdemfilm.de/content/zurschaustellung-und-performanz>.

Netburn, Deborah. "Thomas Edison, Steve Jobs top young adult list of greatest innovators." *Los Angeles Times* 27 Jan. 2012: n.p. Accessed 13 Sept. 2013. <http://archive.is/20120709172336/http://latimesblogs.latimes.com/technology/2012/01/thomas-edison-steve-jobs-top-young-adult-list-of-greatest-innovators.html>.

Neumaier, John. "Jobs, movie review." *New York Daily News* 15 August 2013: n.p. Accessed 27 Aug. 2013. <http://www.nydailynews.com/entertainment/tv-movies/jobs-movie-review-article-1.1426804>.

Newton, Casey. "Review: While Jobs fawns over subject, film falls flat." *CNET* 25 Jan. 2013: n.p. Accessed 18 Sept. 2013. <http://news.cnet.com/8301-13579_3-57565992-37/review-while-jobs-fawns-over-subject-film-falls-flat/>.

Nietzsche, Friedrich. *The Antichrist*. 1985. Trans. Anthony M. Ludovici. New York: Prometheus, 2000.

Nikolow, Sybilla, and Arne Schirrmacher, ed. *Wissenschaft und Öffentlichkeit als Ressourcen füreinander*. Frankfurt: Campus, 2007.

Nisbet, Robert. *History of the Idea of Progress*. New York: Basic, 1980.

Noble, Bobby. "Queer as Box: Boi Spectators and Boy Culture on Showtime's Queer as Folk." *Third Wave Feminism and Television*. Ed. Merri Lis Johnson. London, New York: I.B.Tauris, 2007. 147-165.

Nye, David. *American Technological Sublime*. Cambridge: MIT, 1994.

O'Connell, Sean. "The Day After Tomorrow." *Movie Guide*. Accessed 2. Feb. 2013. <http://movies.amctv.com/movie/2004/The+Day+After+Tomorrow>.

O'Sullivan, Michael. "Jobs: A Film with little Room for Insights." *Buenos Aires Herald* 23 Aug. 2013: n.p. Accessed 24 Aug. 2013. <http://www.buenosairesherald.com/article/139035/jobs-a-film-with-little-room-for-insights>.

OECD. *OECD Guide to Measuring the Information Society 2011*. OECD Publishing, 2011. Accessed 27 January 2012. <http://dx.doi.org/10.1787/10.1787/9789264113541-en>.

"Open and Shut." *CSI: NY*. CBS. Dir. Joe Ann Fogle. Writ. Wendy Battles. 25 Oct. 2006. Television.

Orland, Barbara. "Wissenschaft und Öffentlichkeit." *Werkstattgeschichte* 51 (2009): 108-111.

Overbye, Dennis. "Exploring the Complexities of Nerdiness, for Laughs." *New York Times* 27 April 2010: D4.

Palmer, Jerry. *The Logic of the Absurd*. London: British Film Institute, 1987.

Panse, Silke. "'The Bullets Confirm the Story Told by the Potato': Materials without Motives in *CSI: Crime Scene Investigation*." *Reading CSI: Crime TV Under the Microscope*. Ed. Michael Allen. London: I.B.Tauris, 2007. 153-166.

Pansegrau, Petra. "Stereotypes and Images of Scientists in Fiction Film." *Science Images and Popular Images of the Sciences*. Ed. Bernd Hüppauf and Peter Weingart. New York: Routledge, 2008. 257-266.

Parshall, Peter F. "*Die Hard* and the American Dream Mythos." *Journal of Popular Film and Television* 18.4 (1991): 134-144.

Pearson, Roberta. "Anatomising Gilbert Grissom: The Structure and Function of the Televisual Character." *Reading CSI: Crime TV Under the Microscope*. Ed. Michael Allen. London: I.B.Tauris, 2007. 39-56.

Penley, Constance. *NASA/Trek: Popular Science and Sex in America*. London: Verso, 1997.

Peters, Michael. *Education and the Postmodern Condition*. Westport: Bergin & Garvey, 1995.

Phillips, Adam, ed. "Introduction." *A Philosophical Enquiry into the Origin of our Ideas of the Sublime and the Beautiful*. By Edmund Burke. Oxford: OUP, 1990. ix-xxiii.

Pickering, Andrew. "Book Review: We Have Never Been Modern." *Modernism/Modernity* 1.3 (1994): 257-258.

Pierson, David P. "Evidential Bodies: The Forensic and Abject Gazes in *C.S.I.: Crime Scene Investigation*." *Journal of Communication Inquiry* 34.2 (2010): 184-203.

Pileggi, Mary S., Maria Elizabeth Grabe, Lisa B. Holderman, and Michelle de Montigny. "Business As Usual: The American Dream in Hollywood Business Films." *Mass Communication & Society* 2&3 (2000): 207-228.

"Pilot." *CSI: Crime Scene Investigation*. CBS. Dir. Danny Cannon. Writ. Anthony Zuiker. 6 Oct. 2000. Television.

"Pilot." *The Big Bang Theory*. Dir. Mark Cendrowski. Writ. Robert Cohen and David Litt. CBS. 29 Oct. 2007. Television.

Pirates of the Caribbean: The Curse of the Black Pearl. Dir. Gore Verbinski. Prod. Jerry Bruckheimer. Walt Disney, 2003. Film.

Pirates of the Caribbean: Dead Man's Chest. Dir. Gore Verbinski. Prod. Jerry Bruckheimer. Walt Disney, 2006. Film.

Pirates of the Caribbean: At World's End. Dir. Gore Verbinski. Prod. Jerry Bruckheimer. Walt Disney, 2007. Film.

Pirates of the Caribbean: On Stranger Tides. Dir. Rob Marshall. Prod. Jerry Bruckheimer. Walt Disney, 2011. Film.

Planet of the Apes. Dir. Franklin J. Schaffner. Twentieth Century Fox, 1968. Film.

Podlas, Kimberlianne. "'The CSI Effect': Exposing the Media Myth." *Fordham Intellectual Property, Media, and Entertainment Law Journal* 16.2 (2005): 429-465.

Polar Storm. Dir. Paul Ziller. Sunfilm, 2009. DVD.

Poniewozik, James. "Crimetime Lineup: How the Slick Show Changed Television—in Part by Dragging It into the Past." *Time Magazine* 8 Nov. 2004: n.p. Accessed 22 June 2012. <http://www.time.com/time/magazine/article/0,9171,995590,00.html>.

Popper, Karl. *The Logic of Scientific Discovery*. 1934. New York: Harper, 1959.

Poschardt, Ulf. *Cool*. Hamburg: Roger&Bernhard, 2000.

Pountain, Dick, and David Robins. *Cool Rules: Anatomy of an Attitude*. London: Reaktion, 2000.

"Prey." *CSI: NY*. CBS. Dir. Matt Earl Beesley. Writ. Barry O'Brien. 19 Jan. 2009. Television.

Price, Derek de Solla. *Little Science, Big Science*. New York: Columbia UP, 1963.

—. "The Parallel Structures of Science and Technology." *Science in Context: Readings in the Sociology of Science*. Ed. Barry Barnes and David Edge. Milton Keynes: Open UP, 1982. 164-176.

Prime Suspect. Prod. Gub Neal, Rebecca Eaton, and Lynn Horsford. ITV. 1991. Television.

Prince, Rosa. "Stephen Hawking set for 'geekcom' role in US sitcom *The Big Bang Theory*." *Telegraph* 12 March 2012: n.p. Accessed 21 Sept. 2012. <http://www.telegraph.co.uk/culture/tvandradio/9138730/Stephen-Hawking-set-for-geekcom-role-in-US-sitcom-The-Big-Bang-Theory.html>.

Pynchon, Thomas. *V*. Philadelphia: J.B. Lippincott & Co, 1963.

Quincy, M.E. Prod. Glen A. Larson. NBC. 1976. Television.

Rahilly, Lucia. "The Quintessence of Con: The Las Vegas of *CSI*." *Reading CSI: Crime TV Under the Microscope*. Ed. Michael Allen. London: I.B.Tauris, 2007. 122-125.

Ravetz, Jerome R. "Usable Knowledge, Usable Ignorance." *Knowledge* 9 (1987): 87-116.

Ray. Dir. Taylor Hackford. Universal, 2004. Film.

"Real Life CSI Stories." *The Oprah Winfrey Show*. ABC. 23 October 2003. Television.

Rebel Without a Cause. Dir. Nicholas Ray. Writ. Stewart Stern and Irving Shulman. Warner Bros., 1955.

Reid, Roddey. "Researcher or Smoker? Or, When the Other Isn't Other Enough in Studying 'Across' Tobacco Control." *Doing Science and Culture.* Ed. Roddey Reid and Sharon Traweek. New York: Routledge, 2000. 119-150.

—, and Sharon Traweek, ed. *Doing Science and Culture.* New York: Routledge, 2000.

Reinhardt, Carsten. "Historische Wissenschaftsforschung, heute: Überlegungen zu einer Geschichte der Wissensgesellschaft." *Ber. Wissensschaftsgesch.* 33 (2010): 81-99.

Restivo, Sal. "The Theory Landscape in Science Studies: Sociological Traditions." *Handbook of Science and Technology Studies.* Ed. Sheila Jasanoff, Gerald E. Markle, James C. Petersen, and Trevor Pinch. Thousand Oaks: Sage, 1995. 95-113.

Rice, Jeff. "Writing about Cool: Teaching Hypertext as Juxtaposition." *Computers and Composition: An International Journal for Teachers of Writing* 20.3 (2003): 221-236.

—. "What is Cool? Notes on Intellectualism, Popular Culture, and Writing." *CTheory* 2002: n.p. Accessed 20 May 2009. <http://www.ctheory.net/articles.aspx?id=338.>.

Richta, Radovan, et al. *Civilization at the Crossroads: Social and Human Implications of the Scientific and Technological Revolution.* 1966. White Plains: International Arts and Science Press, 1969.

Rip, Arie. "Legitimations of Science in a Changing World." *Wissenschaftssprache und Gesellschaft.* Ed. Theo Bungarten. Hamburg: Edition Akademion, 1986. 134-148.

Roach, Joseph. *It.* Michigan: U of Michigan P, 2007.

Roddick, Nick. "Only the stars survive: Disaster Movies in the Seventies." *Performance and Politics in Popular Drama: Aspects of Popular Entertainment.* Ed. David Brady. Cambridge: Cambridge UP, 1980. 243-269.

Romein, Jan. *Die Biographie: Einführung in ihre Geschichte und ihre Problematik.* Bern: Francke, 1948.

Rorty, Richard. *The Linguistic Turn: Recent Essays in Philosophical Method.* Chicago: U of Chicago P, 1967.

—. "Habermas and Lyotard on Postmodernity." *Habermas and Modernity.* Ed. Richard J. Bernstein. Cambridge: MIT Press, 1985. 161-175.

—. "Is Natural Science a Natural Kind?" *Objectivity, Relativism, and Truth: Philosophical Papers.* Vol. I. By Rorty. Cambridge: Cambridge UP, 1991. 46-62.

Rosen, David. "Critical Dialogue on Disaster Films: Drugged Popcorn." *Jump Cut: A Review of Contemporary Media* 8 (1975): 19-20.

Rosen, Larry D. *Rewired: Understanding the iGeneration and the Way They Learn.* New York: Palgrave Macmillan, 2010.

—. "Teaching the iGeneration." *Educational Leadership* 68.5 (2011): 10-15.

Ross, Andrew. *No Respect: Intellectuals and Popular Culture.* New York: Routledge, 1989.

—. "The Challenge of Science." *The Cultural Studies Reader.* Ed. Simon During. London: Routledge, 1999. 292-305.

Ross, Art. "What is the True Meaning of the Disaster Film?" *Making Films in New York* 9.2 (April 1975): 28-31.

Rouse, Joseph. *Knowledge and Power: Toward a Political Philosophy of Science.* Ithaca: Cornell UP, 1987.

—. "The Dynamics of Power and Knowledge in the Sciences." *Journal of Philosophy* 88 (1991): 658-665.

—. "What are Cultural Studies of Scientific Knowledge?" *Configurations* 1.1 (1993): 57-94.

Rowe, Kathleen. *The Unruly Woman: Gender and the Genres of Laughter.* Austin: U of Texas P, 1995.

Ryan, Michael, and Douglas Kellner. *Camera Politica: The Politics and Ideology of Contemporary Hollywood Film.* Bloomington: Indiana UP, 1988.

Said, Edward. *Orientalism.* New York: Pantheon, 1978.

Sawday, Jonathan. *The Body Emblazoned: Dissection and the Human Body in Renaissance Culture.* London: Routledge, 1995.

Schelsky, Helmut. *Der Mensch in der wissenschaftlichen Zivilisation.* Köln: Westdeutscher Verlag, 1961.

Schiebinger, Londa. *The Mind Has No Sex? Women in the Origins of Modern Science.* Cambridge: Harvard UP, 1989.

Schiller, Dan. *Digital Capitalism.* Cambridge: MIT Press, 2000.

Schirrmacher, Arne. "Nach der Popularisierung: Zur Relation von Wissenschaft und Öffentlichkeit." *Geschichte und Gesellschaft* 34.1 (2008): 73-95.

Schlesinger, Arthur M. *The Cycles of American History.* 1986. New York: Mariner, 1999.

Schmandt, Jürgen. "Crises and Knowledge." *Science* 174.4006 (1971): 231.

Schröder, Gerald. "'Birth of the Cool': Jazz, Beat und Jackson Pollock." *Coolness: Zur Ästhetik einer kulturellen Strategie und Attitüde.* Ed. Annette Geiger, Gerald Schröder, and Änne Söll. Bielefeld: transcript, 2010. 167-184.

Schröder, Nicole. "Framing Disaster: Images of Nature, Media, and Representational Strategies in Hollywood Disaster Movies." *Local Natures, Global Responsibilities: Ecocritical Perspectives on the New English Literatures.* Ed. Laurenz Volkmann, Nancy Grimm, Ines Detmers, and Katrin Thomson. Amsterdam: Rodopi, 2010. 289-306.

Schummer, Joachim, and Tami I. Spector. "Popular Images versus Self-Image of Science: Visual Representations of Science in Clipart Cartoons and Internet Photographs." *Science Images and Popular Images of the Sciences.* Ed. Bernd Hüppauf and Peter Weingart. New York: Routledge, 2008. 69-95.

Schwartz, John. "The Best Science Show on Television?" *New York Times* 21 Nov. 2006: n.p. Accessed 25 February 2013. <http://www.nytimes.com/2006/11/21/science/21myth.html?n=Top%2FReference%2FTimes%20Topics%2FSubjects%2FA%2FAirships>.

Schweitzer, N.J., and Michael J. Saks. "The *CSI* Effect: Popular Fiction About Forensic Science Affects the Public's Expectations About Real Forensic Science." *Jurimetrics* 47 (2007): 357-364.

Scott, Peter. *The Eye of the Wind*. London: Hodder&Stoughton, 1961.

Scrubs. Prod. Bill Lawrence. NBC. 2001. Television.

Seabrook, John. "Selling the Weather." *The New Yorker* 3 April 2000: 44-53.

Searle, John R. *Speech Acts: An Essay in the Philosophy of Language*. Cambridge: Cambridge UP, 1969.

Sedlmeier, Florian. "The Cool Touch of Things: Libertarian Economics, Complex Simplicity, and the Emergence of the Tactile Erotic." *Is It 'Cause It's Cool? Affective Encounters with American Culture*. Ed. Astrid M. Fellner, Susanne Hamscha, Jennifer Moos, and Klaus Heissenberger. Vienna: LIT, 2014. 275-295.

Seewald, Franziska. *Bilder vom Ende: Zur Affizierung im Kino am Beispiel des Naturkatastrophenfilms Hollywoods*. Marburg: Tectum, 2011.

Seltzer, Mark. "Murder/Media/Modernity." *Canadian Review of American Studies* 38.1 (2008): 11-41.

"Sex, Lies, and Larvae." *CSI: Crime Scene Investigation*. CBS. Dir. Thomas J. Wright. Writ. Josh Berman and Ann Donahue. 22 Dec. 2000. Television.

Shadwell, Thomas. *The Virtuoso*. 1676. London: British Library, 2010.

Shakespeare, William. *Hamlet*. 1599-1602. Ed. George R. Hibbard. Oxford: OUP, 2008.

Shapin, Steven. "History of Science and its Sociological Reconstructions." *History of Science* 20 (1982): 157-211.

—. "Pump and Circumstance: Robert Boyle's Literary Technology." *Social Studies of Science* 14.4 (November 1984): 481-520.

—. "A Scholar and a Gentleman: The Problematic Identity of the Natural Philosophical Practitioner in Early-Modern England." *History of Science* 29 (1991): 279-327.

—, and Simon Schaffer. *Leviathan and the Air Pump: Hobbes, Boyle, and the Experimental Life*. Princeton: Princeton UP, 1985.

Shelton, Donald E., Young S. Kim, and Gregg Barak. "A Study of Juror Expectations and Demands Concerning Scientific Evidence: Does the 'CSI Effect' Exist?" *Vanderbilt Journal of Entertainment & Technology Law* 9 (2006): 331-368.

Shinn, Terry. "The Triple Helix and New Production of Knowledge: Prepackaged Thinking on Science and Technology." *Social Studies of Science* 32 (2002): 599-614.

—, and Richard Whitley, ed. *Expository Science: Forms and Functions of Popularisation.* Sociology of Science Yearbook Vol.9. Dordrecht: Reidel, 1985.

Shortland, Michael. "Screen Memories: Towards a History of Psychiatry and Psychoanalysis in the Movies." *British Journal for the History of Science* 20 (October 1987): 421-452.

—. *Medicine and Film: A Checklist, Survey, and Research Structure.* Oxford: Wellcome Unit for the History of Medicine, 1989.

Slater, Don. "Photography and Modern Vision: The Spectacle of Natural Magic." *Visual Culture.* Ed. Chris Jenks. New York: Routledge, 1995. 218-237.

Slavin, Kevin. "How algorithms shape our world." Public Talk at *TEDGlobal 2011.* 12 July 2011. Accessed 19 January 2012. <http://ted.com/talks/kevin_slavin_how_algorithms_shape_our_world.html>.

Smith, Glenn D. "Love as Redemption: The American Dream Myth and the Celebrity Biopic." *Journal of Communication Inquiry* 33.3 (2009): 222-238.

Smith, John. "A Description of New England." 1616. *The Complete Works of Captain John Smith (1580-1631).* Ed. Philip L. Barbour. Chapel Hill: U of North Carolina P, 1986. 291-370.

Smith, Tracy. "Why TV Crime Shows Are to Die For." *CBS News.* 16 May 2010. Accessed 28 June 2012. <http://www.cbsnews.com/2100-3445_162-6488704.html>.

Söll, Änne. "Raumkälte. Architektur und Distanz in Anton Räderscheidts Porträts der 1920er Jahre." *Coolness: Zur Ästhetik einer kulturellen Strategie und Attitüde.* Ed. Annette Geiger, Gerald Schröder, and Änne Söll. Bielefeld: transcript, 2010. 149-163.

Sommer, Andreas Urs. "Coolness: Zur Geschichte der Distanz." *Zeitschrift für Ideengeschichte* 1.1 (2007): 30-44.

Sontag, Susan. "The Imagination of Disaster." *Against Interpretation.* By Sontag. New York: Farrar, Straus & Giroux, 1966. 209-225.

Soraya, Lynne. "Sheldony or Aspergery?: *The Big Bang Theory.*" *Psychology Today* 20 April 2009: n.p. Accessed 3 August 2012. <http://www.psychologytoday.com/blog/aspergers-diary/200904/sheldony-or-aspergery-the-big-bang-theory>.

South Park. Prod. Trey Parker, Matt Stone, and Anne Garefino. Comedy Central. 1997. Television.

Southgate, Nick. "Coolhunting, Account Planning, and the Ancient Cool of Aristotle." *Marketing Intelligence & Planning* 21.7 (2003): 453-461.

Spencer, Charles D., et al. "Communications from Readers." *Newsletter of the Program on Public Conceptions of Science* 3 (April 1973): 6-13.

Spivak, Gayatri Chakravorty. "The Rani of Simur." *Europe and Its Others Vol 1. Proceedings of the Essex Conference on the Sociology of Literature.* Ed. Francis Barker Francis Barker, Peter Hulme, Margaret Iversen, and Diana Loxley. Colchester: University of Essex P, 1985. 128-151.

Stearns, Peter N. *American Cool: Constructing a Twentieth-Century Emotional Style.* New York: New York UP, 1994.

Stehr, Nico. *Arbeit, Eigentum und Wissen: Zur Theorie von Wissensgesellschaften.* Frankfurt am Main: Suhrkamp, 1994.

Strauss, Gary. "There's a science to CBS' 'Big Bang Theory'." *USA Today* 4 Nov. 2007: n.p. Accessed 18 January 2012. <http://www.usatoday.com/life/television/news/2007-11-04-big-bang_N.htm>.

Street, John. "The Celebrity Politician: Political Style and Popular Culture." *The Celebrity Culture Reader.* Ed. P. David Marshall. London: Routledge, 2006. 359-370.

Sunshine. Dir. Danny Boyle. Fox Searchlight, 2007. Film.

"Swap Meet." *CSI: Crime Scene Investigation.* CBS. Dir. Danny Cannon. Writ. Carol Mendelsohn, David Rambo, and Naren Shankar. 28 Oct. 2004. Television.

Swift, Jonathan. *Gulliver's Travels.* 1726. London: Wordsworth Classics, 1992.

Szöllösi-Janze, Margit. "Wissensgesellschaft in Deutschland: Überlegungen zur Neubestimmung der deutschen Zeitgeschichte über Verwissenschaftlichungsprozesse." *Geschichte und Gesellschaft* 30 (2004): 277-313.

Tagg, Philip. "Analysing Popular Music: Theory, Method and Practice." *Popular Music* 2 (1982): 37-65.

Tait, Sue. "Autoptic Vision and the Necrophilic Imaginary in *CSI.*" *International Journal of Cultural Studies* 9.1 (2006): 45-62.

Tatort. Das Erste. 1970. Television.

Taylor, Henry McKean. *Rolle des Lebens: Die Filmbiographie als narratives System.* Marburg: Schüren, 2002.

Templeton, Neil. "Othering & Writing Back." *The Imperial Archive.* School of English at the Queen's University of Belfast. 17 June 2007. Accessed 25 Sept. 2012. <http://www.qub.ac.uk/schools/SchoolofEnglish/imperial/transnational/Othering.html>.

"The Accused Is Entitled." *CSI: Crime Scene Investigation.* CBS. Dir. Kenneth Fink. Writ. Elizabeth Devine and Ann Donahue. 3 Oct. 2002. Television.

The Big Bang Theory. Prod. Chuck Lorre, Bill Prady, and Steven Molaro. CBS. 2007. Television.

"The Boyfriend Complexity." *The Big Bang Theory.* Dir. Mark Cendrowski. Writ. Bill Prady, Steven Molaro, and David Goetsch. CBS. 18 Nov. 2010. Television.

"The Bus Pants Utilization." *The Big Bang Theory.* Dir. Mark Cendrowski. Writ. Bill Prady, Steven Molaro, and Eric Kaplan. CBS. 6 Jan. 2011. Television.

"The Cooper-Hofstadter Polarization." *The Big Bang Theory.* Dir. Joel Murray. Writ. Chuck Lorre, Lee Aronsohn, and Dave Goetsch. CBS. 17 March 2007. Television.

"The Cooper-Nowitzki Theorem." *The Big Bang Theory.* Dir. Mark Cendrowski. Writ. Tim Doyle and Richard Rosenstock. CBS. 3 Nov. 2008. Television.

The Core. Dir. John Amiel. Paramount, 2003. Film.

"The Cruciferous Vegetable Amplification." *The Big Bang Theory*. Dir. Mark Cendrowski. Writ. Bill Prady, Lee Arohnson, and Steve Holland. CBS. 30 Sept. 2010. Television.

The Day After Tomorrow. Dir. Roland Emmerich. Writ. Roland Emmerich and Jeffrey Nachmanoff. Fox Home Entertainment, 2004. DVD.

"The Dead Hooker Juxtaposition." *The Big Bang Theory*. Dir. Mark Cendrowski. Writ. Steven Molaro. CBS. 30 March 2009. Television.

"The Desperation Emanation." *The Big Bang Theory*. Dir. Mark Cendrowski. Writ. Chuck Lorre, Steven Molaro, and Steve Holland. CBS. 21 Oct. 2010. Television.

"The Electric Can Opener Fluctuation." *The Big Bang Theory*. Dir. Mark Cendrowski. Writ. Steven Molaro. CBS. 21 Sept. 2009. Television.

"The Elegant Universe." *Nova*. Prod. Paula Apsell. PBS. 2003. Television.

"The Excelsior Acquisition." *The Big Bang Theory*. Dir. Peter Chakos. Writ. Bill Prady, Steve Holland, and Maria Ferrari. CBS. 1 March 2010. Television.

The Execution of Mary, Queen of Scots. Dir. Alfred Clark. Edison Manufacturing Company, 1895. Film.

The Fresh Prince of Bel-Air. Prod. Quincy Jones. NBC. 1990. Television.

"The Fuzzy Boots Corollary." *The Big Bang Theory*. Dir. Mark Cendrowski. Writ. Bill Prady and Steven Molaro. CBS. 8 Oct. 2007. Television.

"The Gorilla Experiment." *The Big Bang Theory*. Dir. Mark Cendrowski. Writ. Bill Prady, Steven Molaro, and Maria Ferrari. CBS. 7 Dec. 2009. Television.

"The Hawking Excitation." *The Big Bang Theory*. Dir. Mark Cendrowski. Writ. Chuck Lorre, Eric Kaplan, and Maria Ferrari. CBS. 5 April 2012. Television.

The Iron Lady. Dir. Phyllida Lloyd. Twentieth Century Fox, 2012. Film.

"The Isolation Permutation." *The Big Bang Theory*. Dir. Mark Cendrowski. Writ. Bill Prady, Steven Molaro, and Steve Holland. CBS. 3 Nov. 2011. Television.

"The Justice League Recombination." *The Big Bang Theory*. Dir. Mark Cendrowski. Writ. Robert Cohen and David Litt. CBS. 29 Oct. 2007. Television.

The Kennedys. Prod. Michael Prupas et al. History Channel. 2011. Television.

"The Killer Robot Instability." *The Big Bang Theory*. Dir. Mark Cendrowski. Writ. Robert Cohen and David Litt. CBS. 29 Oct. 2007. Television.

"The Love Car Displacement." *The Big Bang Theory*. Dir. Anthony Rich. Writ. Lee Aronsohn, Steven Molaro, and Steve Holland. CBS. 20 Jan. 2011. Television.

"The Luminous Fish Effect." *The Big Bang Theory*. Dir. Mark Cendrowski. Writ. David Litt and Lee Aronsohn. CBS. 15 Oct. 2007. Television.

"The Maternal Capacitance." *The Big Bang Theory*. Dir. Mark Cendrowski. Writ. Richard Rosenstock and Steven Molaro. CBS. 9 Feb. 2009. Television.

"The Merchants of Cool." *Frontline*. Dir. Barak Goodman. Writ. Rachel Dretzin. *PBS*. 27 Feb. 2001. Television.

"The Middle Earth Paradigm." *The Big Bang Theory*. Dir. Mark Cendrowski. Writ. Robert Cohen and David Litt. CBS. 29 Oct. 2007. Television.

The Nanny. Prod. Prudence Fraser, Robert Sternin, et al. CBS. 1993. Television.

"The Nerdvana Annihilation." *The Big Bang Theory*. Dir. Mark Cendrowski. Writ. Stephen Engel and Steven Molaro. CBS. 28 April 2007. Television.

The New Detectives: Case Studies in Forensic Science. Prod. Mike Sinclair, Peter Koeppen, and Tom Naughton. Discovery. 1996. Television.

The Nutty Professor. Dir. Tom Shadyac. Universal, 1996. Film.

The Office. Prod. Capital United Nations Entertainment and The Identity Company. BBC2. 2001. Television.

"The Panty Piñata Polarization." *The Big Bang Theory*. Dir. Mark Cendrowski. Writ. Jennifer Glickman and Steven Molaro. CBS. 10 Nov. 2008. Television.

The Poseidon Adventure. Dir. Ronald Neame. Twentieth Century Fox, 1972. *Film*.

The Private Lives of Elizabeth and Essex. Dir. Michael Curtiz. Warner Bros., 1939. Film.

"The Robotic Manipulation." *The Big Bang Theory*. Dir. Mark Cendrowski. Writ. Steven Molaro, Eric Kaplan, and Steve Holland. CBS. 23 Sept. 2010. Television.

"The Roommate Transmogrification." *The Big Bang Theory*. Dir. Mark Cendrowski. Writ. Bill Prady, Eric Kaplan, and Jim Reynolds. CBS. 19 May 2011. Television.

"The Rothman Disintegration." *The Big Bang Theory*. Dir. Mark Cendrowski. Writ. Steven Molaro, Eric Kaplan, and Jim Reynolds. CBS. 16 Feb. 2012. Television.

The Scarlet Empress. Dir. Josef von Sternberg. Paramount, 1934. Film.

The Secret of My Success. Dir. Herbert Ross. Universal, 1987. Film.

The Shield. Prod. Michael Chiklis. FX. 2002. Television.

The Social Network. Dir. David Fincher. Writ. Aaron Sorkin. 2010. Sony Home Entertainment, 2011. DVD.

The Story of Alexander Graham Bell. Dir. Irving Cummings. Twentieth Century Fox, 1939. Film.

The Story of Louis Pasteur. Dir. William Dieterle. Warner Bros., 1936. Film.

"The Terminator Decoupling." *The Big Bang Theory*. Dir. Mark Cendrowski. Writ. Tim Doyle and Stephen Engel. CBS. 9 March 2009. Television.

"The Theory of Everything." *CSI: Crime Scene Investigation*. CBS. Dir. Christopher Leitch. Writ. Carol Mendelsohn and David Rambo. 1 May 2008. Television.

The Time Machine. Dir. George Pal. Metro-Goldwyn-Mayer, 1960. Film.

"The Toast Derivation." *The Big Bang Theory*. Dir. Mark Cendrowski. Writ. Chuck Lorre, Steven Molaro, and Jim Reynolds. CBS. 24 Feb. 2011. Television.

The Towering Inferno. Dir. John Guillermin. Twentieth Century Fox, 1974. Film.

"The Vacation Solution." *The Big Bang Theory*. Dir. Mark Cendrowski. Writ. Bill Prady, Steven Molaro, and Maria Ferrari. CBS. 9 Feb. 2012. Television.

The Virgin Queen. Dir. Henry Koster. Twentieth Century Fox, 1955. Film.

The Wild One. Dir. László Benedek. Columbia, 1953. Film.

The Wolf of Wall Street. Dir. Martin Scorsese. Universal, 2013. Film.

Thomas, Ronald. *Detective Fiction and the Rise of Forensic Science.* Cambridge: Cambridge UP, 2003.

Thompson, Robert Farris. "An Aesthetic of the Cool." *African Arts* 7.1 (Autumn 1973): 40-91.

—. *African Art in Motion.* Berkeley: U of California P, 1974.

—. *Flash of the Spirit: African & Afro-American Art & Philosophy.* New York: Random House, 1984.

Thorne, Tony. *Dictionary of Contemporary Slang.* London: A&C Black, 1990.

Thurs, Daniel. *Science Talk: Changing Notions of Science in American Popular Culture.* New York: Rutgers UP, 2007.

Titanic. Dir. James Cameron. Twentieth Century Fox, 1997. Film.

Toffler, Alvin. *Future Shock.* 1970. New York: Bantam, 1984.

—. *The Third Wave.* New York: Bantam, 1980.

"Too Tough to Die." *CSI: Crime Scene Investigation.* CBS. Dir. Richard J. Lewis. Writ. Elizabeth Devine. 1 March 2001. Television.

Touraine, Alain. *Return of the Actor: Social Theory in Postindustrial Society.* Minneapolis. U of Minnesota P, 1988.

Townsend, Mark. "Big Bang Theory fuels physics boom." *Guardian* 6 Nov. 2011: n.p. Accessed 3 August 2012. <http://www.guardian.co.uk/education/2011/nov/06/big-bang-theory-physics-boom>.

Traube, Elizabeth G. "Secrets of Success in Postmodern Society." *Cultural Anthropology* 4.3 (August 1989): 273-300.

Traweek, Sharon. *Beamtimes and Lifetimes.* Cambridge: Harvard UP, 1988.

Turim, Maureen. *Flashbacks in Film: Memory and History.* New York: Routledge, 1989.

Turnbull, Sue. "The Hook and the Look: *CSI* and the Aesthetics of the Television Crime Series." *Reading CSI: Crime TV Under the Microscope.* Ed. Michael Allen. London: I.B.Tauris, 2007. 15-32.

Turow, Scott. "When He Was 17." *New York Times Sunday Book Review* 9 March 2008: n.p. Accessed 15 January 2012. <http://www.nytimes.com/2008/03/09/books/review/Turow-t.html?pagewanted=all>.

Twelve Monkeys. Dir. Terry Gilliam. Universal, 1995. Film.

Twister. Dir. Jan de Bont. Warner Bros., 1998. Film.

Two and a Half Men. Prod. Chuck Lorre. CBS. 2003. Television.

"Two Days Before the Day After Tomorrow." *South Park.* Dir. Trey Parker. Writ. Trey Parker and Kenny Hotz. 19. Oct. 2005. Television.

Ueding, Gert. *Klassische Rhetorik.* München: Beck, 2000.

Uricchio, William. "Television's Next Generation: Technology/Interface Culture/Flow." *Television After TV: Essays on a Medium in Transition.* Ed. Lynn Spigel and Jan Olsson. Durham: Duke UP, 2004. 163-182.

Van Dijk, Jan. *The Network Society: Social Aspects of New Media.* 1991. 2nd ed. London: Sage, 2005.

Vogel, Jakob. "Von der Wissenschafts- zur Wissensgeschichte: Für eine Historisierung der 'Wissensgesellschaft'." *Geschichte und Gesellschaft* 30 (2004): 639-660.

Volcano. Dir. Mick Jackson. Twentieth Century Fox, 1997. Film.

Vonnegut, Kurt. *Cat's Cradle.* 1963. London: Penguin, 2008.

Wachhorst, Wyn. *Thomas Alva Edison: American Myth.* Cambridge: MIT Press, 1981.

Walk the Line. Dir. James Mangold. Twentieth Century Fox, 2005. Film.

Walker, J. Samuel. *Three Mile Island: A Nuclear Crisis in Historical Perspective.* 2004. Berkeley: U of California P, 2006.

Walker, Mark. *Science and Ideology: A Comparative History.* New York: Routledge, 2002.

Wall Street. Dir. Oliver Stone. Twentieth Century Fox, 1987. Film.

Waugh, Patricia. *Practising Postmodernism, Reading Modernism.* London: Edward Arnold, 1992.

Weber, Jutta. *Umkämpfte Bedeutungen: Naturkonzepte im Zeitalter der Technoscience.* Frankfurt: Campus, 2003.

Weber, Max. "Wissenschaft als Beruf." *Gesammelte Aufsätze zur Wissenschaftslehre.* By Weber. Tübingen: J.C.B. Mohr, 1922. 524-555.

—. *The Protestant Ethic and the Spirit of Capitalism.* 1920. Trans. Stephen Kalberg. London: Blackwell, 2002.

Weinberg, Alvin M. "Impact of Large-Scale Science on the United States." *Science* 134.3473 (1961): 161-164.

Weingart, Peter, Claudia Muhl, and Petra Pansegrau. "Of Power Maniacs and Unethical Geniuses: Science and Scientists in Fiction Film." *Public Understanding of Science* 12 (2003): 279-287.

Weingart, Peter. "Science and Technology in a Legitimation Crisis: Hypotheses and Indicators." *Communication and Cognition* 6 (1979): 378-393.

—. *Die Stunde der Wahrheit: Zum Verhältnis der Wissenschaft zu Politik, Wirtschaft und Medien in der Wissensgesellschaft.* Weilerswist: Velbrück, 2001.

—. "The Moment of Truth for Science." *EMBO Reports* 3.8 (2002): 703-706.

—. "Chemists and their Craft in Fiction Film." *HYLE - International Journal for Philosophy of Chemistry* 12.1 (2006): 31-44.

—. "The Ambivalence towards New Knowledge: Science in Fiction." *Science Images and Popular Images of the Sciences.* Ed. Bernd Hüppauf and Peter Weingart. New York: Routledge, 2008. 267-281.

Weir, John. "The Beautiful American Word 'Cool'." *Rolling Stone* April 2001: 67.

Weissmann, Elke, and Karen Boyle. "Evidence of Things Unseen: The Pornographic Aesthetic and the Search for Truth in *CSI*." *Reading CSI: Crime TV Under the Microscope*. Ed. Michael Allen. London: I.B.Tauris, 2007. 90-102.

Wellman, Barry. "Physical Place and Cyber Place: The Rise of Networked Individualism." *International Journal of Urban and Regional Research* 25.2 (June 2001): 227-252.

When Worlds Collide. Dir. Rudolph Maté. Paramount, 1953. Film.

Wiatrowski, Michael. "Abjection, Outsiders and Others in *The Big Bang Theory*." Battleground States Conference. Bowling Green State University Campus, Bowling Green, OH. 25 - 26 Feb. 2011. Conference Paper. Accessed 24 Sept. 2012. <http://bgsu.academia.edu/MycWiatrowski/Papers/444169/Abjection_Outsiders_and_Others_in_The_Big_Bang_Theory>.

Wilkes, John. "Training scientists to be journalists." *EMBO Reports* 3.11 (2002): 1005-1008.

Will and Grace. Prod. Max Mutchnick, David Kohan, et al. NBC. 1998. Television.

Willett, Ralph. *Naked City: Urban Crime Fiction in the USA*. Manchester: Manchester UP, 1996.

Winn, J. Emmett. "Every Dream Has Its Price: Personal Failure and the American Dream in *Wall Street* and *The Firm*." *Southern Communication Journal* 68:4 (2003): 307-318.

—. *The American Dream in Contemporary Hollywood Cinema*. New York: Continuum, 2007.

Wittgenstein, Ludwig. *Philosophical Investigations*. Trans. G.E.M. Anscombe. 1953. New York: Macmillan, 1958.

Workneh, Lilly. "Newspapers use social media to say 'Smart is the New Sexy'." *CNN* online. 27 Oct. 2011. Accessed 15 Oct. 2012. <http://articles.cnn.com/2011-10-27/tech/tech_social-media_newspapers-smart-sexy-campaign_1_newspapers-facebook-and-twitter-social-media?_s=PM:TECH>.

Wozniak, Steve. Interview. *Bloomberg West*. Bloomberg Television. 16. Aug. 2013.

Wynne, Brian. "Public Perception and Communication of Risks: What do we know?" *Journal of NIH Research* 3 (1991): 65-71.

—. "Public Understanding of Science." *Handbook of Science and Technology Studies*. Ed. Sheila Jasanoff, Gerald E. Markle, James C. Petersen, and Trevor Pinch. Thousand Oaks: Sage, 1995. 361-388.

Xanthos, Nicolas. "Wittgenstein's Language Games." *Signo*. Dir. Louis Hébert. Rimouski: Quebec, 2009. n.p. Accessed 21 February 2012. <http://www.signosemio.com/wittgenstein/language-games.asp>.

Yergensen, Brent. "Exploring Science as Salvation in Apocalyptic Films." *Media and the Apocalypse*. Ed. Kylo-Patrick R. Hart and Annette M. Holba. New York: Peter Lang, 2009. 153-165.

Ziman, John. *Real Science: What it Is and What it Means.* Cambridge: Cambridge UP, 2000.
Zuckerberg, Mark. Interview. *World News.* ABC. 21 July 2010.

Index

A

H

I

J

K

L

Q

R

S

T

U